5th DAE-BRNS Workshop on Hadron Physics

(Hadron 2011)

Journal of Physics: Conference Series Volume 374

Mumbai, India
31 October – 4 November 2011

ISBN: 978-1-62276-808-0
ISSN: 1742-6588

Printed from e-media with permission by:

Curran Associates, Inc.
57 Morehouse Lane
Red Hook, NY 12571

Some format issues inherent in the e-media version may also appear in this print version.

Printed by Curran Associates, Inc. (2013)

For permission requests, please contact the Institute of Physics
at the address below.

Institute of Physics
Dirac House, Temple Back
Bristol BS1 6BE UK

Phone: 44 1 17 929 7481
Fax: 44 1 17 920 0979

techtracking@iop.org

Additional copies of this publication are available from:

Curran Associates, Inc.
57 Morehouse Lane
Red Hook, NY 12571 USA
Phone: 845-758-0400
Fax: 845-758-2634
Email: curran@proceedings.com
Web: www.proceedings.com

Papers

Preface

The 5th DAE-BRNS Workshop on Hadron Physics was held at the Bhabha Atomic Research Centre (BARC), Mumbai from 31 October to 4 November 2011. This workshop series, supported by the Board of Research in Nuclear Sciences, Department of Atomic Energy (BRNS, DAE), Govt. of India, began ten years ago with the firs one being held at BARC, Mumbai in October 2002. The second one was held at Puri in 2005, organized jointly by Institute of Physics, Bhubneswar and Saha Institute of Nuclear Physics, Kolkata. The 3rd and 4th ones took place, respectively, at Shantineketan in 2006, organized by Visva Bharati University, and at Aligarh in 2008, organized by Aligarh Muslim University, Aligarh.

The aim of the present workshop was to bring together the experts and young researchers in the fiel of hadron physics (both experiment and theory) and to have in-depth discussions on the current research activities in this field The format of the workshop was: a series of review lectures by various experts from India and abroad, the presentation of advanced research results by researchers in the field and a review of major experimental programs being planned and pursued in major laboratories in the fiel of hadron physics, with the aim of providing a platform for the young participants for interaction with their peers.

The upcoming international FAIR facility at GSI is a unique future facility for studies of hadron physics in the charm sector and hyper nuclear physics. The Indian hadron physics community is involved in this mega science project and is working with the PANDA collaboration on the development of detectors, simulation and software tools for the hadron physics programme with antiprotons at FAIR. A one-day discussion session was held at this workshop to discuss India–PANDA activities, the current collaboration status and the work plan.

This volume presents the workshop proceedings consisting of lectures and seminars which were delivered during the workshop. We are thankful to the authors for giving us the manuscripts in good time.

The workshop was financiall supported by BRNS, DAE, GoI. We also received partial funding support from the India-FAIR coordination centre, Kolkata, for the organization of the India–PANDA discussion meeting. We acknowledge the financia support received from BRNS and DST (Department of Science and Technology). The assistance from various departments of BARC and the Homi Bhabha Centre for Science Education (HBCSE), TIFR is gratefully acknowledged. We also thank the members of the advisory committee and organizing committee and colleagues from NPD and Physics Group, BARC for their contributions.

May 2012, Mumbai

Bidyut Jyoti Roy
A Chatterjee
S Kailas
Bhabha Atomic Research Centre

Journal of Physics: Conference Series **374** (2012) 011001 doi:10.1088/1742-6596/374/1/011001

Committees and Sponsors

Advisory Committee

R S Bhalerao	TIFR, Mumbai
R K Chaudhary	BARC, Mumbai
Y K Gambhir	IIT, Mumbai
R Godbole	IISc, Bangalore
Wasi Haider	AMU, Aligarh
B K Jain	Bombay University, Mumbai
S Kailas	BARC, Mumbai
M V N Murthy	IMSc, Chennai
S C Phatak	IOP, Bhubaneswar
Radhey Shyam	SINP, Kolkata
S K Singh	HNB Garhwal University
B C Sinha	VECC, Kolkata
D K Srivastava	VECC, Kolkata

Organizing Committee

B Ananthanarayan	IISc, Bangalore
A Chatterjee	BARC, Mumbai
Pankaj Jain	IIT, Kanpur
N Mathur	TIFR, Mumbai
H Mishra	PRL, Ahemadabad
A K Mohanty	BARC, Mumbai
P K Rath	Lucknow University
B J Roy	BARC, Mumbai *(Convener)*
P K Sahu	IOP, Bhubaneswar
A B Santra	BARC, Mumbai
M K Saran	SINP, Kolkata
S Sarkar	VECC, Kolkata
M Shoeb	JMI, New Delhi
Raghav Varma	IIT, Mumbai *(Convener)*
K B Vijaykumar	Mangalore University
P C Vinodkumar	Sardar Patel University, Gujrat
M Zafar	AMU, Aligarh

Sponsors

Board of Research in Nuclear Sciences
Department of Atomic Energy, Government of India

Photographs

Hadron Workshop 2011, Bhabha Atomic Research Centre, Mumbai, India

Journal of Physics: Conference Series **374** (2012) 011001 doi:10.1088/1742-6596/374/1/011001

Inauguration

from left: Ambar Chatterjee, Bikass Sinha

from left: Bidyut Jyoti Roy, Bikash Sinha, Ambar Chatterjee

Journal of Physics: Conference Series **374** (2012) 011001 doi:10.1088/1742-6596/374/1/011001

Academic Programme

<u>October 31, Monday</u>

10.00am-10.45am — Inauguration:
Welcome address: A.Chatterjee, R.Varma, B.J.Roy
Vote of thanks: V.Jha
Inaugural Talk: Bikas Sinha

10.45am-11.15am — *High Tea*

11.15am-12.05pm — (L1) B K Jain, "Question of η and K^- Nuclear Bound States"
12.05pm-12.55pm — (L2) S.Wycech, "In search of baryonium"

12.55pm-2.30pm — *Lunch Break*

2.30pm-3.20pm — (L3) C. Schwarz, "The PANDA Experiment at FAIR"
3.20pm- 4.10pm — (L4) A.B.Santra, "Hadron physics in holographic QCD-Part I"

4.10pm-4.40pm — *Tea Break*
4.40pm-5.30pm — (L5) W. Haider, "Nuclear Matter, Neutron Matter and Nucleon Optical potential in Brueckner-Hartree-Fock Theory"

8.00pm — *Dinner*

<u>Nov. 01, Tuesday</u>

9.45am-10.35am — (L6) S.Wycech, "On the nuclear states of Kbar mesons"
10.35am-11.25am — (L7) Susan Schadmand, "Hadron Physics via Meson Decays"

11.25am-11.45am — *Tea Break*

11.45am-12.35pm — (L8) S. Afsar Abbas, "Colour Confinement and Deformed Baryons in Quantum Chromodynamics"

1.05pm-2.30pm — *Lunch Break*

2.30pm-3.20pm — (L9) K. Khemchandani, "Hadron dynamics at intermediate energies"
3.20pm-4.10pm — (L10) A. Martinez Torres, "Novel interest in three-hadron system"

4.10pm-4.40pm — *Tea Break*

4.40pm-5.30pm — (L11) N. Mathur, "Exotic Hadrons from lattice QCD"
5.30pm-6.00pm — (S1) A.K.Jain, "N-N t-matrix effective interaction for (p,2p) reactions"

8.00pm — *Dinner*

<u>Nov. 02, Wednesday</u>

9.45am-10.35am — (L12) A.B.Santra, "Hadron physics in holographic QCD- Part II"
10.35am-11.25am — (L13) S.Sarkar, "Low mass lepton pair production in heavy ion collisions"

11.25am-11.45am — *Tea Break*
11.45am-13.10pm
(L14) B. Ananthanarayan, "The theory of unitarity bounds for form factors"
(S2) G.Abbas, "Improving the phenomenology of K_{l3} form factors with analyticity and unitarity"

(S3) S.Imsong, "Implications of unitarity and analyticity for the D\pi form factor"

1.00pm-2.30pm *Lunch Break*

2.30pm-6.30pm Discussion on India-PANDA programme

2:30pm-4:00pm

Introductory remarks	: S.Kailas and A.Chatterjee
An introduction to DIRC and SciTil detectors	: C.Schwarz
Status report from IIT-Bombay	: R.Varma
Status report from BARC	: B.J. Roy
Status from FAIR	: Susan Schadmand
Discussion	

4.00pm-4.30pm *Tea Break*

4:30pm – 6:00pm

On Luminosity detector	: M.Jadhav
Physics simulation in PANDAroot	: V.Jha
On SiPM	: Anita Topkar
Discussion	

8.00pm *Dinner*

Nov. 03, Thursday

9.45am-10.35am (L15) SAWADA, Shin'ya. "Hadron Physics at J-PARC"

10.35am-11.25am (L16) H.Fujioka, "Experimental search for eta-mesic nuclei at J-PARC"

11.25am-11.45am *Tea Break*

11.45am-12.35pm (L17) P C Vinod Kumar, "Heavy Flavour Hadron Spectroscopy: Challenges and Future Prospects"

12.35pm – 1.05pm (S4) Ajai kr. Rai, "Properties of heavy flavoured meson using NRQCD Formalism"

1.05pm-2.30pm *Lunch Break*

2.30pm-3.20pm (L18) N. Mathur, "Lattice QCD and Hadron Spectra"

3.20pm-4.10pm (L19) Y. Ilieva, "Studies of Baryon Structure and Baryon Interactions at Thomas Jefferson National Accelerator Facility"

4.10pm-4.40pm *Tea Break*

4.40pm-5.30pm (L20) M.Shoeb, "Variational Monte Carlo Calculations for the energies of multi strange α-cluster hypernuclei"

5.30pm-6.00pm (S5) Sonika Gupta, "Energy of partial *A*-body problem for ${}^{9}_{\Lambda}$Be and ${}^{10}_{\Lambda\Lambda}$Be

8.00pm *Dinner*

Nov. 04, Friday

9.45am –10.35am (L21) R. Raja, "Main Injector Particle production Experiment (MIPP) at Fermilab-- status and plans"

10.35am-11.25am (L22) M. Sajjad Athar, "Nuclear medium effects in structure functions"

11.25am-11.45am *Tea Break*

Journal of Physics: Conference Series **374** (2012) 011001 doi:10.1088/1742-6596/374/1/011001

11.45am-12.35pm	(L23) C.Samanta, "STRANGE TO SUPERSTRANGE NUCLEI"
12.35pm-2.00pm	*Lunch Break*
2.00pm-2.30pm	(S6) P.Shukla, "Probing Strongly Interacting Matter Produced in Pb+Pb Collisions at LHC"
2.30 pm	Concluding Session Summary Talk : B.K.Jain and S.Wojech Concluding remarks : A.Chatterjee

Journal of Physics: Conference Series **374** (2012) 011002 doi:10.1088/1742-6596/374/1/011002

Peer review statement

All papers published in this volume of *Journal of Physics: Conference Series* have been peer reviewed through processes administered by the proceedings Editors. Reviews were conducted by expert referees to the professional and scientific standards expected of a proceedings journal published by IOP Publishing.

Question of η- and K^-- Nucleus Bound States

B. K. Jain[1], N. J. Upadhyay[2] and Mohammad Shoeb[3]

[1] MU-DAE Centre for Excellence in Basic Sciences, Mumbai University, Kalina Campus, Mumbai-400098, INDIA

[2] National Superconducting Cyclotron Laboratory, Michigan State University, East Lansing, Michigan 48824-1321, USA

[3] Physics Department, Aligarh Muslim University, Aligarh, INDIA

E-mail: `brajeshk@gmail.com`

Abstract. Interaction of the η-meson and that of the K^--meson with nucleons is of special interest because both of them are strongly attractive near threshold. This raises the strong possibility that we may find in nature the bound η- and K^--nuclear (quasi) bound states. This led to experimental programs to hunt for the existence of these states and theoretical studies to keep pace with them. The efforts had positive results. The η-meson studies had been there for several years by now, while those with the K^--meson had been relatively recent. The talk gives a brief critical overview of the η-nuclear interaction studies, especially in context with the η-mesic state explorations.

For the K^--meson we give a brief summary of the efforts in understanding the basic K^--nucleon interaction and theoretical explorations for the existence of K^--nucleus bound states. We critically examine the FINUDA measurements as a signal for the existence of the K^--nucleus bound states, especially in context with the contribution of the single nucleon knock-out final state interaction in the $(K^-, p\Lambda)$ reaction.

1. Introduction

It is known that most of the visible mass of cosmos is accounted for by the hadrons. Therefore, the study of hadrons and their interaction is of fundamental importance in understanding certain aspects of nature. The origin of hadron masses from the practically massless quarks is believed to lie in the non-perturbative domain of QCD. Therefore, the hadrons provide a "laboratory" where one can pursue the study of not-so-easy-to-explore non-perturbative QCD domain.

Equally important is the understanding of the interaction between hadrons, where an impressive progress has been made in describing it using chiral perturbation theories (χPT). Of special interest in these studies are the η-N and K^--p interactions near threshold, where both of these interactions are found to be attractive and mediated by certain "doorway" states. This raises a tantalizing possibility that, if these elementary interactions generate sufficiently strong nuclear interaction when they are put in the nucleus, we might see the existence of (quasi-)bound η-/K^--mesic states in nature. This has generated great interest in the study of the nuclear interaction of these mesons near threshold, theoretically as well as experimentally. In the present talk we will discuss briefly the overall situation in the field of the η-meson interaction and the current state of the K^--mesic studies.

2. η- Meson Interaction

The existence of η-mesic states depends sensitively on the precise quantitative knowledge of the strength of the η-N interaction, which, unfortunately, is not easily forthcoming. This is because, η-meson being unstable (decays quickly in 10^{-18} sec), we can not have eta beams. This makes the investigation of η-meson dynamics possible only through the η-meson producing reactions, which, unfortunately, complicates the matter, and also renders the extracted information uncertain. A large set of phenomenologically extracted η-N scattering amplitudes, therefore, exist in the literature [1, 2]. They differ in details, but all of them accept that the η-N interaction near threshold is attractive and proceeds through an intermediate S_{11} (1535) $J^P = 1/2^-$ nucleon resonance.

Theoretically, the existence of the η-nucleus bound states in nuclei have been explored in several ways. For nuclei beyond few-nucleon systems, calculations have been done by finding complex energy solutions of the momentum-space relativistic three-dimensional integral equation for some η-nucleus optical potential [3]. This potential is essentially constructed following the "$t\rho$" approximation with the choice of the η-N amplitude guided by the phenomenologically determined on-shell amplitude extrapolated off-shell using the separable model ansatz. Existence and the position of bound states in such models depend strongly on the sub-threshold behaviour of the η-N amplitude, which it seems can not be constrained in any way. These calculations predict bound states for nuclei with mass number greater than 12. Another approach used is the QCD based quark-meson-coupling approach (QMC), where the η-meson is embedded in the nuclear medium, and couples to quarks and mixes with $\eta\prime$ [4]. The η self energy obtained from such calculations is used in the local density approximation in the Klein-Gordon equation to obtain the complex energy solutions. This approach, apart from predicting the energy and width of the bound states, shows important role of $\eta - \eta\prime$ mixing in increasing the value of the η-N scattering length.

For few-nucleon systems existence of η binding is explored by calculating η-nucleus scattering lengths for them. For the existence of bound states the scattering lengths need to be large and negative. For η-NN system it is calculated using relativistic Faddeev equations and some choice for potentials for binary sub-systems in it [5]. This calculation finds at best the existence of a quasivirtual state only. For an η-d system one uses a certain version of the multiple scattering formalism [6] and for ^{2}He, ^{3}H, ^{3}He and ^{4}He nuclei scattering lengths are calculated in the Finite Rank Approximation (FRA) [7]. Both these calculations need elementary η-N amplitude with a prescription for off-shell extrapolation. The outcome of these calculations obviously depends on the input quantities, which as we mentioned above are not known so well. Using the one term Yamaguchi extrapolation, these calculations suggest that if $a_{\eta N}$ is large ($\sim$0.6 fm) it is possible to get η-nucleus bound states even for A=4 nucleus.

Experimentally, anticipated η-nucleus bound states should get reflected in a straight forward way in the measured eta producing nuclear cross sections. Such observations indeed have been made in the proton induced η production reaction on nuclei where the measured amplitude on the deuteron target shows a sharp rise as we reach the threshold [8]. The η-^{3}He (or η-d) scattering amplitude from data has been extracted usually by writing the cross section as a product of the plane wave result and an enhancement factor, which has been parametrized in terms of the scattering length [9]. Large values of the extracted scattering lengths from such analyses are normally taken as an indicator for the existence of bound states.

This procedure of extracting the scattering lengths, however, ignores reference to any reaction mechanism for η production and lacks totally a detailed treatment of the final state interaction. Therefore, while the extracted scattering lengths by this method may at best be indicative, their actual values may not be fully reliable for determining the bound states. A proper procedure should be to first analyze these data using a certain η production mechanism and incorporating in detail the final state interaction as accurately as possible. Once such a procedure, using the available ηN t-matrix, reproduces the measured cross sections, one can then infer from them the resultant η-nucleus scattering amplitudes and use them to investigate the possibility of the existence of unstable bound states.

Based on the above concept, our group carried out a systematic analysis in this direction over about the last ten years [10] to explore theoretically the possibility of the existence of the η-meson bound state.

Experimentally two measurements which give positive indication of the existence of η-mesic states have been reported in literature. One experiment is by the TAPS collaboration [11] on photo-production of η on ^{3}He, $\gamma\, ^3\text{He} \rightarrow \pi^0$ p X, where one essentially sees the decay of bound η in ^{3}He through the S_{11} resonance. The other is a bit more recent measurement from COSY on the ^{27}Al (p, ^{3}He)X reaction in a recoil free kinematic set-up [12], where one sees in coincidence with ^{3}He, the decay of a possible bound η-^{25}Mg bound state, again through the S_{11} resonance. This experiment was done at COSY in collaboration with the Mumbai experimental group who also designed and fabricated the Enstar detector. The COSY collaboration recently also reported the measurement of the p $^6\text{Li} \rightarrow ^7\text{Be}\, \eta$ reaction cross section at an excess energy around 11 MeV [13].

We proceeded to study the case of η-^{3}He system.

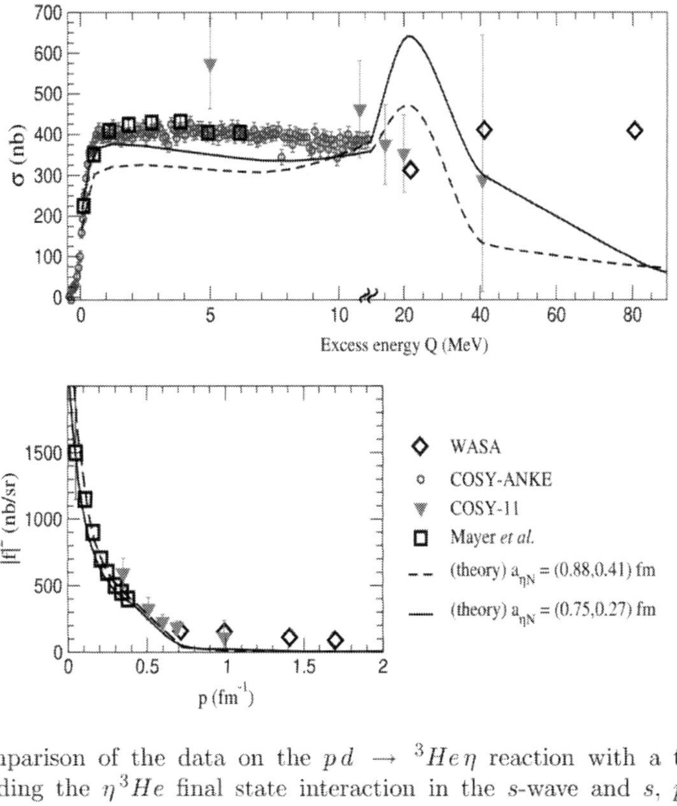

Figure 1. Comparison of the data on the $pd \rightarrow\, ^3He\eta$ reaction with a two step model calculation including the $\eta\, ^3He$ final state interaction in the s-wave and s, p and d partial waves in the reaction mechanism. The scale in the upper plot has been broken at 11.2 MeV for clarity.

We analysed the pd$\rightarrow \eta^3$He reaction for which the good quality data were available. The η production was described by a two step process, where the beam proton interacts with the proton (or neutron) in the deuteron in the first step. This produces a deuteron and a π^+ (or π^0).

The pion in the second step interacts with another nucleon in the target deuteron and produces an η. The final state interaction (FSI) between η and ^{3}He was incorporated by solving few body Faddeev type equations using the finite-range-approximation. These studies reproduced the available eta production data in p d fusion near threshold for η-^{3}He exit channel [Fig. 1] very well [14]. The η-^{3}He t-matrix in these calculations entered through the description of the η-^{3}He scattering wave function, ψ^- in the final state,

$$\langle\psi^-| = \langle\vec{k}| + \int \frac{d\vec{q}}{(2\pi)^3} \frac{\langle\vec{k}|T_{\eta He}(\vec{q})|\vec{q}\rangle}{E(k) - E(q) + i\epsilon} \langle\vec{q}|. \tag{1}$$

Having determined the η-^{3}He t-matrix which fit the pd$\rightarrow \eta^3$He data well we proceeded to explore if this t-matrix admits any bound state. For this we used the method of Wigner's time delay [15]. Naively, this method means that in a scattering process $ab \rightarrow ab$, if there is a resonance, the scattering process will take longer time compared to the situation if there is no resonance. Equivalently, the scattering phase shift, δ will show a large variation with energy across the resonance. This method has been used in the literature for the analysis of hadron resonances successfully [16]. For our purpose a modification of Wigner's method was done by Kelkar in [17] and we obtained an eta-mesic state in the η-^{3}He system in agreement with experiments [17, 18].

3. K^--Nuclear Interaction: Experimental

First indication of the existence of the K^- bound nuclear states came in 2005 in the FINUDA measurements [19] of the stopped K^- absorption in Li, C and other target nuclei. These experiments using the FINUDA spectrometer installed at the DAΦNE collider detected a Λ hyperon and a proton pair in coincidence following K^- absorption at rest in several nuclei. The emitted Λ-p pair was found to emerge, predominantly back to back in all target nuclei, and had their invariant mass distributions peaking significantly below the sum of a kaon and two proton mass in free state (2.370 GeV) [Fig. 2]. If it is assumed that the Λ-p pair is emitted from a "$K^- pp$" system in the nucleus, this mass shift implies a bound $K^- pp$ system in nuclei with the binding energy above 100 MeV. In a more elaborate second run of these experiments carried out recently [20] it is further reported that these mass shifts occur only for the $K^- pp$ module, and not for the $K^- np$ cluster. The absorption on an $n\,p$ pair gives Λ-n and Σ^--p pairs in the final state.

Recent analysis of the old DISTO data from the Saturne accelerator on $p\,p \rightarrow p\,\Lambda\,K^+$ reaction too suggests the existence of a $K^- pp$ cluster with the binding energy around 100 MeV [21].

As the K^-N interaction in s-state is attractive, attractive behaviour of the K^--nucleus interaction was expected but the existence of such deeply bound modules came as a big surprise. The occurrence of only "$K^- pp$" modules is understandable because the K^-N interaction is such that it is much stronger in T=0 state than in T=1 state. Since the K^-p system is an admixture of T=0 and T=1 states, while the K^-n system is a pure T=1 state, strong interaction between a K^- and a proton is understandable. In the presence of another proton, *a la* Heitler-London theory of hydrogen molecules, K^-, by continuously shuttling between two protons, can understandably cause strong attraction between them, producing thereby bound "$K^- pp$" modules.

4. K^--Nuclear Interaction: Theoretical

4.1. $\bar{K}N$ Interaction

Theoretically, the K^--p issue has been investigated in the framework of the SU(3) chiral perturbation theories, first developed and applied in Ref. [22]. Subsequently it was expanded in Ref. [23]. The fact that, unlike pion-nucleon and K^+-N cases, the basic interaction here

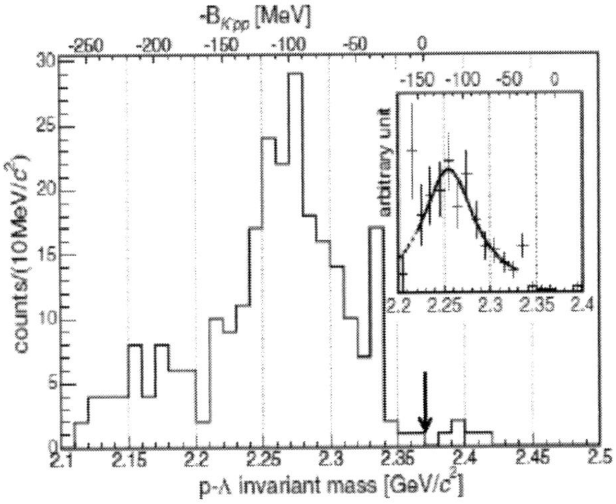

Figure 2. Invariant mass of a Λ and a proton in back-to-back correlation ($\cos\theta^{\rm Lab}$ < -0.8) from light targets before the acceptance correction. The inset shows the result after the acceptance correction for the events which have two protons with well-defined good tracks. Only the bins between 2.22 and 2.33 GeV/c^2 are used for the fitting. The plot is taken from Ref. [19].

is relatively strong and is incorporated in these studies by including terms up to order q^2 in the chiral Lagrangian expansion. The chiral Lagrangian approach is further connected in these studies to the potential model approach. A meson-baryon potential is constructed for this purpose by constraining that in the Born approximation, this potential has the same s-wave scattering length as the effective chiral Lagrangian , up to order q^2. Such a potential in local representation between channel i and j with a Yukawa form comes out as,

$$V_{ij}(r) = \frac{C_{ij}\alpha_{ij}^2}{8\pi f^2}\sqrt{\frac{m_i m_j}{s\omega_i\omega_j}}\frac{e^{-\alpha_{ij}r}}{r}, \tag{2}$$

where $m_i(m_j)$ is the mass of the baryon in channel i(j), $\omega_i(\omega_j)$ is the reduced energy in channel i(j) in the centre-of-mass. C_{ij} and α_{ij} are constants. They are obtained by iterating this potential in a set of coupled channel Lippmann-Schwinger equations and fitting the variety of low energy data of the kaon-baryon system. Such a system for S = -1 has six channels, $\pi^+\Sigma^-$, $\pi^0\Sigma^0$, $\pi^-\Sigma^+$, $\pi^0\Lambda$, K^-p, K^0n. This pseudo-potential for K^-p and K^-n systems turns out as

$$V_{K^-p}(r) = -360\frac{e^{-\alpha r}}{r}MeV, \tag{3}$$

and

$$V_{K^-n}(r) = -30\frac{e^{-\alpha r}}{r}MeV, \tag{4}$$

with $1/\alpha$=0.5 fm. This clearly shows that for the s-wave

$$V_{K^-p} >> V_{K^-n}. \tag{5}$$

The amplitude resulting from these calculations for K^-N system is shown in Fig. 3. Below K^-N threshold it shows a characteristic resonance behaviour for the K^-p system. Its imaginary part has a maximum around the resonance energy of the $\Lambda(1405)$, and the real part goes through

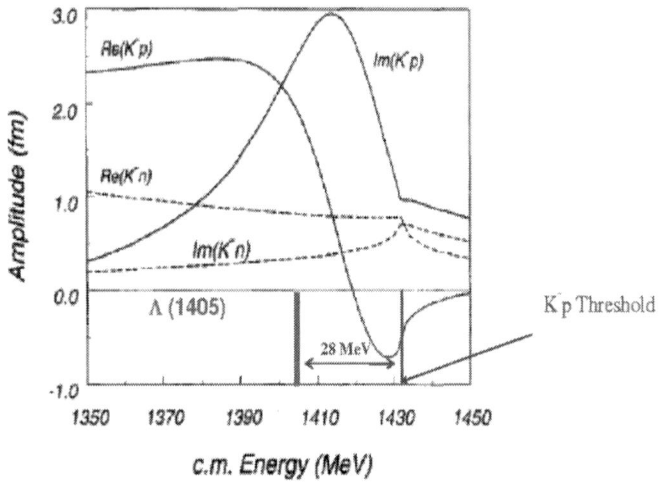

Figure 3. Plot of real and imaginary parts of the K$^-$N scattering amplitude below K$^-$N threshold energy is taken from Ref. [22].

zero around the same energy. The K^-n amplitude does not have any such behaviour. That's the reason why the K^-p potential is much stronger than the K^-n potential.

Identifying that the experimentally observed $\Lambda(1405)$ is a bound state of the K^-p system, the K^-p potential is alternatively obtained by reproducing this bound state at the (experimental) binding energy of around 28 MeV (1405 MeV-K^-p free mass) by varying the parameters of an assumed potential.

Because of the coupling amongst different S = -1 channels, the $\Lambda(1405)$ also acquires a width of about 50 MeV. This width is mainly for the $\Lambda(1405) \to \Sigma\pi$ channel.

4.2. K^--Nucleus Binding

Having learnt about the K^-N potential, theoretical search for the antikaon-nucleus bound state has been made in the literature for K^- plus 2-3 nucleons and heavier nuclei . These studies follow the variational approach [24, 25, 26, 27] and the Faddeev method [28]. All these calculations need, as input, realistic choice for the NN and the K^--N potentials. For the nucleon-nucleon potential, following extensive work over the years on this subject, it is always possible to make a correct choice. For the K^--N potential some calculations use a phenomenological pseudo-potential, while others use a leading order chiral interaction. All these calculations, in line with the experimental observations, found K^-pp bound states with about 50 MeV or more binding energies.

The variational calculations in [24, 25, 27] are done in the framework of Brueckner-Hartree-Fock or anti-symmetric molecular dynamics and use a phenomenologically constructed baryon-meson potential. We, in our calculations, use Variational Monte Carlo (VMC) method and use the leading order chiral interaction for the K^-N potential. In the following we give a brief description of our calculations for a K^-pp and K^-pn systems.

4.2.1. Variational Monte Carlo (VMC) Calculations The VMC method, in brief, consists of constructing the general Hamiltonian for K^--nucleus system, which for the K^-pp system, for example, takes the form

$$H^{K^-}_{N_2N_3} = T_{K^-}(1) + \sum_i (T_{N_i}(i) + V_{K^-N_i}(r_{1i})) + V_{N_2N_3}, \qquad (6)$$

Table 1. The calculated binding energy for K$^-$pp and K$^-$np systems for Urbana V14 potential and $\langle R_{NN}^{rms}\rangle$ of NN pair and root mean square radius $\langle R_{(NN)}^{K}\rangle$ that of K$^-$ from c.m. of NN pair.

	K$^-$pp	K$^-$np
BE (in MeV)	-124.84	-99.72
$\langle R_{NN}^{rms}\rangle$	1.53	1.85
$\langle R_{(NN)}^{K}\rangle$	0.89	1.24

where T_Y is the kinetic energy operator for the particle Y and V_{hh} denotes the hadron-hadron interaction for the pair hh (=K^-N and NN). The translationally invariant trial wave function for the system is taken as the product of two-body correlation functions, f_{hh} (hh=K^-N and NN) and the appropriate spin function, χ_s^σ, i.e.

$$\Psi_{NN}^{K^-} = [\Pi_i f_{K^-N}(r_{1i})] f_{NN}(r_{23}) \chi_s^\sigma. \tag{7}$$

For the K^-pp system anti-symmetry requires that $s = \sigma = 0$. The correlation functions $f_{hh}(r)$ have the asymptotic behaviour, $f_{hh}(r) \sim r^{-\alpha} exp(-\kappa_{hh} r)$ and are obtained from the solution of the Schrodinger-type equation, a procedure developed by the Urbana group [29]. The energy $-B_{NN}^{K^-}$ for K^-NN system is evaluated using the relation:

$$-B_{NN}^{K^-} = \frac{\langle \Psi_{N_2N_3}^{K^-} | H_{N_2N_3}^{K^-} | \Psi_{N_2N_3}^{K^-} \rangle}{\langle \Psi_{N_2N_3}^{K^-} | \Psi_{N_2N_3}^{K^-} \rangle}. \tag{8}$$

For NN pair we use very well known potential V14 of Urbana group which has repulsive soft core. It is constrained by the NN scattering data up to 350 MeV and the ground state properties of the deuteron.

Our estimates of the energy were made for 100,000 points. The statistical errors in the energies are of the order of 1%. The calculated binding energy for K^-pp and K^-np systems and r.m.s. radii for NN, $\langle R_{NN}^{rms}\rangle$ and that of K^- w.r.t. NN centre-of-mass $\langle R_{(NN)}^{K}\rangle$ are listed in Table 1. We find that both the systems are very strongly bound, and the separation of K^- w.r.t. NN is very small. The NN nucleon core also shrinks considerably. In the K^-np system, for example, np system shrinks by about 50% compared to deuteron ($\langle R_d^{rms}\rangle$=3.8 fm). Our calculations include the Coulomb interaction. Ignoring it lowers the binding energy by about 2 MeV. These results are almost of the same order as found by the Japanese group earlier [24, 25].

Various variational calculations thus predict affirmatively about the existence of deeply bound and compact K^-NN modules. This is in accord with the experimental findings of the FINUDA group at Frascati, Rome.

However, consequent to above accord between experiments and theoretical calculations, doubts have been raised about the conclusions of the χPT calculations as well as the interpretation of the experimental findings. The χPT calculations are questioned for the legitimacy of the use of the phenomenological or the chiral leading order prescription for the $\bar{K}N$ potential. The correlation of the mass-shift seen in the FINUDA data to the binding of the K^--pp module in nuclei is questioned for ignoring in this interpretation the final state interaction (FSI) in the $(K^-, p\Lambda)$ reaction.

5. More on K^--Nucleus Binding

What is argued is that, though the correct K^-N potential is the one which predicts or reproduces the $\Lambda(1405)$ binding, the binding energy of this state should not be identified with the observed

Table 2. Binding energy B and mean distance $\sqrt{(r^2)}$ for $\bar{K}N$ two-body system for the local $\bar{K}N$ potentials in I=0 channel derived from four variants of the chiral SU(3) coupled channel approach. The range parameter b of the potential is given in the first row.

	ORB [32]	HNJH [33]	BNW [34]	BMN [31]
		Two-body $\bar{K}p$ potential with	Coulomb potential	
b	0.52	0.47	0.51	0.41
B [MeV]	12.78	12.25	11.42	13.51
$\sqrt{(r^2)}$ [fm]	1.84	1.84	1.93	1.74
		Two-body $\bar{K}p$ potential without	Coulomb potential	
B [MeV]	11.1	10.54	9.60	11.73
$\sqrt{(r^2)}$ [fm]	1.92	1.94	2.04	1.82

peak position (around 1405 MeV, i.e. B.E.∼27 MeV) in the $\pi\Sigma$ invariant mass distribution in the $\bar{K}N$ scattering. This is because, in the chiral $SU(3)$ $\bar{K}N$ dynamics there is a strong $\bar{K}N \leftrightarrow \pi\Sigma$ coupled-channel dynamics, which results in a two pole structure in the $\bar{K}N$ t-matrix. One pole corresponds to the $\bar{K}N$ quasi-bound state and another to the $\Sigma\pi$ resonance. Because of this, the observed peak in the $\pi\Sigma$ invariant mass distribution seen in the $\bar{K}N$ scattering needs to be fitted by the superposition of both these poles. This, as a thorough investigation of this issue in Ref. [30] argues results in the mass of the $\Lambda(1405)$ quasi-bound state to be 1420 MeV, and not the normally used 1405 MeV value, making thereby the $\Lambda(1405)$ bound only by 12 MeV below $\bar{K}N$ threshold, not 27 MeV assumed in other studies.

Recent measurements on the $pp \rightarrow pK^+Y^0$ reaction at COSY [31], however, seem to present experimental evidence which do not support the above two pole model for the $\Lambda(1405)$. The shape and position of the $\Lambda(1405)$ distribution in these measurements is reconstructed cleanly in the $\Sigma^0\pi^0$ channel using invariant- and missing-mass techniques. The mass of the $\Lambda(1405)$ is found to be $\sim$ 1400 MeV with width $\sim$ 60 MeV.

The 12 MeV binding for the $\Lambda(1405)$ from the "two-pole" model, however, reduces the strength of the $\bar{K}N$ potential considerably. The use of such a weaker potential in variational calculations then reduces the binding of the K^--nuclear systems. The potential itself has the form

$$V_{K^-N}(\sqrt{s}, r) = -V_{K^-N}(\sqrt{s}, 0)\frac{e^{-r^2/b^2}}{\pi^{3/2}b^3}, \tag{9}$$

where the strength of the potential is complex and is $\bar{K}N$ invariant mass dependent. $\sqrt{s}$ is the total centre-of-mass energy of the system. We consider only real part of this potential. The calculated binding energy and rms distance for the two-body $\bar{K}N$ system are given in Table 2. The use of the effective $\bar{K}N$ potential derived from four variants of the chiral SU(3) coupled channel approach [30] in our VMC calculations gives the results shown Table 3 for the K^-pp system. The binding energy in our calculations has come down to 35-40 MeV, while in calculation of Dote et al. [25] it drops to around 18-23 MeV for the same $\bar{K}N$.

6. Final State Interaction

As regards the interpretation of the FINUDA measurements a suggestion was advanced in Ref. [35]. According to it the down-shift observed in the invariant mass of the p and Λ in the FINUDA experiment could be the result of the final state interaction of these particles with the recoiling nucleus. This seems quite plausible because, due to Q-value of the $K^- + pp \rightarrow \Lambda p$ reaction

Table 3. The detailed composition of the total K^-pp binding energy (BE) using VMC calculation for $\bar{K}N$ interaction in I=0 channel without coulomb potential. Auxiliary antikaon "binding energy" (B_K) and the centre-of -mass energy $\sqrt{s}$ of $\bar{K}N$ subsystem within K^-pp are related as $\sqrt{s} = M_N + m_K - B_K/2$, energy of two-nucleon $E_N = \langle\Psi^{K^-}_{N_2N_3}|\sum_i T_{N_i}(i) + V_{N_2N_3}|\Psi^{K^-}_{N_2N_3}\rangle$, kinetic energy $T_{nuc} = \langle\Psi^{K^-}_{N_2N_3}|\sum_i T_{N_i}(i)|\Psi^{K^-}_{N_2N_3}\rangle$ of the two-nucleon subsystem, total kinetic energy of the three-particle $T_{kin} = \langle\Psi^{K^-}_{N_2N_3}|T_{K^-}(1) + \sum_i T_{N_i}(i)|\Psi^{K^-}_{N_2N_3}\rangle$, potential energy of two-proton $V(NN) = \langle\Psi|V_{N_2N_3}|\Psi\rangle$, and potential energy of three-particle $V(KN) = \langle\sum_i V_{K^-N_i}(r_{1i})) + V_{N_2N_3}\rangle$. R_{pp} is the rms distance between the two protons, $R_{\bar{K}p}$ between antikaon and nucleon and $R_{(pp)\bar{K}}$ between cm of pp and $\bar{K}$ (all energies in MeV and distances in fm).

Effective potential	BE	B_K	E_N	T_{nuc}	T_{kin}	V_{NN}	$V_{\bar{K}N}$	R_{pp}	$R_{\bar{K}p}$	$R_{(pp)\bar{K}}$
ORB [32]	39.80	22.82	62.61	83.20	196.33	-20.6	-215.52	1.84	1.58	1.28
HNJH [33]	36.10	26.32	62.64	81.59	191.69	-19.18	-208.61	1.99	1.81	1.20
BNW [34]	35.96	18.84	54.81	74.31	173.37	-19.46	-189.87	2.01	1.78	1.17
BMN [31]	35.34	38.96	74.35	92.96	224.30	-18.61	-241.03	1.86	1.63	1.34

being around 317 MeV, the kinetic energies of outgoing p and Λ are 160 MeV or so. At these energies, it is well known that in the nucleon-nucleus scattering the reactive cross section mainly consists of the single nucleon knock-out channel [36]. Thus, the knock-out of one nucleon in the nucleus by the out going p or Λ can shift their energies considerably. The calculations in Ref. [35] indeed reproduce the observed mass shifts in the FINUDA experiments. However, these calculations are the computer simulations of the inter-nuclear cascade model for the nucleon-nucleus scattering. This approach describes the sequence of nucleon-nucleon collisions of the outgoing nucleon while passing through the residual nucleus in the final state in the framework of classical physics. The trajectory of each nucleon is followed. After a mean free path a N-N collision takes place and its results are computed by Monte Carlo or some similar method. Apart from the Pauli principle there are no quantum mechanical effects in this approach of describing the FSI. The nucleus too is described by the Fermi gas model, thus being totally devoid of any nuclear structure effect. In view of the crucial role played by the FSI in interpreting the FINUDA Λ-p measurements for K^--nuclear bound states, it is absolutely necessary that the FSI in this reaction is described using quantum mechanical scattering theory and nuclear wave functions. Such a calculation has been done by us recently [37]. These calculations, while confirm the mass shift due to single nucleon knock-out channel, they also predict another peak in the $p\Lambda$ invariant mass distribution due to elastic channel in the FSI. The latter peak is not seen in the FINUDA measurements. Briefly, our calculations go as follows.

The K^- meson capture by the nucleus is considered to occur from the 3d and 2p atomic orbits. Capture from the 1s orbit is insignificant. The capture itself is assumed to occur on two nucleons in the nucleus as $K^-NN \rightarrow \Lambda N$. For the measurements involving p-Λ detection in coincidence, the quasi-free process $K^-N \rightarrow \Lambda\pi$ does not contribute. The absorption on more than two nucleons is also expected to be weak because the probability of finding three or more nucleons together in the nucleus is small. Prescription to describe the absorption vertex is not clear and also not simple. One thing, which is definite about it is that, it involves large momentum transfer, hence, spatially it can not be localized over any extended volume.

Therefore, in our calculations we disentangle it from the shell model dynamics of the absorbing nucleons in the nucleus. We, thus factorize the absorption probability as

$$\omega_{abs}(\vec{p_p}, \vec{p_\Lambda}) = g_{abs}(q)\, G(Q), \tag{10}$$

where $\vec{q} = (m_p \vec{p_\Lambda} - m_\Lambda \vec{p_p})/(m_p + m_\Lambda)$ and $\vec{Q} = \vec{p_p} + \vec{p_\Lambda}$ are the centre-of-mass and the total momenta of the p and Λ respectively. g and G are respectively the absorption strength for $K^- pp \to p\Lambda$ process in their centre of mass and the momentum probability distribution of nuclear wave functions corresponding to the total momentum, Q. Because of the back-to-back emission of p and Λ, obviously the magnitude of q is very large and that of Q is small. Due to these vastly different momentum scales for q and Q, over most of the variables measured in the $(K^-, p\Lambda)$ reaction, while $G(Q)$ can go through a large variation, the factor $g(q)$ should not change much.

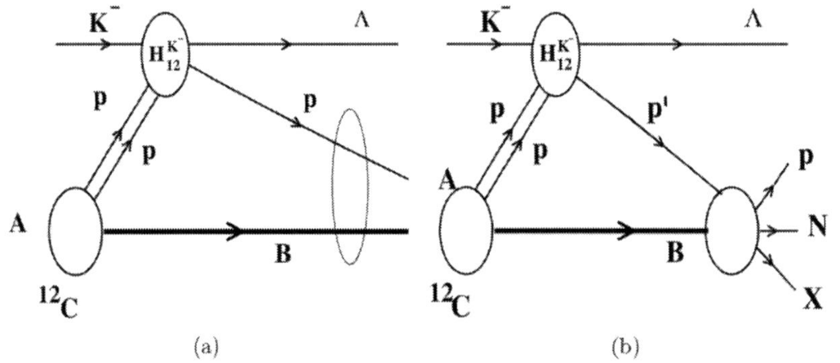

Figure 4. Reaction mechanism of (a) Elastic process and (b) Knock-out process.

For the final state interaction (FSI) of p and Λ with the nucleus we consider the diagrams (a) and (b) in Fig. 4. They represent respectively the elastic scattering and the single-nucleon knock-out on the recoiling nucleus. The inclusive probability for a process like

$$K^- + A \to p + \Lambda + X, \tag{11}$$

is therefore written as

$$d\omega = d\omega_{elas} + d\omega_{KO}. \tag{12}$$

The sum of two terms in the above is incoherent because in principle (by making exclusive measurements) we can distinguish between elastic scattering and single scattering with one target nucleon being knocked out. We calculate the contribution of both these diagrams using fully quantum mechanical scattering theory.

6.1. Elastic Scattering

For the absorption of K^- on a pair of nucleons in the shell model orbitals $n_1 l_1, n_2 l_2$, $G(Q)$ for the elastic channel comes out, (for details see [37])

$$G(Q) = \frac{[PS]\, N_{l_1 l_2}}{(2l_1+1)(2l_2+1)} \times \sum_{LM} \left| \int d\vec{r_1}\, d\vec{r_2}\, \chi^{-*}(\vec{r_1}, \vec{r_2})\, \phi_K(\vec{r_1})\, \phi_{LM}(\vec{r_1}, \vec{r_2})\, \delta(\vec{r_1} - \vec{r_2}) \right|^2, \tag{13}$$

where $\chi^{-*}(\vec{r_1}, \vec{r_2})$ describes the elastically scattered waves in the final state. $\phi_{LM}(\vec{r_1}, \vec{r_2})$ is the bound state wave function of the absorbing nucleons in the target nucleus.

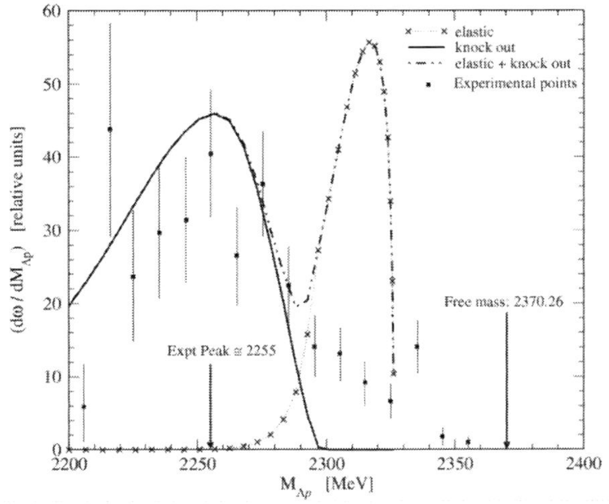

Figure 5. Invariant $p\Lambda$ mass distribution from both the vertices along with the experimental points [19].

6.2. Single Nucleon Knock-out

The knock-out contribution for a nucleon in the shell "nl" is written as

$$d\omega_{KO} = f(\vec{p_p'})\,\frac{\sigma_{KO}(\vec{p_p},\vec{p_p'})}{\sigma_T^{pB}(p_p')}, \tag{14}$$

where $\sigma_T^{pB}(p_p')$ is the total proton-nucleus cross section at the momentum $|\vec{p_p'}|$. Momentum $|\vec{p_p'}|$ is given in terms of the variables at the first vertex. Function $f(\vec{p_p'})$, which describes the distribution of this momentum is also determined from the dynamics at the first vertex. The inclusive knock-out cross section is obtained using the proper scattering theory. It is given by

$$d\sigma_{KO} = d\vec{p_p}\left[\frac{2}{qp_p'}\right]\left[\frac{d\sigma}{d\Omega}(\epsilon)\right]_{cm}^{pN}\left[S(l)\int_{Q_R^{min}}^{\infty} Q_R\, dQ_R\left|\frac{1}{(2\pi)^{3/2}}\,\phi_{nl}(Q_R)\right|^2\right]. \tag{15}$$

Here Q_R^{min} is that minimum momentum which a nucleon of binding energy B_N must have in the nucleus for the scattered proton to be observed at a scattering angle θ with a momentum $|\vec{p_p}|$. Its value is given by

$$Q_R^{min} = \frac{[p_p^2 - p_p' p_p \cos(\theta) + m B_N]}{[p_{p'}^2 - p_p' p_p \cos(\theta) + m B_N]^{1/2}}. \tag{16}$$

$[\frac{d\sigma}{d\Omega}(\epsilon)]_{cm}^{pN}$ is the pN elastic scattering cross section in its c.m. at an appropriate c.m. energy. $\phi_{nl}(Q_R)$ is the single particle wave function in the nucleus.

Using this formalism in Fig. 5 we show the calculated $p\Lambda$ invariant mass distribution along with the FINUDA measured distribution. In calculated distribution we see two peaks. The one around 2320 MeV is due to elastic scattering, while the one at 2250 MeV is due to the knock-out FSI. The latter peak agrees both in position and width with the measured distribution. The position of the K^-pp mass in the free state (2370 MeV) is indicated by the arrow in the figure.

7. Conclusions

In an overall summary, to the K^-pp cluster interpretation of the observed downshift of about 100 MeV in the FINUDA measurements of the Λp invariant mass compared to its free value, the knock-out reaction in the final state seems to be a definitive alternative for this shift. The only discomfiture in this conclusion comes from the absence of the elastic scattering peak in the observed invariant mass distribution in the FINUDA measurements.

Existence of two pole structure in the $\bar{K}N$ scattering dynamics in the χPT studies is also not fully accepted. Hence the value of the $\Lambda(1405)$ binding energy is also not settled.

A way forward to make further unambiguous progress in this subject could be to have experimental and theoretical studies on the following topics:

(i) Study of the $pp \to K^+(K^-pp) \to K^+(p\Lambda)$ reaction. As the K^+N interaction is weak, this reaction can give a direct handle on the study of the $K^-pp \to p\Lambda$ process. This might help in settling the binding energy of the "K^-pp" module.

(ii) Study of the $^3He(K^-, n)p\Lambda$ reaction. This may avoid the single nucleon knock-out FSI, as it might proceed as

$$K^- +^3 He(ppn) \to n + [K^-pp] \to n + (p\Lambda).$$

(iii) Availability of absolute magnitudes of the FINUDA measurements should be very useful.

References

[1] A. M. Green and S. Beech, Phys. Rev. C 71, 014001 (2005); A. Fix and H. Arenhovel, Nucl. Phys. A 697, 277 (2002); Eur. Phys. J. A. 19, 275 (2004).
[2] R. S. Bhalerao and L. C. Liu, Phys. Rev. Lett. 54, 865 (1985).
[3] Q. Haider and L. C. Liu, Phys. Lett. B 172, 257 (1986); Phys. Rev. C 66, 045208 (2002); J. Phys. G 37, 125104 (2010); L. C. Liu and Q. Haider, Phys. Rev. C 34, 1845 (1986).
[4] K. Tsushima et al., Phys. Lett. B 443, 26 (1998); S. Bass and A. W. Thomas, Phys. Lett. B 634, 368 (2006).
[5] H. Garcilazo, Phys. Rev. C 71, 048201 (2005).
[6] S. Wycech and A. M. Green, Phys. Rev. C 64, 045206 (2001); S. Wycech, Acta Phys. Polon. B 41, 2201 (2010); S. Wycech et al., Phys. Rev. C 52, 544 (1995); A. Sibirtsev et al., Eur. Phys. J. A 22, 495 (2004); J. A. Niskanen et al., Int. J. Mod. Phys. A 20, 634 (2005).
[7] S. A. Rakityansky et al., Phys. Lett. B 359, 33 (1995); S. A. Rakityansky et al., Phys. Rev. C 53, R2043 (1996); S. A. Rakityansky et al., Chin. J. Phys. 34, 998 (1996).
[8] B. Mayer et al. Phys. Rev. C 53, 2068 (1996): J. Berger et al. Phys. Lett. 61, 919 (1988).
[9] C. Wilkin, Phys. Rev. C 47, R938 (1993); G. Faldt and C. Wilkin, Nucl. Phys. A587, 769 (1995).
[10] K. P. Khemchandani, N. G. Kelkar and B. K. Jain, Nucl. Phys. A 708, 312 (2002); N. J. Upadhyay, K. P. Khemchandani, B. K. Jain and N. G. Kelkar, Phys. Rev. C 75, 054002 (2007); *ibid*, Mod. Phys. Lett. A 24, 2319 (2009); N. J. Upadhyay, N. G. Kelkar and B. K. Jain, Nucl. Phys. A 824, 70 (2009); K. P. Khemchandani, N. G. Kelkar and B. K. Jain, Phys. Rev. C 68, 064610 (2003); *ibid* Phys. Rev. C 76, 069801 (2007).
[11] M. Pfeiffer et al., Phys. Rev. Lett. 92, 252001 (2004)
[12] A. Budzanowski *et al.*, Phys. Rev. C 79, 012201 (2009).
[13] A. Budzanowski *et al.*, Phys. Rev. C 82, 041001 (2010).
[14] B. K. Jain, N. J. Upadhyay, K. P. Khemchandani and N. G. Kelkar, Int. J. Mod. Phys. E 18, 1271 (2009).
[15] E. P. Wigner, Phys. Rev. 98, 145 (1955).
[16] N. G. Kelkar et al., Nucl. Phys. A730, 121 (2004).
[17] N. G. Kelkar, Phys. Rev. Lett. 99, 210403 (2007).
[18] N. G. Kelkar, K. P. Khemchandani and B. K. Jain, J. Phys. G 32, 1157 (2006); *ibid*, L19 (2006).
[19] M. Agnello *et al.*, Phys. Rev. Lett. **94**, 212303 (2005).
[20] H. Fujioka *et al.*, Nucl. Phys. **A827**, 303c (2009).
[21] T. Yamazaki *et al.*, Proceedings of the International Conference on Exotic Atoms and Related Topics and International Conference on Low Energy Antiproton Physics (EXA/LEAP 2008), Vienna, Austria, 2008; arXiv:0810.5182v1.
[22] N. Kaiser, P. B. Siegel, and W. Weise, Nucl. Phys. **A594**, 325 (1995).
[23] E. Oset and A. Ramos, Nucl. Phys. **A635**, 99 (1998); J. A. Oller and U. G. Meißner, Phys. Lett. **B500**, 263 (2001); M. F. M. Lutz and E. E. Kolomeitsev, Nucl. Phys. **A700**, 193 (2002).
[24] Y. Akaishi and T. Yamazaki, Phys. Rev. C **65**, 044005 (2002).
[25] A. Dote, T. Hyodo, and W. Weise, Nucl. Phys. **A804**, 197 (2008); Phys. Rev. C **79** (2009).
[26] Mohammad Shoeb and B. K. Jain, DAE Symposium on Nuclear Physics, Roorkee, 53, 561 (2008).
[27] S. Wycech and A. M. Green, Phys. Rev. C 79, 014001 (2009).

[28] N. V. Shevchenko, A. Gal, and J. Mares, Phys. Rev. Lett. **98**, 082301 (2007); J. Revai, Phys. Rev. C **76**, 044004 (2007).
[29] I. E. Lagaris and V. R. Pandharipande, Nucl. Phys. **A359**, 331 (1981).
[30] T. Hyodo and W. Weise, Phys. Rev C **77**, 035204 (2008).
[31] I. Zychor *et al.*, Phys. Lett. **B660**, 167 (2008).
[32] E. Oset *et al.*, Phys. Lett. **B527**, 99 (2002).
[33] T. Hyodo *et al.*, Phys. Rev. **C68**, 018201 (2003).
[34] B. Borasoy *et al.*, Eur. Phys. Journ. **A25**, 79 (2005).
[35] V. K. Magas, E. Oset, A. Ramos, and H. Toki, Phys. Rev. C **74**, 025206 (2006).
[36] N. S. Wall and P. R. Roos, Phys. Rev. C **150**, 811 (1966).
[37] G. M. Pandejee, N. J. Upadhyay, and B. K. Jain, Phys. Rev. C **82**, 034608 (2010).

In search of baryonium

S.Wycech
National Centre for Nuclear Studies, Warsaw, Poland
E-mail: wycech@fuw.edu.pl

J.P. Dedonder
Laboratoire de Physique Nucléaire et de Hautes Énergies, Groupe Théorie, IN2P3-CNRS, Universités Pierre & Marie Curie et Paris-Diderot, France
E-mail: Jean-pierre.dedonder@univ-paris-diderot.fr

B. Loiseau
Laboratoire de Physique Nucléaire et de Hautes Énergies, Groupe Théorie, IN2P3-CNRS, Universités Pierre & Marie Curie et Paris-Diderot, France
E-mail: loiseau@lpnhe.in2p3.fr

Abstract. Recent experimental hints of the existence of nucleon-antinucleon quasi-bound states are presented and discussed. One, very broad and weakly bound S wave state is indicated in two decay modes of the J/Ψ meson. It finds additional support in the level widths measured in light antiprotonic atoms. Another, a narrow P-wave state, is found instrumental in the understanding of several results obtained in heavy antiprotonic atoms. Both states have a natural explanation in a fairly traditional Paris potential model of the $N\bar{N}$ interactions.

1. Introduction

The concept of baryonium - a state coupled strongly to a nucleon antinucleon system - emerged when it was realized that the attraction due to pion exchange may be particularly strong in this system. The special role of the pion is due to the G-parity transformation which changes the sign of the pion mediated interaction in $N\bar{N}$ systems as compared to NN systems [1]. There is no Pauli principle in the $N\bar{N}$ system and strong attraction may exist in the states which are not allowed in the NN systems. In some of such states there arises strong coherence of pion and heavier meson exchange forces. Initial expectations indicated possibilities of very deeply bound states. Later on baryonium was also understood as any meson system coupled to $N\bar{N}$.

Experimental attempts to find these exotic objects started some 30 years ago. However, there has been two obstacles in the way. First, the detection in scattering or formation experiments is difficult as the measurements involve large number of partial waves. Second, the annihilation process is very fast and if such bound states exists they are expected to be very broad. No evidence for such states exists as the result of a long series of CERN experiments performed in the last century. Let us mention two results which offer definite limits. A measurement [2] of $p\bar{p} \to \pi + X$ performed at 1.3 GeV/c antiproton momentum found no signal of baryonium in the 1.6-2 GeV energy range. The possibility of a narrow 4 MeV wide baryonium is limited to

8 μb/sr cross section. This work contains references to earlier experiments claiming discoveries in particular findings of baryonia in the radiative transitions. The latest of these radiative transition studies involving antiprotons stopped in hydrogen performed at LEAR [3] excludes baryonium of widths $\Gamma < 25$ MeV and energies $E < 1770$ MeV.

Apparently, broad baryonia close to the $N\bar{N}$ threshold or states weakly coupled to $N\bar{N}$ have not been excluded. To test these possibilities new experiments are required which select specific partial waves. Recently, such selectivity, has to a certain extent become accessible due to the fine structure resolution of antiprotonic atomic levels. It has also been obtained by the BES Collaboration [4, 5] in the studies of radiative J/ψ decays

$$J/\psi \to \gamma p\bar{p}, \tag{1}$$

$$\text{and } J/\psi \to \gamma\pi^+\pi^-\eta' \; . \tag{2}$$

Reaction (1) indicates an enhancement at the $p\bar{p}$ threshold, which may be related to a strong attraction or to a quasi-bound state. The allowed $p\bar{p}$ states are restricted by CP conservation to three partial waves. An analysis [6] performed in terms of Paris $N\bar{N}$ potential model [7] reaches consistency of the data with a 52 MeV wide $^{11}S_0$ quasi-bound state located at 4.8 MeV below threshold. On the other hand, there is no quasi-bound state in the Jülich potential model [8] which also explains the data. To resolve the difference, one has to look below the $N\bar{N}$ threshold. Two ways are open nowadays: the selective mesonic decay mode (2) and non-selective widths of the $\bar{p}$ atomic levels. The latter are determined by the absorptive $N\bar{N}$ amplitudes at negative energies and indicate absorption increasing with the increasing nucleon binding energies. It is consistent with the $^{11}S_0$ quasi-bound state. The mesonic decay mode shows a distinct peak at 40 MeV below threshold, the X(1835), which apparently is also related to the $^{11}S_0$ quasi-bound state. The actual shape of the X(1835) may be roughly reproduced by an interference of the $^{11}S_0$ state with a background amplitude [9].

The exotic states indicated by the BES and the atomic results find a natural explanation in a fairly traditional Paris model of the $N\bar{N}$ interactions based on dispersion relations, G-parity transformation, correlated two-pion exchanges and semi-phenomenological absorptive and short-range potentials.

The text that follows is divided into four sections:

• Section II, the J/Ψ evidence, examines and discusses reactions (1) and (2).

• In section III, we present the atomic evidence which contains indications of a broad S-wave state, and a narrow P-wave one.

• Section IV summarizes the description in terms of the Paris potential.

• Some perspectives are outlined in a brief section V."

At this moment, the experimental and model indications suggest existence of two quasi-bound states: a broad $^{11}S_0$ state, and a narrow $^{33}P_1$ one.

2. The J/Ψ evidence

The measurements by the BES collaboration [4] of the J/ψ $\to$ $\gamma p\bar{p}$ decays indicated an enhancement at $p\bar{p}$ threshold which however is not present in the J/ψ decays into $\pi p\bar{p}$. The explanation is due in part to the CP invariance and in part to the properties of the final state interactions in the $p\bar{p}$ system. First the CP invariance limits the radiative transitions to three possible final $p\bar{p}$ states. These are given in table 1.

Models of final state nucleon-antinucleon interaction allow to pinpoint the proper final state. The standard way of reaction parametrization in multichannel systems is a K-matrix which

Table 1. Low partial wave states of $p\bar{p}$ allowed in the $J/\psi \to \gamma p\bar{p}$ and $J/\psi \to \pi^0 p\bar{p}$ decays. The first column gives decay modes to the specified internal state of the $p\bar{p}$ pair. Spectroscopic notation $^{2T+1\ 2S+1}L_J$ is used with isospin T, spin S, angular momenta L and total spin J. The notation $^{2S+1}L_J$ implies all possible isospin components contribute. Two particle analogs are indicated in the second column [10]. The third column gives J^{PC} for the light spectator particles, photons or π mesons. The fourth column gives J^{PC} for the internal $p\bar{p}$ system, the last column gives the relative angular momentum of the light particle vs. the pair. $J^{PC} = 1^{(--)}$ for J/ψ.

decay mode	analogue	$J^{PC}[\gamma \text{ or } \pi^0]$	$J^{PC}[p\bar{p}]$	relative l
$\gamma p\bar{p}(^1S_0)$	$\gamma\eta(1405)$	1^{--}	0^{-+}	1
$\gamma p\bar{p}(^3P_0)$	$\gamma f_0(1710)$	1^{--}	0^{++}	0
$\gamma p\bar{p}(^3P_1)$	$\gamma f_1(1285)$	1^{--}	1^{++}	0
$\pi^0 p\bar{p}(^{31}P_1)$		0^{-+}	1^{+-}	0
$\pi^0 p\bar{p}(^{33}S_1)$	$\pi^0\rho$	0^{-+}	1^{--}	1

guarantees unitarity of the description. The transition amplitude from a channel i to a two-body channel f may be presented in the form

$$T_{if} = \frac{A_{if}}{1 + iqA_{ff}} \tag{3}$$

where A_{if} is a transition length, A_{ff} is the scattering length in the channel f, and q is the momentum in this channel (see e.g. [11]). Both lengths can be expressed in terms of energy dependent K matrix elements. The same formalism describes the scattering amplitude in the channel f as

$$T_{ff} = \frac{A_{ff}}{1 + iqA_{ff}}. \tag{4}$$

In the decay of interest the final state interaction is assumed to occur in the proton-antiproton state f which emerges after the boson emission. For slow $p\bar{p}$ pairs the final state interactions in the $\pi^0 p\bar{p}$ and $\gamma p\bar{p}$ systems are dominated by interactions in the $p\bar{p}$ sub-system. The formation amplitude A_{if} is unknown, but A_{ff} is calculable in terms of $N\bar{N}$ interaction models constrained by other experiments. The results obtained with a plausible form $A_{if} = C/(1 + (r_o q)^2)^2$, where C is a constant and $r_o = .55$ fm is a best fit length, are shown in Fig. 1. Only one partial wave the $^1S_0$ is consistent with the data and this result seems fairly independent on the latest versions of the potential. In particular the enhancement generated in both the Paris and the Jülich -Bonn potential models is due to strong attraction in the isospin zero $^{11}S_0$ component of the proton-antiproton wave function. In this partial wave the Paris potential [7] generates a 52 MeV broad quasi-bound state at 4.8 MeV below threshold. A similar conclusion has been reached by the Jülich group although the Bonn-Jülich potential does not generate a bound state in the $p\bar{p}$ $^{11}S_0$ partial wave [8].

To understand better the nature of the $p\bar{p}$ states involved,one should look directly into the subthreshold energy region. This may be achieved in the antiproton-deuteron or the antiproton-helium atoms or very low energy scattering. Another way is the detection of $N\bar{N}$ decay products. The decay mode (2) has been studied by the BES collaboration [5]. This reaction is attributed by BES to an intermediate $p\bar{p}$ configuration in the $J^{PC}(p\bar{p}) = 0^{-+}$ state which corresponds to spin singlet S wave. A peak in the invariant mass of the mesons is observed. It has been interpreted as a new baryon state and named X(1835).

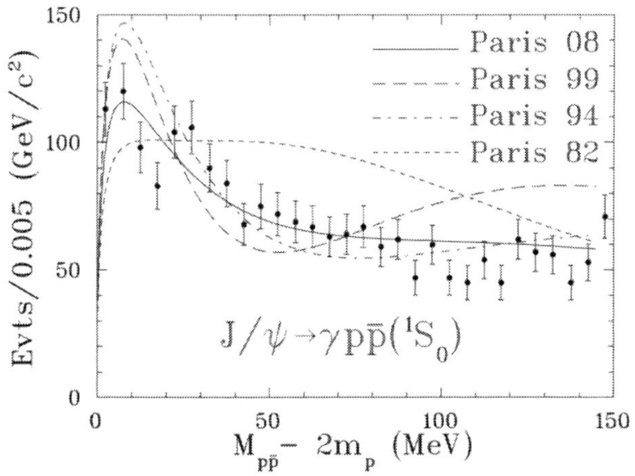

Figure 1. The $\gamma p\bar{p}(^1S_0)$ decays. The final state factor $q \mid T_{if} \mid^2$. The latest Paris model [7] offers the best fit to the BES data with an $^{11}S_0$ wave involving a 4.8 MeV quasi-bound state of 52 MeV width.

Under the assumption that all mesons are produced in relative S waves, reaction (2), if attributed to an intermediate $p\bar{p}$ is even more restrictive than reaction (1). It allows only one intermediate state the $p\bar{p}\ ^1S_0$, which coincides with the previous findings. The intermediate state of $p\bar{p}$ in reaction (2) is possible but not warranted. In Ref. [9] a more consistent interpretation is obtained with the dominance of the $^{11}S_0$ state which is a mixture of $p\bar{p}$ and $n\bar{n}$ pairs. It has been argued there that the peak is due to an interference of the quasi-bound, isospin 0, $N\bar{N}$ state with a background amplitude. The results are plotted in Fig. 2. Both Figs. 2 and 1 indicate that the agreement is semi-quantitative and a more involved decay model is necessary to clarify the understanding.

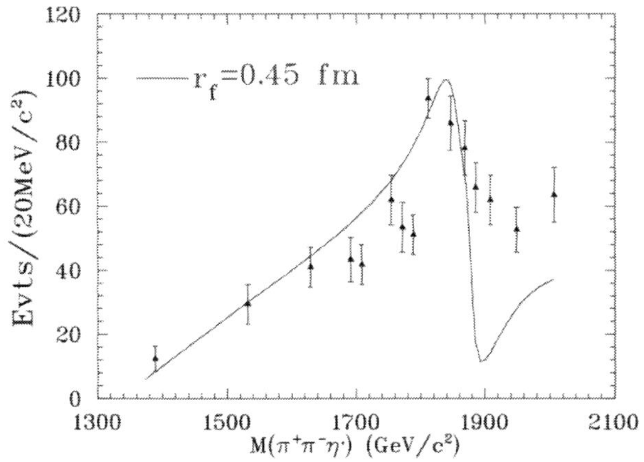

Figure 2. The distribution of the invariant meson mass from reaction (2) measured by BES. Calculations are performed with Paris potential and the curve is due to interference od direct decay and an intermediate $N\bar{N}$ interaction amplitude. The latter is dominated by the $^{11}S_0$ quasi-bound state, r_f is the size of the initial $N\bar{N}$ source.

3. The evidence from antiprotonic atoms

Antiprotonic atom X ray measurements may provide shifts and widths of atomic levels which are generated by strong interactions (see review [12]). In the lightest atoms Hydrogen and Helium experiments test the $1S$, $2P$ and $3D$ levels. The widths and shifts are related to low energy antiproton-nucleon scattering amplitudes

$$f(E) = a + 3\, b\, \mathbf{q} \cdot \mathbf{q}' \tag{5}$$

where a are scattering lengths and b are scattering volumes. The atomic physics opens a method to test the subthreshold $N\bar{N}$ energy region as both nucleons and antiprotons are bound. The scattering amplitudes relevant to atomic states $f(-E_{binding} - E_{recoil})$ involve binding as well as recoil energies. In light atoms the accessible energies span the region from - few keV (Hydrogen) to about - 40 MeV (Helium). As fine structure is not fully resolved one can extract only spin and isospin averaged absorptive parts of a and b. The results are given in Fig. 3.

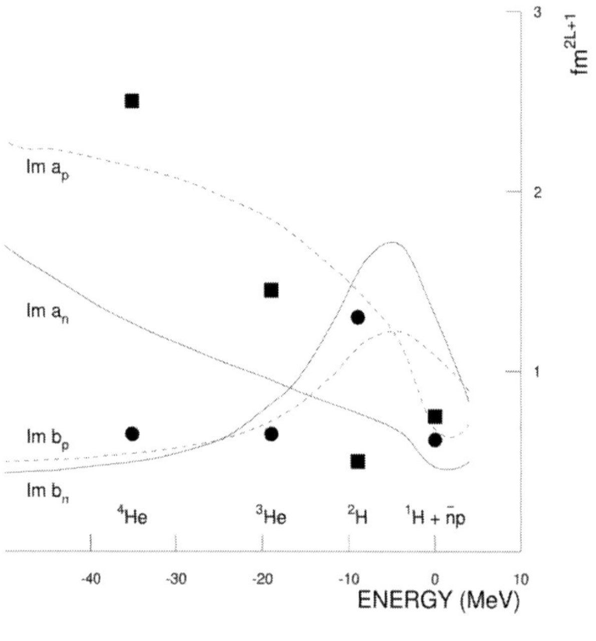

Figure 3. The spin and isospin averaged, $N\bar{p}$ amplitudes in the sub-threshold region extracted from light antiprotonic atoms : circles- absorptive parts of length, squares - absorptive parts of volumes. The curves present Im a, Im b calculated with the Paris potential.

The energy dependence extracted in this way indicates a sizable increase of the S wave absorption in the subthreshold region consistent with the existence of an S wave $N\bar{N}$ quasi-bound state [13]. In addition, the deuteron atomic level widths indicate enhanced absorption in a P wave which occurs at energies very close to the threshold. The result obtained in deuteron coincides with certain anomalies observed in antiproton annihilation on a single nucleon studied in heavy nuclei by radiochemical methods (so called cold capture). The method discussed in Ref. [14] is devoted to studies of the neutron distribution at the surface of nuclei. One can test the frequency of antiproton capture on neutrons relative to the frequency of capture on protons and in this way infer the ratio of neutron to proton densities at the site of capture. The capture site is different in two atomic levels (a lower one and an upper one) tested by X-ray method. The site tested by cold capture is much more peripheral. Nuclear densities involved are $\sim 10\%$, $\sim 3\%$ and $\sim 0.1\%$ of central density correspondingly. Nuclear physics expects the neutron/proton ratio to increase along the way out of the nucleus. This behavior is observed in almost a hundred of tested nuclei. Three examples are given in the upper part of table 2. However, there are anomalous cases indicated in lower part of the table. These are characterized by a low ~ 7 MeV proton separation energies and larger ~ 11 MeV neutron separation energies. In the cold capture one tests essentially the valence nucleons and these respond differently to the narrow resonance. Thus the anomalous cases are not related to any "proton halos" in these nuclei but to the different chances to form a narrow $N\bar{N}$ state very close to the threshold.

Table 2. Ratios of $(n\bar{p})$ and $(p\bar{p})$ capture rates from atomic states. The last column shows experimental numbers from radiochemical experiments. Other columns give ratios calculated with optical potential and plausible nuclear densities [13] based on experimental results from [14].

atom	lower	upper	radiochemistry
^{96}Zr	0.95(9)	1.53(29)	2.6(3)
^{116}Cd	1.63(49)	2.65(61)	5.00(21)
^{124}Sn	1.79(10)	2.44(39)	5.0(6)
^{106}Cd	1.64(80)	2.10(80)	0.5(1)
^{112}Sn	1.90(13)	2.43(49)	0.79(14)

4. The Paris potential model

The 1982 Paris $N\bar{N}$ optical potential was itself readjusted in 1994, 1999 and 2008 [7]. For all these potentials the $N\bar{N}$ interaction is described by an energy dependent optical potential

$$V_{N\bar{N}}(\mathbf{r}, T_{Lab}) = U_{N\bar{N}}(\mathbf{r}, T_{Lab}) - i\, W_{N\bar{N}}(\mathbf{r}, T_{Lab}), \tag{6}$$

where the non-locality of the real $U_{N\bar{N}}$ and imaginary $W_{N\bar{N}}$ potentials are accounted for by a linear energy dependence in the kinetic energy T_{Lab}. Meson exchanges explain in a satisfactory way the NN force for large and medium distances between the nucleons. Therefore the long and intermediate range real parts, i.e., those for inter $N\bar{N}$ distances $r \geq 1$ fm, are obtained by the G-parity transformation of the corresponding parts of the Paris NN potential. These real potentials contain, besides the one-pion exchange, the two-pion exchange and the ω and A_1 meson exchanges as parts of the three-pion exchange. For $r < 1$ fm heavier meson exchanges and/or other degrees of freedom, such as quarks and gluons take place but the available theoretical calculations are not free from phenomenological parameters (see for instance Ref. [12]. Consequently, following the choice made in the case of the Paris NN potential an empirical short range real potential is used.

For $r < 1$ fm the phenomenological radial potentials are expanded in powers of r and matched to the theoretical ones in the vicinity of 1 fm. Then, above 1 fm the theoretical potentials are entirely preserved. For each isospin state, the spin structure of the $N\bar{N}$ interaction requires five independent invariant with five radial potentials. All together it leads to nine parameters representing the strength of the different empirical potentials. These parameters are fitted to about 4000 scattering data and four complex scattering lengths obtained from hydrogen atoms.

The central potentials essential to generate the S state are plotted in Figs. 4 and 5, and the two quasi-bound states given in this model are shown in table 3. The values of the recent model are consistent with the experimental data discussed above.

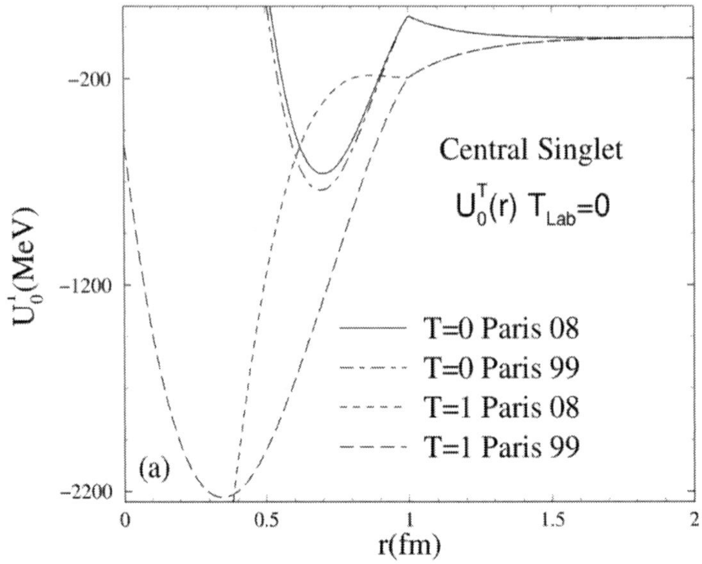

Figure 4. The real part of the central potential which dominates the attraction in the singlet wave.

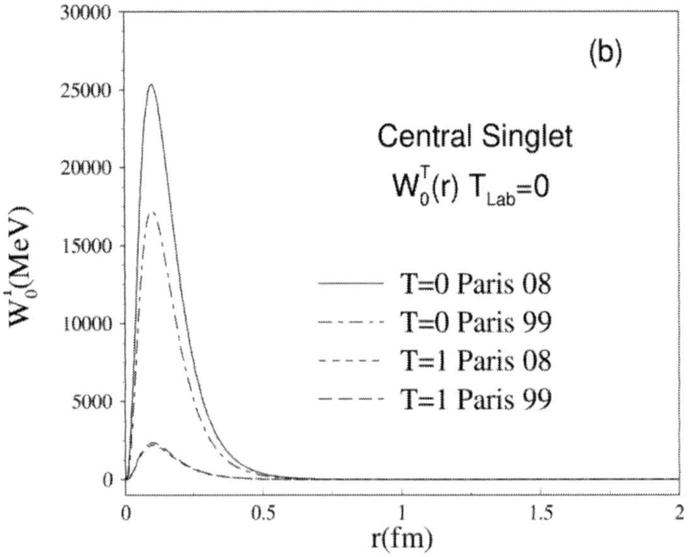

Figure 5. Absorptive part of central potential .

Table 3. Binding energy in MeV of the close to threshold quasi-bound states of the Paris 08 [7] and of the Paris 99 potential [15].

${}^{2T+1\ 2S+1}L_J$	Paris 08	Paris 99
${}^{11}S_0$	-4.8-i26	
${}^{33}P_1$	-4.5-i9.0	-17-i6.5

5. Conclusions

The arguments for the existence of two nucleon-nucleon quasi-bound states were presented. These are based on experimental measurements performed with antiprotonic atoms and J/Ψ decay studies as well as potential interaction models. We believe that a further confirmation is necessary. It may be achieved with several experiments feasible with the new FAIR / PANDA project:

- resolution of the fine structure in lightest antiprotonic atoms, in particular in the antiprotonic deuterium.
- repetition of the CERN $(\bar{p}, \pi)$ and $(\bar{p}, \gamma)$ experiments performed in flight with polarized antiproton and polarized targets.
- repetition of the CERN studies of the atomic radiative transition from the atomic to baryonium states.

The quasi-bound nuclear states of antiproton constitute a separate branch of this research. In general such states are too broad to be detected easily. However, certain states of antiprotons bound loosely in the region of nuclear surface are likely to exist.

Acknowledgements

Part of this research was performed in the framework of the IN2P3-Polish Laboratory Convention (collaboration N° 05-115). Financial support from Hadron 3 - LEANNIS project is acknowledged.

References

[1] R.A. Bryan and R.J. Phillips, Nucleon-antinucleon potential from single meson exchanges, Nucl. Phys. **B 5** (1968) 201; Nucl. Phys. **B 7** (1968) 481 (erratum).
[2] R. Bertini *et al*, Search for narrow baryonium states near the $N - \bar{N}$ threshold, Nucl Phys **B 209**(1982) 269.
[3] L. Adiels *et al* , Search for narrow signals in the γ-spectrum from $p\bar{p}$ annihilation at rest, Phys Lett **B 182** (1982) 405.
[4] J.Z.Bai for BES Collaboration. Observation of a Near-Threshold Enhancement in the $p\bar{p}$ Mass Spectrum from Radiative $J/\psi$to $\gamma p\bar{p}$ Decays , Phys.Rev.Lett. **91**(2003) 02200.
[5] M. Ablikim for BES Collaboration, Observation of a resonance X(1835) in $J/\psi \to \gamma\pi^+\pi^-\eta'$. Phys.Rev.Lett. **95** (2005) 262001.
[6] B. Loiseau and S. Wycech, Antiproton-Proton Channels in J/ψ Decays., Phys.Rev. C **72** (2005) 011001.
[7] M. Lacombe B. El-Bennich, B. Loiseau and S.Wycech, The Paris $N\bar{N}$ potential constrained by recent antiprotonic-atom data and $\bar{n}p$ total cross section, Phys. Rev. C **79** (2009) 054001.
[8] J. Haidenbauer, Ulf-G. Meißner and A. Sibirtsev, Near threshold $p\bar{p}$ enhancement in B and J/ψ decay. Phys.Rev.D **74** (2006) 017501.
[9] J-P. Dedonder, B. Loiseau, B. El-Bennich and S. Wycech, Structure of X(1835) baryonium, Phys.Rev. C **80** (2009) 0145207.
[10] K. Nakamura for Particle Data Group, J.Phys. **G 37** (2010) 07021.
[11] H. Pilkuhn, *The Interactions of Hadrons*, North Holland P.C., 1967, p 19.
[12] E. Klempt, F. Bradamante, A. Martin and J.-M. Richard, Antinucleon-nucleon interaction at low energy: scattering and protonium, Phys. Rep. **368**, 119 (2002).
[13] S. Wycech, F.J. Hartmann, J. Jastrzebski, B. Klos, A. Trzcinska, and T. von Egidy , Nuclear surface studies with antiprotonic atom X-rays, Phys. Rev.C **76** (2007) 034316.
[14] A. Trzcinska *et al.*, Information on antiprotonic atoms and nuclear periphery from the PS 209 experiment, Nucl. Phys. **A 692**, (2001) 176c.
[15] B. El-Bennich, M. Lacombe, B. Loiseau and R. Vinh Mau, Refining the inner core of the Paris $N\bar{N}$ potential, Phys. Rev. C **59** (1999) 2313.

Journal of Physics: Conference Series **374** (2012) 012003 doi:10.1088/1742-6596/374/1/012003

The PANDA Experiment at FAIR

C Schwarz[1] for the PANDA collaboration

[1] GSI Helmholtzzentrum GmbH, Planckstraße 1, 64291 Darmstadt, Germany

E-mail: C.Schwarz@gsi.de

Abstract.

The PANDA experiment represents the central part of the hadron physics branch of the antiproton program of the new Facility for Antiproton and Ion Research (FAIR) at Darmstadt, Germany. It will investigate antiproton annihilations with almost 4π acceptance from two subsequent spectrometers in the momentum range from 1.5 GeV/c to 15 GeV/c. The gluon rich reaction product of antiproton-proton and antiproton-nucleus collisions on one hand and the high intensity and phase space cooled antiproton beam of the HESR on the other hand constitutes an ideal environment to study non-conventional hadronic states. It allows the high precision spectroscopy of the charmonium system and the copious production of hypernuclei.

1. Introduction

The Facility for Antiproton and Ion Research (FAIR) will be an accelerator facility leading the European research in nuclear and hadron physics in the coming decade. It will address a wide range of physics topics in the fields of nuclear structure, nuclear matter, studies using high energy and very slow antiprotons, atomic and plasma physics. Several topics in applied science and accelerator development will be addressed as well. FAIR builds on the experience and technological developments from the existing GSI facility, and incorporates new technological concepts [1, 2].

The existing GSI accelerators will be upgraded by the addition of a proton-linac and used as injectors for the newly built complex of storage rings to form the FAIR facility. An overview of the layout is given in Fig. 1. To the left, the existing GSI facility is shown. New proposed beam lines and accelerator structures are shown in red. This facility will provide intense secondary beams of antiprotons and rare isotopes which will be used for research at the main experimental setups labelled in green.

The PANDA experiment will make use of antiprotons, which will be produced as follows. Protons will be accelerated to 70 MeV in a new linear accelerator and injected into the existing synchrotron SIS18 and accelerated to 2 GeV. After the subsequent transfers to the SIS100 synchrotron, the protons will be accelerated to 29 GeV. Up to 2×10^{13} protons will be compressed into a single short bunch of less than 50 ns length in order to minimize the heating in the antiproton production target. The proton bunch will be directed onto a nickel target of about 60 mm length followed by a magnetic horn. The cycle of proton acceleration will be repeated every 10 s. This scheme is expected to produce a bunch of at least 1×10^8 antiprotons in the phase space volume which can be accepted by the magnetic separator and the Collector Ring (CR). The CR will be used for pre-cooling of the antiprotons. Thereafter, the antiprotons will be moderately compressed to a single bunch and transferred to the RESR storage ring.

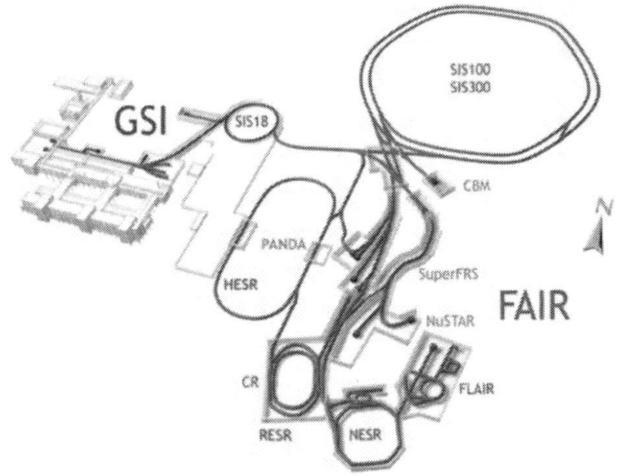

Figure 1. Layout of the FAIR facility. Shown are: synchrotrons with bending powers up to 18, 100 and 300 Tm the SIS 18, 100 and 300, respectively; the Collector Ring CR; the Accumulator Ring RESR; the low and high energy experimental storage rings NESR and HESR, respectively; and the Fragment Separator FRS.

The antiprotons will be accumulated to high intensities in the RESR by a dedicated stochastic cooling system with a rate that matches the speed of cooling in the CR. The antiprotons will then be transferred at a momentum of 3.8 GeV/c to the High Energy Storage Ring (HESR) and decelerated or accelerated to momenta in the range of 1.5 to 15 GeV/c.

2. The Physics Case

The full quantitative understanding of the strong interaction is the subject of hadron physics. New experimental results like the discovery of the new X,Y, and Z resonances have stimulated a very intense theoretical activity and a refinement of theoretical tools. Still there are many fundamental questions to be addressed like the confinement of quarks, the existence of hybrids and glueballs, the origin of the mass of hadrons.

Four major physical research topics will be addressed:

i. High precision spectroscopy of resonances in the energy region of charmonium and above. The exact knowledge about masses, widths, and branching ratios provides information about the mechanisms of quark confinement.

ii. In-medium effects of open and hidden charm and their relationship to the chiral symmetry breaking addressing especially the origin of hadron mass.

iii. So-called glueballs, predicted by QCD, need a firm experimental establishment in our picture of strong interactions. Also hybrid meson states, so-called excited glue, in the mass range between 3 and 5 GeV/c^2 are predicted to be rather narrow without mixing into neighboring resonances.

iv. Hypernuclei, when abundantly produced, and examined with the next generation of γ-detectors, will revive the physics with single and double hypernuclei. This will improve our modest knowledge on their structure and give information on the interaction of hyperons with nucleons as well as hyperon-hyperon interaction.

GSI has a distinguished history of having made important contributions to the physics of strong interaction, in particular nuclear physics. The proposed PANDA experiments play a significant role in strong interaction physics, providing a link between nuclear and hadron physics.

3. The High Energy Storage Ring

An important feature of the new anti-proton facility is the combination of phase-space cooled beams and dense internal targets, comprising challenging beam parameter in two operation modes: high-luminosity mode (HL) with beam intensities up to 10^{11}, and high-resolution mode (HR) with a momentum spread down to a few times 10^{-5}, respectively. Powerful electron and stochastic cooling systems are necessary to meet the experimental requirements.

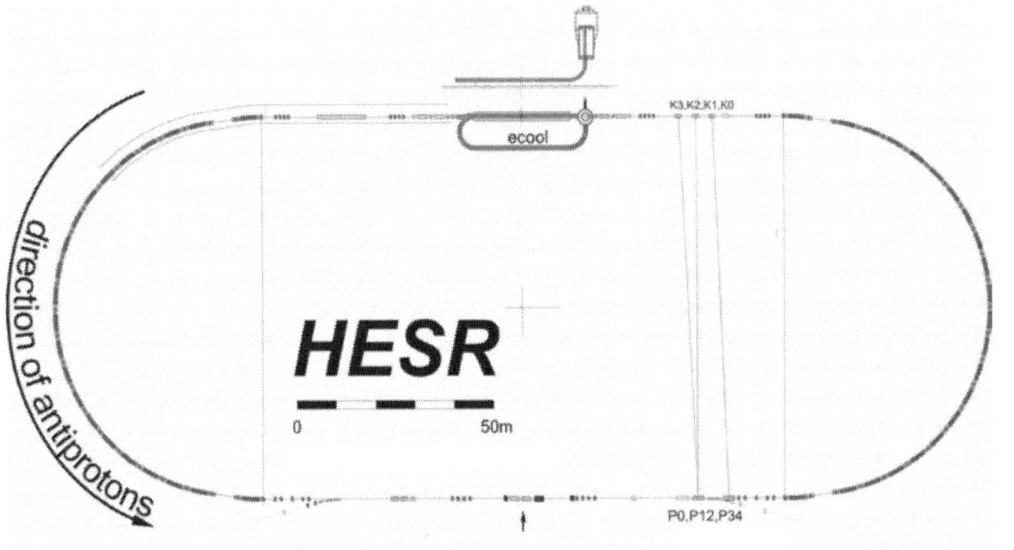

Figure 2. Schematic view of the HESR. Positions for injection, cooling devices and experimental installations are indicated.

The HESR lattice is designed as a racetrack shaped ring, consisting of two 180° arc sections connected by two long straight sections. One straight section will mainly be occupied by the electron cooler. The other section will host the experimental installation with internal H_2 pellet target, RF cavities, injection kickers and septa (see Fig. 2). For stochastic cooling pickup and kicker tanks are also located in the straight sections, opposite to each other. To improve longitudinal stochastic cooling a third pickup location in the arc is presently being investigated. Special requirements for the lattice are dispersion free straight sections and small betatron amplitudes in the range of a few meters at the internal interaction point. In addition the betatron amplitudes at the electron cooler are adjustable within a large range.

Table 1 summarizes the specified injection parameters, experimental requirements and operation modes.

4. The PANDA Experiment

The main objectives of the design of the PANDA experiment are to achieve 4π acceptance, high resolution for tracking, particle identification and calorimetry, high rate capabilities and a versatile readout and event selection. To obtain a good momentum resolution the detector

Injection Parameters	
Transverse emittance	1 $mm \cdot mrad$ (normalized, RMS) for $3.5 \cdot 10^{10}$ particles, scaling with number of accumulated particles: $\epsilon_\perp \sim N^{4/5}$
Relative momentum spread	$1 \cdot 10^{-3}$ (normalized, RMS) for $3.5 \cdot 10^{10}$ particles, scaling with number of accumulated particles: $\sigma_p/p \sim N^{2/5}$
Bunch length	Below 200 m
Injection Momentum	3.8 GeV/c
Injection	Kicker injection using multi-harmonic RF cavities
Experimental Requirements	
Ion species	Antiprotons
$\bar{p}$ production rate	$2 \cdot 10^7$/s ($1.2 \cdot 10^{10}$ per 10 min)
Momentum / Kinetic energy range	1.5 to 15 GeV/c / 0.83 to 14.1 GeV
Number of particles stored in HESR	10^{10} to 10^{11}
Target thickness	$4 \cdot 10^{15}$atoms/cm^2
Transverse emittance	< 1 mm$\cdot$ mrad
Betatron amplitude E-Cooler	25–200 m
Betatron amplitude at IP	1–15 m
Operation Modes	
High resolution (HR)	Luminosity of $2 \cdot 10^{31} cm^{-2} s^{-1}$ for 10^{10} $\bar{p}$ RMS momentum spread $\sigma_p/p \leq 4 \cdot 10^{-5}$, 1.5 to 9 GeV/c, electron cooling up to 9 GeV/c
High luminosity (HL)	Luminosity of $2 \cdot 10^{32}$cm^{-2} s^{-1} for 10^{11} $\bar{p}$ RMS momentum spread $\sigma_p/p \sim 10^{-4}$, 1.5 to 15 GeV/c, stochastic cooling above 3.8 GeV/c

Table 1. Injection parameters, experimental requirements and operation modes.

will be composed of two magnetic spectrometers: the *Target Spectrometer (TS)*, based on a superconducting solenoid magnet surrounding the interaction point, which will be used to measure at large angles and the *Forward Spectrometer (FS)*, based on a dipole magnet, for small angle tracks (Fig. 3). A silicon vertex detector will surround the interaction point. In both spectrometer parts, tracking, charged particle identification, electromagnetic calorimetry and muon identification will be available to allow to detect the complete spectrum of final states relevant for the PANDA physics objectives.

4.1. Target Spectrometer

The Target Spectrometer will surround the interaction point and measure charged tracks in a highly homogeneous (better than ±2%) solenoidal field of 2 T. In the manner of a collider detector it will contain detectors in an onion shell like configuration. Pipes for the injection of target material will have to cross the spectrometer perpendicular to the beam pipe.

The Target Spectrometer will be arranged in three parts: the barrel covering angles between 22° and 140°, the forward end cap extending the angles down to 5° and 10° in the vertical and horizontal planes, respectively, and the backward end cap covering the region between about 145° and 170°. Please refer to Fig. 4 for an overview.

The compact geometry of the detector leaves very restricted space for the target installations. In order to reach the design luminosity of $2 \cdot 10^{32}$ s^{-1}cm^{-2} a target thickness of about $4 \cdot 10^{15}$ hydrogen atoms per cm^2 is required assuming 10^{11} stored anti-protons in the HESR ring. At

Figure 3. The PANDA detector consists of a central and a forward magnetic spectrometer. The figure shows the subdetectors within the solenoid and subdetectors inside and behind the dipole magnet of the forward spectrometer.

present, two different, complementary techniques for the internal target are developed further: the cluster-jet target and the pellet target. Both techniques are capable of providing sufficient densities for hydrogen at the interaction point, but exhibit different properties concerning their effect on the beam quality and the definition of the interaction point. In addition, internal targets also of heavier gases, like deuterium, nitrogen or argon can be made available. For non-gaseous nuclear targets the situation is different in particular in case of the planned hyper-nuclear experiment. In these studies the whole upstream end cap and part of the inner detector geometry will be modified.

The tracking system has to provide efficient reconstruction of charged particle tracks by meeting the following requirements: almost full solid angle coverage around the interaction area; resolving complex pattern of multiple tracks; high spatial resolution for secondary vertices, $\sigma_{r\phi} \sim 150\ \mu$m, $z \sim 1$ mm; high momentum resolution of reconstructed trajectories, $\delta p/p \sim 1$ %; minimal detector material, $X_0 \sim 1$ % for low multiple scattering and energy loss of passing tracks and suppression of secondary background production by interactions with the detector material; high rate capability of the detector and especially of its components in the forward region reaching values of $3 \cdot 10^4$ cm^{-2}s^{-1}, fast electronic readout; resistance against ageing effects with the expected accumulated charge density on the level of 0.1-1 C/cm/year; operation in non-uniform magnetic field of the PANDA dipole magnet varying by up to 0.3 T along the wires of the forward tracking detectors. This will be achieved by surrounding the vertex detector with a cylindrical low-mass device, and placing two sets of Multiwire Drift Chambers (MDCs) in the forward direction.

The design of the micro-vertex detector (MVD) for the target spectrometer is optimized for the detection of secondary vertices from D and hyperon decays and maximum acceptance close to the interaction point. It will also strongly improve the transverse momentum resolution. The

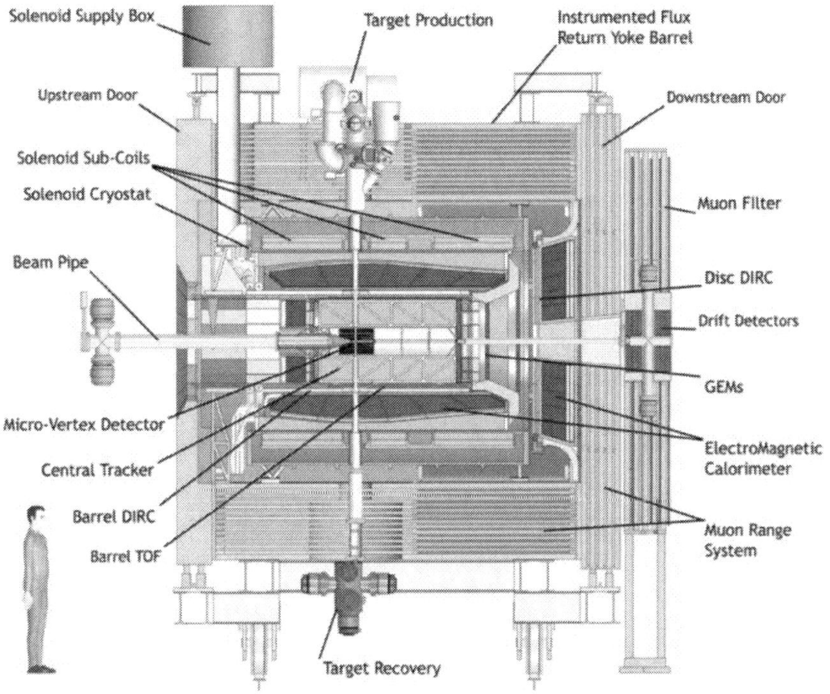

Figure 4. Artistic side view of the Target Spectrometer (TS) of PANDA. To the right of this the Forward Spectrometer (FS) follows, which is illustrated in Fig. 7.

setup is depicted in Fig. 5.

The concept of the MVD is based on radiation hard silicon pixel detectors with fast individual pixel readout circuits and silicon strip detectors. The layout foresees a four layer barrel detector with an inner radius of 2.5 cm and an outer radius of 13 cm. The two innermost layers will consist of pixel detectors while the outer two layers are considered to consist of double sided silicon strip detectors.

Eight detector wheels arranged perpendicular to the beam will achieve the best acceptance for the forward part of the particle spectrum. Here again, the inner two layers are made entirely of pixel detectors, the following four are a combination of strip detectors on the outer radius and pixel detectors closer to the beam pipe. Finally the last two wheels, made entirely of silicon strip detectors, are placed further downstream to achieve a better acceptance of hyperon cascades.

The present design of the pixel detectors comprises detector wafers which are 200 μm thick ($0.25\% X_0$). The readout via bump-bonded wafers with ASICs as it is used in ATLAS and CMS [3, 4] is foreseen as the default solution. It is highly parallelized and allows zero suppression as well as the transfer of analog information at the same time. The readout wafer has a thickness of 300 μm ($0.37\% X_0$). A pixel readout chip based on a 0.13 μm CMOS technology is under development for PANDA. This chip allows smaller pixels, lower power consumption and a continuously sampling readout without external trigger.

The next tracking system, the Straw Tube Tracker (STT), is covering a large area and is designed as a barrel around the MVD. This detector consists of aluminized mylar tubes called *straws*, which are self supporting by the operation at 1 bar overpressure. The straws are arranged in planar layers which are mounted in a hexagonal shape around the MVD. In total there are 24

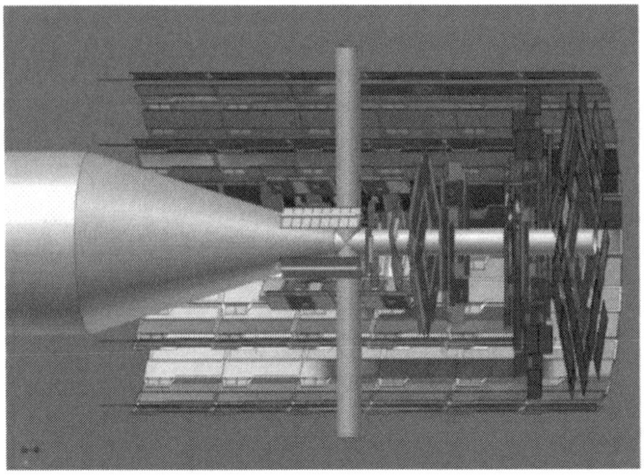

Figure 5. The Microvertex detector of PANDA

layers of which the 8 central ones are tilted to achieve an acceptable resolution of 3 mm also in z (parallel to the beam). The gap to the surrounding detectors is filled with further individual straws. In total there are 4200 straws around the beam pipe at radial distances between 15 cm and 42 cm with an overall length of 150 cm. All straws have a diameter of 10 mm. A thin and light space frame will hold the straws in place, the force of the wire however is kept solely by the straw itself. The mylar foil is 30 μm thick, the wire is made of 20 μm thick gold plated tungsten. This design results in a material budget of 1.3 % of a radiation length.

The gas mixture used will be Argon based with CO_2 as quencher. It is foreseen to have a gas gain no greater than 10^5 in order to warrant long term operation. With these parameters, a resolution in x and y coordinates of about 150 μm is expected.

Particles emitted at angles below 22° which are not covered fully by the Straw Tube Tracker will be tracked by three stations of GEM detectors placed 1.1 m, 1.4 m and 1.9 m downstream of the target.

Charged particle identification of hadrons and leptons over a large range of angles and momenta is an essential requirement for meeting the physics objectives of PANDA. There will be several dedicated systems which, complementary to the other detectors, will provide means to identify particles. The main part of the momentum spectrum above 1 GeV/c will be covered by Cherenkov detectors. Below the Cherenkov threshold of kaons several other processes have to be employed for particle identification. In addition a time-of-flight barrel will identify slow particles.

For slow particles at large polar angles particle identification will be provided by a time-of-flight detector. In the Target Spectrometer the flight path is only in the order of 50 – 100 cm. Therefore the detector must have a very good time resolution of $\approx$ 100 ps.

One of the main concerns with the barrel Time-of-Flight detectors is the influence on the electromagnetic calorimeter (EMC) due to its material causing photon conversions. This effect is reduced to a minimum if one places the detector material as close as possible to the EMC and tries to reduce it to the absolute minimum. Taking all these concerns into account one arrives at a new timing detector concept based on small 5 mm thin plastic scintillator tiles (SciTil) coupled directly to Silicon Photomultipliers (SiPM) with excellent time resolution to read out the scintillation light. A minimum ionizing particle generates approximately 10000 photons across the tile thickness of which about 100 are directly measured by the SiPMs. This allows a threshold well above the intrinsic noise of the SiPMs.

The basic unit of the SciTil detector is the scintillator tile to which two SiPMs are attached on different sides to allow the collection of photons from all locations within the scintillator volume. The size of the tile is 28.5 mm in width and length and 5 mm in thickness. With these dimensions the barrel section will house 5670 SciTil and the endcap section about 2000 SciTil.

At polar angles between 22° and 140°, particle identification will be performed by the detection of internally reflected Cherenkov (DIRC) light as realized in the BaBar detector [5]. It will consist of 1.7 cm thick synthetic fused silica slabs surrounding the beam line at a radial distance of 45 - 54 cm. At BaBar the light was imaged across a large stand-off volume filled with water onto 11 000 photomultiplier tubes. At PANDA, it is intended to focus the images by lenses onto micro-channel plate photomultiplier tubes (MCP PMTs) which are insensitive to magnet fields. This fast light detector type allows a more compact design and the readout of two spatial coordinates and the time of arrival of the photons. The DIRC design with its compact radiator mounted close to the EMC will minimize the conversions.

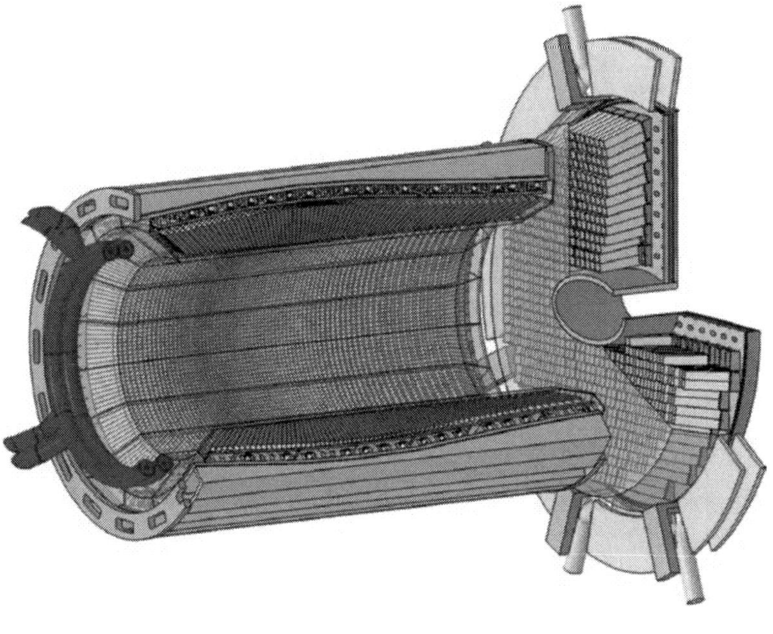

Figure 6. Barrel and forward end cap of the Electro-Magnetic Calorimeter (EMC) with its mounting structures and cooling pipes. These structures will be mounted directly inside the cryostat and forward end of the flux return yoke.

A similar concept can be employed in the forward direction for particles between 5° and 22°. The same radiator, fused silica, is to be employed however in shape of a disk. At the rim around the disk the Cherenkov angle will be measured either directly by focusing elements or indirectly using time-of-propagation (ToP). In the first option focusing will be done by mirroring quartz elements reflecting onto MCP PMTs. Once again two spatial coordinates plus the propagation time for corrections will be read. In the ToP design this angle is reconstructed by using the known path of the photon and its time of propagation. This will be achieved by the use of alternating dichroic mirrors transmitting and reflecting different parts of the light spectrum. The disk will be 2 cm thick and will have a radius of 110 cm. It will be placed directly upstream of the forward end cap calorimeter.

Expected high count rates and a geometrically compact design of the Target Spectrometer require a fast scintillator material with a short radiation length and Molière radius for the construction of the electromagnetic calorimeter (EMC). Lead tungsten ($PbWO_4$) is a high density inorganic scintillator with sufficient energy and time resolution for photon, electron, and hadron detection even at intermediate energies [6, 7, 8].

The crystals will be 20 cm long, i.e. approximately $22\,X_0$, in order to achieve an energy resolution below 2% at 1 GeV at a tolerable energy loss due to longitudinal leakage of the shower. Tapered crystals with a front size of $2.1 \times 2.1\,\text{cm}^2$ will be mounted in the barrel EMC with an inner radius of 57 cm. This implies 11360 crystals for the barrel part of the calorimeter. The forward end cap EMC will be a planar arrangement of 3600 tapered crystals with roughly the same dimensions as in the barrel part, and the backward end cap EMC comprises of 592 crystals. The readout of the crystals will be accomplished by large area avalanche photo diodes in the barrel and vacuum photo-triodes in the forward and backward end caps. The light yield can be increased by a factor of about 4 compared to room temperature by cooling the crystals down to -25° C. The arrangement of the barrel and forward end cap calorimeters is shown in Fig. 6.

The EMC will allow to achieve an e/π ratio of 10^3 for momenta above 0.5 GeV. Therefore, e-π-separation will not require an additional gas Cherenkov detector in favour of a very compact geometry of the EMC. For further details please refer to Ref. [9].

The muons will be identified by a stack of alternating detection and absorption layers. In the barrel region the yoke is segmented and will consist of 13 layers in total: the innermost layer will have a thickness of 6 cm equal to the outermost one, in between 11 layers of 3 cm thickness will be located. The gaps for the detectors will be 3 cm wide. This is enough material for the absorption of pions in the momentum range in PANDA at these angles. In the forward end cap more material is needed. Since the downstream door of the return yoke has to fulfil constraints for space and accessibility, the muon system will split in several layers. Six detection layers will be placed around five iron layers of 6 cm each within the door, and a removable muon filter with additional five layers of 6 cm iron is located in the space between the solenoid and the dipole.

As detector within the absorber layers rectangular aluminum drift tubes will be used as they were constructed for the COMPASS muon detection system. They are essentially drift tubes with additional capacitive coupled strips read out on both ends to obtain the longitudinal coordinate.

The hypernuclei study will make use of the modular structure of PANDA. Removing the backward end cap calorimeter will allow to add a dedicated nuclear target station and the required additional detectors for γ spectroscopy close to the entrance of PANDA. While the detection of anti-hyperons and low momentum K^+ can be ensured by the universal detector and its PID system, a specific target system and a γ-detector are additional components required for the hypernuclear studies.

The production of hypernuclei proceeds as a two-stage process. First hyperons, in particular $\Xi\bar{\Xi}$, are produced on a nuclear target. In some cases the Ξ will be slow enough to be captured in a secondary target, where it reacts in a nucleus to form a double hypernucleus.

The geometry of this secondary target is determined by the short mean life of the Ξ^- of only 0.164 ns. This limits the required thickness of the active secondary target to about 25–30 mm. It will consist of a compact sandwich structure of silicon micro strip detectors and absorbing material. In this way the weak decay cascade of the hypernucleus can be detected in the sandwich structure.

An existing germanium-array with refurbished readout will be used for the γ-spectroscopy of the nuclear decay cascades of hypernuclei. The main limitation will be the load due to neutral or charged particles traversing the germanium detectors. Therefore, readout schemes and tracking algorithms are presently being developed which will enable high resolution γ-spectroscopy in an environment of high particle flux.

4.2. Forward Spectrometer

The Forward Spectrometer (FS) will cover all particles emitted in vertical and horizontal angles below $\pm 5^\circ$ and $\pm 10^\circ$, respectively. Charged particles will be deflected by an integral dipole field of 2 Tm. Cherenkov detectors, calorimeters and muon counters ensure the detection of all particle types. Please refer to Fig. 7 for an overview.

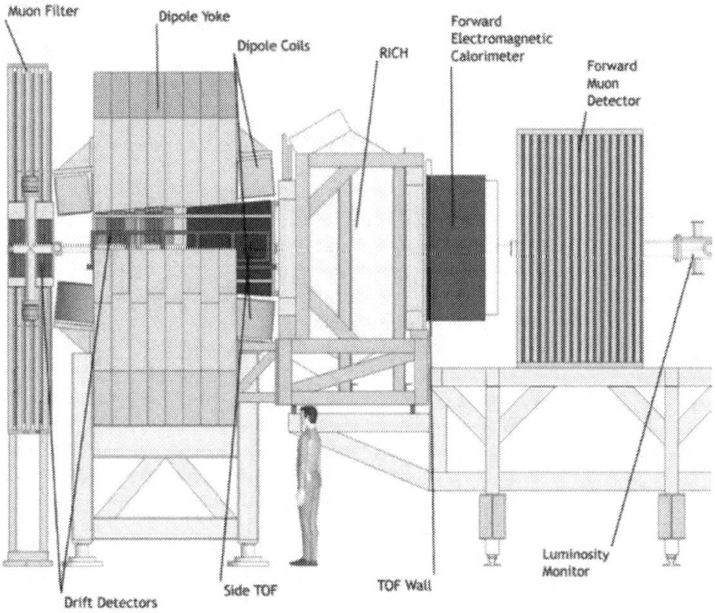

Figure 7. Artistic side view of the Forward Spectrometer (FS) of PANDA. It is preceded on the left by the Target Spectrometer (TS), which is illustrated in Fig. 4.

The deflection of particle trajectories in the field of the dipole magnet will be measured with three pairs of tracking drift detectors. The first pair will be placed in front, the second within and the third behind the dipole magnet. Each pair will contain two autonomous detectors as described below. Thus, in total, 6 independent detectors will be mounted. Each tracking detector will consist of four double-layers of straw tubes, two with vertical wires and two with wires inclined by a few degrees. The optimal angle of inclination with respect to vertical direction will be chosen on the basis of ongoing simulations. The planned configuration of double-layers of straws will allow to reconstruct tracks in each pair of tracking detectors separately, also in case of multi-track events.

A wall of slabs made of plastic scintillator and read out on both ends by fast photo-tubes will serve as time-of-flight stop counter placed at about 7 m from the target. In addition, similar detectors will be placed inside the dipole magnet opening, to detect low momentum particles which do not exit the dipole magnet. With the expected time resolution of $\sigma = 50\,\mathrm{ps}$ π/K and K/p separation on a $3\,\sigma$ level will be possible up to momenta of 2.8 GeV/c and 4.7 GeV/c, respectively.

For the detection of photons and electrons a Shashlyk-type calorimeter with high resolution and efficiency will be employed. The detection is based on lead-scintillator sandwiches read out

Journal of Physics: Conference Series **374** (2012) 012003 doi:10.1088/1742-6596/374/1/012003

with wave-length shifting fibers passing through the block and coupled to photo-multipliers. The technique has already been successfully used in the E865 experiment [10] and it has been adopted for various other experiments like PHENIX and LHCb. An energy resolution of $4\%/\sqrt{E}$ [11] has been achieved. To cover the forward acceptance, 351 modules arranged in 13 rows and 27 columns at a distance of 7.5 m from the target are required.

For the very forward part of the muon spectrum, a further range tracking system consisting of interleaved absorber layers and rectangular aluminum drift-tubes is being designed, similar to the muon system of the Target Spectrometer, but laid out for higher momenta. The system allows discrimination of pions from muons, detection of pion decays and, with moderate resolution, also the energy determination of neutrons and anti-neutrons. The forward muon system will be placed at about 9 m from the target.

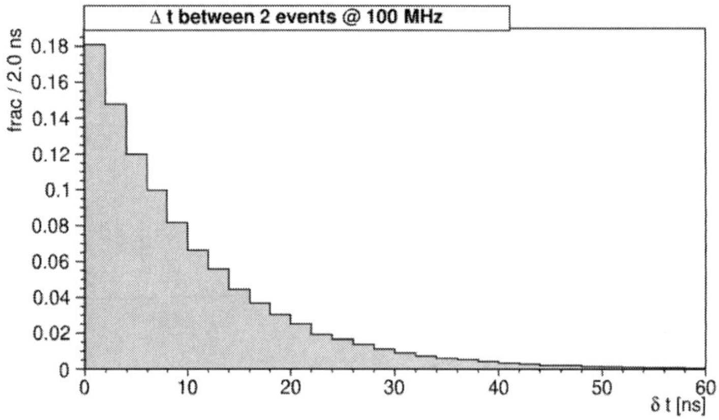

Figure 8. Distribution of the time difference δt between two sequent reactions at an instantaneous event rate of 100 MHz.

4.3. DAQ

In PANDA, a data acquisition concept is being developed to be as much as possible matched to the high data rates, to the complexity of the experiment and the diversity of physics objectives.

In our approach, every sub-detector system is a self-triggering entity. Signals are detected autonomously by the sub-systems and are preprocessed. Only the physically relevant information is extracted and transmitted. This requires hit-detection, noise-suppression and clusterisation at the readout level. The data related to a particle hit, with a substantially reduced rate in the preprocessing step, is marked by a precise time stamp and buffered for further processing.

The trigger selection finally occurs in computing nodes which access the buffers via a high-bandwidth network fabric. The new concept provides a high degree of flexibility in the choice of trigger algorithms. It makes trigger conditions available which are outside the capabilities of the standard approach.

One of the key features of the PANDA experiment will be a high luminosity mode with sufficiently high event rates to tackle physical questions with adequate statistical power. In this mode the average event rate given by the product of the average luminosity of $\mathcal{L} = 2 \cdot 10^{32}$ cm^{2}s$^-1$ and the total cross section $\sigma_{tot} \approx 100$ mb of $\bar{p}p$ scattering will be of the order of 20 MHz.

Since the luminosity is a time dependent quantity influenced by both the beam current in the HESR as well as the spatial distribution of the target, significant higher instantaneous event rates up to 100 MHz have to be taken into account. The average time difference between two events in that case is given by already challenging $\delta t_{avg} = 10$ ns, i. e. on that time scale all detector signals have to be collected to form event data packages. This technical challenge finally is intensified by the fact that the events are not distributed equidistant in time but exhibit an exponential behaviour concerning the time difference between sequent events as illustrated in Fig. 8.

This means that smaller time differences are in general more likely; in fact the probability density function has a maximum at $\delta t = 0$ ns. In particular 63% of all reactions have a δt smaller than the average value. At this point it seems clear that the high resolution timing of the SciTil detector is desirable to disentangle signals from different events.

5. Conclusions

For experiments with cooled antiproton beams in a storage ring on an internal target the PANDA experiment will be realized within the FAIR project at GSI/Darmstadt. It consists of a solenoid as target spectrometer, containing tracking detectors and Cherenkov detectors for particle identification for charged particles as well as an electromagnetic calorimeter. It exceeds with its high antiproton beam quality, its almost 4π acceptance and its high rate capability all previous and planned experiments. New detector concepts are developed for this experiment, technologies as well as detector materials optimized together with a novel trigger and data read out system to constitute a programmable physics machine in the charmonium mass region. If the performances will match the expectations, PANDA will have the opportunity to contribute to the understanding of the hadron structure and dynamics. A pre-installation of the detector including most subdetectors and the solenoid will take place in 2015 in Julich, the installation on the FAIR site will be in 2017 and first physics is to expect for 2018.

[1] FAIR Project, Baseline Technical Report, Technical report, GSI, Darmstadt, 2006, ISBN: 3-9811298-0-6, EAN: 978-3-9811298-0-9, http://www.gsi.de/fair/reports/btr.html.
[2] FAIR Project, FAIR Brochure, 2008, http://www.gsi.de/documents/DOC-2008-May-86-1.pdf.
[3] Technical report, ATLAS Technical Design Report 11, CERN/LHCC 98-13.
[4] Technical report, CMS Technical Design Report 5, CERN/LHCC 98-6.
[5] R. Aleksan et al., NIM A538 281 (2005).
[6] K. Mengel et al., IEEE Trans. Nucl. Sci. **45**, 681 (1998).
[7] R. Novotny et al., IEEE Trans. Nucl. Sci. **47**, 1499 (2000).
[8] M. Hoek et al., Nucl. Instrum. Meth. **A486**, 136 (2002).
[9] PANDA Collaboration, Technical Design Report, PANDA Electromagnetic Calorimeter (EMC), Technical report, FAIR-GSI, 2008, http://arxiv.org/abs/0810.1216.
[10] G. S. Atoyan et al., Nucl. Instrum. Meth. **A320**, 144 (1992).
[11] I.-H. Chiang et al., (1999), KOPIO Proposal.

Journal of Physics: Conference Series **374** (2012) 012004 doi:10.1088/1742-6596/374/1/012004

Hadron physics in holographic QCD

A. B. Santra[a], **U. Lombardo**[b], **A. Bonanno**[c]

[a] Nuclear Physics Division, Bhabha Atomic Research Centre, Mumbai 400085, India.
[b] INFN-LNS and University of Catania, 44 Via S.-Sofia, I-95125 Catania, Italy.
[c] INAF- Catania Astrophysical Observatory, Via S.Sofia 78, 95123 Catania, Italy.

E-mail: santra@barc.gov.in

Abstract.
Hadron physics deals with the study of strongly interacting subatomic particles such as neutrons, protons, pions and others, collectively known as baryons and mesons. Physics of strong interaction is difficult. There are several approaches to understand it. However, in the recent years, an approach called, holographic QCD, based on string theory (or gauge-gravity duality) is becoming popular providing an alternative description of strong interaction physics. In this article, we aim to discuss development of strong interaction physics through QCD and string theory, leading to holographic QCD.

1. Introduction

The accepted theory of strong nuclear forces, or strong interactions, is Quantum Chromodynamics (QCD) [1], an Yang-Mills gauge theory with $SU(3)$ gauge group. It can be studied at high energies using perturbation theory with enormous success, but, it is very hard, known to be intractable for the analytical analysis at low energies because of the strong coupling problem. The low energy regime of QCD contains the most interesting phenomena related to hadron physics, thus it is of great theoretical interest. Due to unavailability of suitable analytical tools for understanding QCD in the non-perturbative regime, QCD inspired heuristic models and approximation schemes, ranging from the bag models [2, 3] to chiral perturbation theory [4, 5] have been used to get partial information about the dynamics of QCD in the strongly coupled regime. These models use simple frameworks for the analysis of some aspects of non-perturbative QCD, but give impressive results. Though the simplicity of the effective models make them very interesting and useful for the phenomenology of strong interactions, no rigorous relation has been established with these models to QCD in spite of much efforts. Another approach of studying non-perturbative QCD is by numerical simulation using lattice gauge theory or lattice QCD [6, 7]. This procedure, though computationally intensive, appears to be successful for the calculation of static quantities such as the vacuum structure and the spectrum.

Almost immediately after the discovery of asymptotic freedom in QCD, existence of a new phase of matter called Quark-Gluon-Plasma (QGP) was predicted [8, 9] at very high temperature and density. QGP was expected to be made of de-confined free quarks and gluons. The programme of heavy ion collision at relativistic energies started to realize QGP experimentally. The experimental observations, over the last decade, in relativistic heavy ion collisions have revealed that QGP is, contrary to expectations, also a strongly coupled system instead of a gas of free quarks and gluons [10, 11]. Lattice QCD has been an effective tool for understanding the static properties related to equilibrium thermodynamics of QGP, but for the real time dynamic properties of strongly coupled plasma, the transport coefficients particularly, it is much more difficult [12]. This is mainly because lattice QCD methods are inherently Euclidean, and it is very difficult to extrapolate numerical results to Minkowski signature. Thus, unfortunately we do not have appropriate theoretical methods that would be applicable to the problems of QCD at strong coupling.

Another attempt of understanding strong interactions had been through S-matrix formalism [13, 14, 15]. It eventually led to the development of string theory [16, 17, 18]. The string description reproduces the Regee trajectories of hadrons very well, however this model was unable to reproduce the experimental results for hard and deep inelastic scatterings. Such a process described by string theory in flat space-time is soft, that is the amplitudes decay exponentially with energy, while both experimental data and QCD theoretical predictions [19, 20] indicate a hard behavior in which the amplitudes decay with a power of energy. So, the interest in string theory shifted towards making it a description of quantum gravity [21, 22]. This happened primarily due to the huge success of QCD at high energies, and consistency problem of string theory as well. Though terms like string tension, flux tubes are frequently used in hadron physics, still it was generally believed that string theory had very little to do as description of hadron physics.

A new way of thinking about strongly coupled gauge theories has emerged now, due to progress in string theory over the past decade, in the form of gauge-gravity duality [23, 24, 25] which connects a gauge theory in d-dimensional space-time to a gravity theory in $(d+1)$-dimensional spactime. Thus, the duality is holographic. The most widely studied example of the duality is known as AdS/CFT correspondence. AdS stands for anti de Sitter space and CFT for confromal field theory. The AdS/CFT correspondence connects a maximally supersymmetric Yang-Mills theory in the gauge theory side to a string theory in a particular ten-dimensional space-time, $AdS_5 \times S^5$, in the gravity side. This duality, between strongly-coupled Yang-Mills theories and weakly-coupled gravity is very attractive for its potential application in understanding QCD in strongly coupled regime.

Direct application of the AdS/CFT correspondence to QCD is not possible, since QCD is neither supersymmetric nor it is fully conformal, though there are some energy regime in which it is 'quasi conformal.' So, gravity dual of QCD is not known yet. However, inspired by AdS/CFT correspondence, and assuming that dual gravity theory to QCD exists, Witten [26] first proposed to deform the AdS/CFT correspondence to describe strong interactions. A large amount of activity started since then in building holographic models of QCD or AdS/QCD models.

The idea behind Ads/QCD is to introduce an additional spatial direction to the four dimensional space-time, which roughly corresponds to the energy scale of the field theory, and try to construct a model that captures important non-perturbative aspects of the original 4-dimensional field theory, such as confinement and chiral symmetry breaking. These models have been proved to be successful in understanding these non-perturbative effects in simple terms.

The gravity dual of gauge theory at finite temperature has also been constructed introducing black hole horizon in AdS_5 [26]. Using the dual theory it becomes relatively easy to compute [27] transport coefficients, such as viscosity and diffusion constants, for strongly coupled plasma. The analysis of scattering amplitudes in the AdS black hole background led to the universal viscosity bound [28], which plays an important role in understanding the physics of the elliptic flow of QGP observed in relativistic heavy ion collisions.

In this article we aim to describe pedagogically the evolution of the understanding of strong interaction from QCD to holographic QCD. In section (2), we give an introduction to strong interaction in terms of QCD and S-matrix formulism. In section (3), we give very elementary description of string theory as least as required to appreciate the terms used in AdS/CFT correspondence. In section (4), we introduce gauge gravity duality and holographic principle. In section (5), we discuss AdS/CFT correspondence through Maldacena conjecture introducing anti de Sitter space and conformal field theory. In section (6), we discuss different approaches of AdS/QCD models. In section (7), we discuss application of AdS/CFT correspondence to gauge theories at finite temperature. Finally, section (8) contains the summary and conclusions.

2. Strong interactions

Microscopic understanding of strong interaction (nuclear force) started with Yukawa [29] postulating that protons and neutrons (nucleons) interact by exchanging pions, pseudo scalar particles (0^-) of mass about 140 MeV. This idea got experimental approval after pions were discovered [30] in cosmic rays. However, it was soon realized that, (a) nucleons and pions show excitations (resonances), *i.e* these are not point like elementary particles, and (b) exchange of pions alone cannot explain the nuclear force. Strong interaction appears to be far more complicated.

With the availability of higher and higher energy accelerators, more and more particles other than the protons, neutrons and pions were produced. Fermions among them were named baryons, and bosons were named mesons. Some of these have strange properties, namely copious production and slow decay. A new quantum number called strangeness (S) was assigned to these particles. Strangeness has been found to be conserved in strong interactions.

To face this situation of understanding a large number of particles and their interactions, two proposals came by almost concurrently, (1) approach based on symmetry, that gave rise to the quark model and QCD, (2) approach based on S-matrix, that led to hadronic strings and string theory.

2.1. Symmetry approach of strong interaction

Patterns were found, when mesons and baryons of same spin and parity were placed on a two dimensional plot with charge (Q) and strangeness (S) as axes [31]. These patterns were identified with higher dimensional representations of $SU(3)$ group. Strong interaction must have $SU(3)$ symmetry. As higher dimensional representations of $SU(3)$ can be constructed taking direct product of fundamental representations, it was thus suggested that the particles corresponding to higher dimensional representations of $SU(3)$ could be made up of the particles corresponding to its fundamental (3-dimensional) representation.

Fundamental representation of $SU(3)$ constitutes three spin-$\frac{1}{2}$ particle states with fractional charge, and baryon number ($B = \frac{1}{3}$), namely, up (u) with charge, $Q = \frac{2}{3}$, strangeness, $S = 0$, isospin, $I = \frac{1}{2}$, third component of isospin, $I_z = \frac{1}{2}$, down (d) with charge, $Q = \frac{-1}{3}$, strangeness, $S = 0$, isospin, $I = \frac{1}{2}$, third component of isospin, $I_z = \frac{-1}{2}$, and strange (s) with charge, $Q = \frac{-1}{3}$, strangeness, $S = -1$, isospin, $I = 0$. These are collectively called quarks [32, 33], and are not observed in free state. Quark Model was developed to understand baryons as bound states of three quarks (qqq), ($p \equiv (uud), n \equiv (udd)$ etc) and mesons as bound state of quark and antiquark ($q\bar{q}$), ($\pi^+ \equiv (u\bar{d}), \pi^- \equiv (d\bar{u})$ etc.)

However, immediately, quark model was found to suffer from spin-statistics problem. In ($J = \frac{3}{2}, J_z = \frac{3}{2}$) state of Δ^{++} there are three u quarks with spin aligned in the same direction. Identical quarks in the same quantum state, giving rise to symmetric wavefunction for a fermion, violates Pauli principle. The problem disappears if quarks are not identical. A new quantum number, called color, was introduced for quarks [34, 35]. Each quark must appear in three colours (say, red q_r, green q_g and blue q_b). Quark model became extremely successful book-keeping model for strongly interacting particles. However, quarks were considered at that time as just "mathematical entities" having little to do with physical reality.

It was known quite early that quantum field theory is the appropriate framework to understand the dynamics of a system obeying principles of quantum mechanics and relativity. It was very successful in describing electromagnetic interactions. However, for strong interactions, there were some basic difficulties in constructing a quantum field theory, namely, (1) there are too many hadrons and too many forces to construct a sensible interaction lagrangian to build a useful field theory wit predictive power, (2) perturbative expansion of field theory becomes meaningless as the coupling constant in strong interaction is large. So, at the end of the 1960's, a fundamental theory of strong interactions did not exist.

Finally, conclusive evidences of point like constituents (quarks) of nucleons were found in deep inelastic scattering of electrons off nucleons [36]. However, isolated quarks with color charge were not "seen". Quarks are found to be, (1) almost non-interacting at high energy or short distances (asymptotic freedom), and, (2) very strongly interacting at low energy or large distances (infrared slavery).

Quarks, instead of baryons and mesons, were identified as the fundamental degrees of freedom of strong interaction. There are 6 flavors of quarks, as shown in table 1 and each flavor is a three component vector in color space.

Flavor	$Q(e)$	I	I_3	S	C	B	T	Mass
u	2/3	1/2	1/2	0	0	0	0	2 MeV
d	-1/3	1/2	-1/2	0	0	0	0	5 MeV
s	-1/3	0	0	-1	0	0	0	100 MeV
c	2/3	0	0	0	1	0	0	1.3 GeV
b	-1/3	0	0	0	0	1	0	4.2 GeV
t	2/3	0	0	0	0	0	1	175 GeV

Table 1. Different flavors of quarks with respective quantum numbers and mass.

After quarks were identified as the fundamental degrees of freedom of strong interaction, complete understanding of strong interaction reduced to, (1) knowing interaction of quarks that explains observations of deep inelastic scattering experiment, (2) constructing bound states of three quarks for baryons, and of quark and anti-quark for mesons with this interaction, and (3) deriving the interaction between baryons and mesons. The methodology to achieve the above objective is quantum field theory with perturbative techniques that worked extremely well for electromagnetic interactions.

Interaction of quarks is modeled in the same way as interaction of electrons in electromagnetism. Quarks interact through its color charge by exchanging gluons, G^a_μ, which are, like photon, mass-less spin 1 particle having two polarizations (μ), and unlike photon, carry color charge ($a = 1, \cdots 8$). Quantum field theoretic description of the dynamics of quarks and gluons is called Quantum Chromodynamics (QCD) [37] in which the form of the interaction lagrangian is determined by the principle of local gauge invariance; that is, by demanding the total lagrangian to be invariant under the transformations belonging to the local (color) $SU(3)$ group. Such theories are called gauge theories or Yang-Mills theories [38]. The QCD lagrangian is written as,

$$\mathcal{L}_{QCD} = \bar{q}^f i\gamma^\mu \left(D_\mu - m_f\right) q^f - \frac{1}{2} Tr\left[G_{\mu\nu} G^{\mu\nu}\right] , \tag{1}$$

where the covariant derivative, $D_\mu = \partial_\mu - igT^a G^a_\mu$, the gluon field tensor, $G_{\mu\nu} = \frac{i}{g}\left[D_\mu, D_\nu\right]$, m_f is the mass of quark of flavor f, T^a are SU(3) generators and g is quark-gluon coupling constant.

Expanding $\mathcal{L}_{QCD}$ of (1) we get,

$$\begin{aligned}\mathcal{L}_{QCD} &= \underbrace{\bar{q}^f\,(i\gamma_\mu\partial^\mu - m)\,q^f - \frac{1}{4}\left(\partial_\mu G^a_\nu - \partial_\nu G^a_\mu\right)^2}_{\mathcal{L}_0} \\ &+ \underbrace{g\,\bar{q}^f\,\gamma_\mu T^a G^{a\,\mu}\,q^f}_{\mathcal{L}_I} \\ &- \underbrace{\left(g\,f^{abc}\,\partial_\mu\,G^a_\nu\,G^{b\,\mu}\,G^{c\,\nu} + \frac{1}{4}g^2\,f^{eab}f^{ecd}G^a_\nu\,G^b_\mu G^{c\,\nu}G^{d\,\mu}\right)}_{\mathcal{L}_{II}},\end{aligned} \tag{2}$$

where $\mathcal{L}_0$ represents the lagrangian for free quarks and gluons, $\mathcal{L}_I$ represents quark-gluon interaction, and $\mathcal{L}_{II}$ represents gluon self interaction of gluons. Self interactions of gluons, a feature of non-abelian gauge theory, occurs because gluons have color charges. Such interactions make QCD non-linear and thus complicated. The quark-gluon interaction as well as the gluon self interaction strength is governed by the same coupling constant, g.

The QCD lagrangian, $\mathcal{L}_{QCD}$, possess, beside Poincare and local SU(3) (color) gauge symmetry, chiral symmetry and scale symmetry. In the limit, $m_u\,, m_d \approx 0$, ($m_u \simeq 1.7 - 3.3\,MeV\,, m_d \simeq 4.1{-}5.8\,MeV$), $\mathcal{L}_{QCD}$ in the (u, d) sector is invariant under $SU(2)\times SU(2)$ chiral symmetry. However, due to absense of parity doublets in hadronic states the chiral symmetry is spontaneously broken. The QCD action,

$$S_{QCD} = \int d^4x\,\mathcal{L}_{QCD}\,, \tag{3}$$

is invariant under the scale (conformal) transformations,

$$x^\mu \to \lambda x^\mu\,, \quad q \to \frac{1}{\lambda^{\frac{3}{2}}}q\,, \quad G^a_\mu \to \frac{1}{\lambda}G^a_\mu\,, \tag{4}$$

in absence of the quark mass term in QCD lagrangian. This is called scale symmetry (or conformal symmetry), however, this symmetry is also broken due to quantum effects.

After quantization and renormalization, the QCD lagrangian, $\mathcal{L}_{QCD}$, uniquely defines the algorithm (or Feynman rules) for calculating amplitudes for all physical processes of strong interaction in perturbative expansion. The quadratic terms give the propagators, whereas the cubic and quartic terms give the interaction vertices.

The wisdom of mixing quantum mechanics and relativity is that the vacuum or the ground state of a relativistic quantum mechanical system is not trivial, *i.e.* it is not just the empty space. Actually it can be thought of as a medium of virtual particles. Because of this, a general feature of quantum field theory is that the coupling constant, g, depends on the energy scale, μ at which observations are made. This dependence is given by the renormalization group (RG) equation, relating the beta function,

$$\beta(\alpha) = \mu\frac{d\alpha}{d\mu} \tag{5}$$

with the coupling constant. In case of QCD the RG equation is given as,

$$\mu\frac{d\alpha}{d\mu} = -\beta_0\,\frac{\alpha^2}{2\pi} - \beta_1\,\frac{\alpha^3}{4\pi^2} - \cdots\,, \tag{6}$$

where the strong coupling constant, $\alpha = \frac{g^2}{4\pi}$. In the lowest order, solution of (6) is,

$$\alpha(\mu) = \frac{2\pi}{\beta_0 \ln\left(\frac{\mu}{\Lambda}\right)}, \tag{7}$$

with $\beta_0 = 11 - \frac{2}{3}n_f$. The integration constant, Λ, is an intrinsic energy scale of QCD, which is introduced dynamically. Several measurements of the strong coupling constant at the Z-boson mass scale fix [41] the QCD scale parameter, Λ, as,

$$\alpha(M_Z) = 0.117, \; M_z = 91\,\text{GeV}, \quad \Rightarrow \Lambda = 217\,\text{MeV}. \tag{8}$$

It is seen from (7) that the coupling constant decreases with increasing energy, thus explaining asymptotic freedom, and the growth of the coupling at small energy is consistent with the non-observation of isolated quarks [39, 40]. The quarks are always confined to form hadrons. It is also seen from (7) that $\alpha(\mu)$ diverges at Λ. This is known as strong coupling problem of QCD. Moreover, the appearance of an intrinsic energy scale (Λ) in QCD breaks the scale symmetry (conformal symmetry).

Observationally to test the correctness of QCD, we require to compute the correlation functions, $\langle \mathcal{O}(x_1)\,\mathcal{O}(x_2)\rangle$, of gauge invariant operators, $\mathcal{O}(x_1)$. At high energy when the strong coupling constant, α, is small ($\alpha < 1$, weak coupling domain) the correlation functions are expressed as a perturbative expansion in α, and a few terms in the expansion give correct result. However, at low energies, below Λ (≈ 217 MeV), the strong coupling constant, α, is large ($\alpha > 1$, strong coupling domain), the perturbative expansion is invalid, consequently interesting domain of strong interaction remains intractable.

At low energy QCD possesses no expansion parameter. 't Hooft proposed to consider a generalization of QCD obtained by replacing the gauge group $SU(3)$ by $SU(N_c)$ and to perform an expansion in $\frac{1}{N_c}$ [42, 43] in the limit $N_c \to \infty$. It was found that $SU(N_c)$ Yang Mills gauge theory has a well defined perturbative expansion in the parameter, $\frac{1}{N_c}$, only if the number of colors, N_c, is large, while keeping the 't Hooft coupling, $\lambda = g^2 N_c$ fixed. This is what is called large N_c limit of QCD. Representing gluon lines in double line notation of color and anti-color, it can be shown that a certain class of Feynman diagrams called planer diagrams survive in this limit. The diagrams look like two-dimensional surfaces. The topological classification of the Feynman diagrams can be made precise by associating a Riemann surface to each Feynman diagram. It turns out that contribution of a given Feynman diagram to the amplitude is governed by the factor N_c^χ, where χ is the Euler number of the Riemann surface corresponding to that Feynman diagram. Thus, the expansion of any amplitude, $\mathcal{A}$, of QCD in terms of Feynman diagrams can be expressed as,

$$\mathcal{A} = \sum_\chi N_c^\chi A^{(\chi)}(\lambda) \tag{9}$$

where the coefficients $A^{(\chi)}(\lambda)$ are to be computed separately for the diagrams. The large N_c QCD is clearly not the same as QCD, however, in many cases the results for large N_c are almost the same as those for $N_c = 3$ theories. There are two regimes in large N_c QCD when parameterized by the coupling λ. For $\lambda \ll 1$ the theory is perturbative in λ and Feynman diagram summation can be used to calculate amplitudes, however, for $\lambda \gg 1$ the theory is in the non-perturbative domain, the diagrams become discretized two dimensional surfaces which are too difficult to compute.

In this situation, when QCD is known to be highly difficult for the analytical analysis at low energies because of the strong coupling problem, two paths were followed, namely, (1) QCD inspired models and effective theories, (2) numerical calculations using lattice QCD.

Journal of Physics: Conference Series **374** (2012) 012004 doi:10.1088/1742-6596/374/1/012004

QCD inspired models aim to understand structure of baryons and mesons and their interactions incorporating some features of QCD, mainly, asymptotic freedom, confinement and chiral symmetry. These models include several versions of phenomenological bag models [2]. In the (u, d) sector, QCD lagrangian (1) has chiral symmetry which is spontaneously broken as parity doublets are not found in the hadronic ground states. The chiral symmetry and its spontaneous breaking was identified to be the key feature of low energy QCD. Taking pseudoscalar mesons, the Goldstone bosons of spontaneous chiral symmetry breaking, as the fundamental degrees of freedom of strong interaction at low energy, an effective theory called chiral perturbation theory was developed to replace QCD for low energy hadron physics [44]. Though the QCD inspired models and effective theories describe low energy hadron physics phenomena quite well, but these are not rigorously derivable from QCD.

Lattice QCD appears to be the only known way to study QCD without approximations [6, 7, 45]. It aims to solve QCD numerically by discretizing space-time on a four dimensional lattice, so that it can be simulated on a computer using methods analogous to those used for systems of statistical mechanics. Fields representing quarks are defined at lattice sites while the gluon fields are defined on the links connecting neighboring sites. The input parameters in these simulations are the strong coupling constant, α_s, and the quark masses, m_q. The discrete space-time lattice can be thought of as a nonperturbative regularization scheme. At finite values of the lattice spacing a, (providing an ultraviolet cutoff of the order of $\frac{1}{a}$) there are no infinities, and in the limit, $a \to 0$, finite renormalized physical quantities are recovered. These simulations allow to calculate correlation functions of hadronic operators and matrix elements of any operator between hadronic states in terms of the fundamental quark and gluon degrees of freedom.

These calculations being computationally demanding require the use of largest available supercomputers. It seems to be the only hope for understanding non-perturbative QCD. In the past several years, it has produced quantitatively impressive results, however, there exists conceptual and technical problems to apply it in all situations of strongly coupled QCD.

One of the major successes of $SU(3)$ lattice QCD calculation is to demonstrate that a confining potential naturally emerges between a static, infinitely heavy quark (q) and antiquark $(\bar{q})$ by considering the Wilson loop [46, 47]. The $(q\bar{q})$ potential can be well reproduced by a sum of the Coulomb term due to the perturbative one-gluon-exchange process, a linear confinement term and a constant,

$$V_{q\bar{q}}(r) = -\frac{A}{r} + \sigma r + V_0 \,, \tag{10}$$

where $\sigma \simeq 0.89\,\mathrm{GeV/fm}$ of the confining term is like string tension. The linear potential at the long distance can be physically interpreted with the flux-tube picture or the string picture for hadrons, in which the quark and the antiquark are linked with a one-dimensional flux-tube with the string tension σ. This flux-tube picture (or the string picture) in the infrared region is supported by the Regge trajectory of hadrons.

Therefore, the starting problem of hadron physics, that is, how to construct the spectrum of mesons and baryons from the fundamental constituents, quarks and gluons, remains one of the most challenging unsolved problems of strong interaction dynamics. In addition, data from RHIC experiments also have convincingly suggested, contrary to expectations, that the quark gluon plasma (QGP) is strongly coupled [10]. Actually, every experimental datum of hadron physics contains at least a part that depends on strong coupling physics. It is ironical that though QCD is a beautiful physical theory, but in the low energy domain, that contains interesting phenomena of hadron physics, QCD remains inaccessible. No escape from strongly coupled regime !

2.2. S-matrix approach to strong interaction

In the 1960's, when list of strongly interacting particles, mesons and baryons, was ever increasing, it was clear that none of these particles were elementary. The idea of "nuclear democracy" was invoked, in which the notion that some particles are more elementary than others was rejected. All hadrons have similar status and the concept of elementary constituent was thought to be unnecessary. The form of a quantum theory to describe the strongly interacting particles can be determined by using very general consistency criteria, and features of the spectrum of particles.

It was observed that the resonances (very short lived strongly interacting particles) can be arranged in straight lines in two dimensional plots with spin (J) and square of mass (m) as axes. These are called Chew-Frautschi plots [48], and the straight lines are called Regge trajectories which follow the relation,

$$J\,(=\alpha(s)) = \alpha\, s + \alpha(0)\,, \tag{11}$$

with $s = m^2$. The Regge trajectories can be derived from a simple assumption that the hadrons are described by rotating relativistic strings. Energy (E) and the angular momentum (J) of a rigidly rotating string of length L with tension T, the endpoints of which move at near the speed of light c, can be written as,

$$E = T\int_{-L/2}^{L/2}\frac{dr}{\sqrt{1-v(r)^2}} = \frac{\pi}{2}TL\,, \qquad J = T\int_{-L/2}^{L/2}\frac{v(r)\,r\,dr}{\sqrt{1-v(r)^2}} = \frac{\pi}{8}TL^2\,. \tag{12}$$

that reproduces the Regge trajectory as,

$$J = \frac{1}{2\pi T}E^2 = \alpha\, s\,, \tag{13}$$

taking $E = m$. Understanding Regge trajectories in terms of rotating string marks the beginning of string description of hadrons.

A self consistent theory of strong interaction in which there are no elementary constituents, all particles are composites (lying on Regge trajectories), and they scatter by exchanging composite particles was formulated using S-matrix (scattering matrix) that describe what happens when particles collide. The S-matrix transforms the initial state, $|i\rangle$, at $t \to -\infty$ into the evolved state, $|f\rangle$, at $t \to +\infty$,

$$|f\rangle = S\,|i\rangle\,, \tag{14}$$

satisfying general principles of unitarity, analyticity, crossing symmetry, and asymptotic behavior. The transition matrix T is defined as,

$$S = I + iT\,, \tag{15}$$

and the scattering amplitude, M_{fi}, for a scattering process $|i\rangle \to |f\rangle$, is defines as,

$$\langle f|T|i\rangle = (2\pi)^4\delta(p_f - p_i)M_{fi}\,. \tag{16}$$

The scattering amplitude, M_{fi}, can be expressed as product of two factors,

$$M_{fi} = K_{fi}\,A(s,t,u)\,. \tag{17}$$

where K_{fi} is a covariant factor depending on spin, isotopic spin, and momentum components of the external particles, and the invariant amplitudes $A(s,t,u)$ is the dynamical part depending only on the Mandelstam variables, s, t and u of which only two are independent. The scattering in the "t-channel" and "s-channel" must produce the same amplitudes, because same set of poles (resonances) participate in scattering in both channels. Veneziano [15] constructed the amplitude of a scattering process as,

$$A(s,t,u) = A(s,t) + A(s,u) + A(t,u)\,, \tag{18}$$

with

$$\begin{aligned} A(x,y) &= \frac{\Gamma(-\alpha(x))\Gamma(-\alpha(x))}{\Gamma(-\alpha(x)-\alpha(y))} \\ &= \int_0^1 dz\, z^{-\alpha(x)-1}(1-z)^{-\alpha(y)-1}\,, \end{aligned} \tag{19}$$

where $\alpha(x)$ is a linear function of x, satisfying Regge trajectory. Veneziano formula was extremely successful in understanding low energy data of meson scattering.

The 4-particle Veneziano formula was generalized to an N-particle amplitudes which were factorized using operators consisting of an infinite number of harmonic oscillators. The complete spectrum of mesons was determined. It turned out that the degeneracy of states grew up exponentially with the mass that agrees with Hagedorn observation [49]. This suggested that Veneziano formula might be indicating something more than just approximate phenomenological descriptions of hadron scattering. These could be regarded as the lowest order approximation of amplitudes from some new quantum theory.

The Veneziano amplitude grew out of S-matrix theory where the scattering amplitude is the only observable object, the action or the lagrangian does not play any role. Nambu [16], Nielsen [17] and Susskind [18] found that the infinite number of oscillators, that one gets through the factorization of N-point Veneziano amplitude, naturally comes out from quantization of the lagrangian for a one-dimensional relativistic string, and Veneziano formula can be obtained as the tree-level string interaction amplitude. This is the beginning of string theory to understand hadronic physics. In the following section we give a very brief introduction to string theory.

3. Elements of string theory

We discuss some basic concepts of string theory mostly following the text books [50, 51, 52, 53, 54, 55, 56, 57]. Some of these concepts will be required to appreciate the connection between gauge theories and string theories through Ads/CFT correspondence.

3.1. Bosnic strings

String are objects extended in one spatial dimension, having no thickness but only length, typically of the order of Planck length $\left(l_P = \sqrt{\frac{\hbar G}{c^3}} \approx 10^{-33}\,\text{cm}\right)$, thus practically look like point particles in the hadronic length scale. The strings could be open ended or closed. As the string moves, it sweeps out a two-dimensional surface (sheet), called the worldsheet, similar to the world-line generated by the motion of a point particle. We need two coordinates, (σ, τ), to describe the worldsheet, where σ is a coordinate along the string, and τ a timelike coordinate, with the limits, $0 \le \sigma \le l_s$ and $-\infty \le \tau \le \infty$, where l_s is the length of the string. The world sheet is embedded in a d-dimensional space-time, $\mathcal{M}$. Thus, any point P having worldsheet coordinates (σ, τ) can also be specified by d-coordinates, $X^\mu(\sigma,\tau)$ $(\mu = 0, 1, \cdots, d-1)$, of $\mathcal{M}$.

The action to describe the dynamics of string can be taken as proportional to the invariant area of the worldsheet,

$$S \propto \int dA\,. \tag{20}$$

The infinitesimal area, dA, can be written as,

$$dA = d\sigma d\tau \sqrt{-\det(h_{ab})}\,, \tag{21}$$

where

$$h_{ab} = G_{\mu\nu}\partial_a X^\mu \partial_b X^\nu\,, \quad (a,b) \in (\sigma,\tau)\,, \tag{22}$$

is the induced metric on the worldsheet. $G_{\mu\nu}$ is the metric of $\mathcal{M}$, it governs all the geometric properties of $\mathcal{M}$. The Nambu-Goto action is written as,

$$S_{NG} \;=\; -T\int d\sigma d\tau \sqrt{-\det(h_{ab})} = -T\int d\sigma d\tau \sqrt{-\det(\partial_a X^\mu \partial_b X_\mu)}\,, \tag{23}$$

where T, called 'string tension', has dimension of force, and can be related to the Regge slope, α, as,

$$T = \frac{1}{2\pi\alpha}\,, \quad \alpha \equiv l_s^2\,, \tag{24}$$

with fundamental length scale, l_s, (string length). This action is difficult to quantize, because of the square root. An alternate action, equivalent to the Nambu-Goto action, S_{NG}, the so called 'Polyakov action' is constructed by introducing an auxiliary field, γ_{ab}, as,

$$S_P \;=\; -\frac{T}{2}\int d\sigma d\tau \sqrt{-\det(\gamma_{ab})}\,\gamma^{ab} G_{\mu\nu}\partial_a X^\mu \partial_b X_\mu\,. \tag{25}$$

The polyakov action is similar to the action of d bosonic fields, $X^\mu(\sigma,\tau)$ in 2-dimensions (one space and one time). It is invariant under Lorentz transformations and translations of $X^\mu(\sigma,\tau)$, in d-dimensions, thus exhibits Poincare symmetry. In addition, S_P is invariant under 2-dimensional diffeomorphism or reparametrization, $\tau \to \tau'(\sigma,\tau)$, $\sigma \to \sigma'(\sigma,\tau)$, and the Weyl rescaling of the auxiliary field, $\gamma_{ab} \to \gamma_{ab}\, e^{\phi(\tau,\sigma)}$.

The equations of motion of $X^\mu(\sigma,\tau)$ can be obtained by varying S_P with respect to $X^\mu(\sigma,\tau)$ as,

$$\frac{\delta S_p}{\delta X^\mu} = 0 \quad\Longrightarrow\quad \partial_a\left(\sqrt{-\gamma}\gamma^{ab}\partial_b X_\mu\right) = 0\,, \tag{26}$$

which are constrained by the equation of motion for γ^{ab}, obtained by varying S_P with respect to γ^{ab} as,

$$\frac{\delta S_p}{\delta \gamma^{ab}} = 0 \quad\Longrightarrow\quad \partial_a X^\mu \partial_b X_\mu - \frac{\gamma_{ab}}{2}\gamma^{cd}\partial_c X^\mu \partial_d X_\mu = 0\,. \tag{27}$$

These constraints are called the Virasoro constraints. The equation of motion for X^μ are supplemented with boundary conditions that make the boundary terms vanish in variation. The closed string satisfies the periodic boundary conditions, $X^\mu(\sigma,\tau) = X^\mu(\sigma + l_s,\tau)$, and the open string satisfies either Neumann boundary conditions, $\partial_\sigma X^\mu(0,\tau) = \partial_\sigma X^\mu(l_s,\tau) = 0$, or Dirichlet boundary conditions, $\delta X^\mu(0,\tau) = \delta X^\mu(l_s,\tau) = 0$, at the end points of the string. For an open string there could be Neumann boundary conditions in p directions, and Dirichlet boundary conditions in $d-p$ directions.

The re-parametrization and Weyl symmetry of S_P are local gauge symmetry on the worldsheet. These can be utilized to fix the auxiliary field, γ^{ab}. However, even after gauge fixing, S_P is left with a residual 2-dimensional conformal symmetry.

The solution of (27) can be written as an expansion in terms of different oscillation modes of string. The string action, S_P, can be quantized using canonical quantization procedure, by making the dynamical variable, $X^\mu(\sigma, \tau)$, and its conjugate momenta as operators and employing commutation relations between them. The expansion coefficients of oscillation modes become creation and annihilation operators. A Fock space of quantum states corresponding to different oscillation modes are generated by applying these creation operators on the vacuum. The residual symmetry is implemented on quantum states using Gupta-Bleuler like methodology of quantum electrodynamics. The particles corresponding to the excitations of the open string propagate in the world-volume on which the end points of the string is fixed, while particles corresponding to the excitations of the closed string propagate in the full d-dimensional space-time. The main results of bosonic string theory can be listed as,

(i) The bosonic string can be quantized consistently only if the dimension of $\mathcal{M}$, $d = 26$,

(ii) Each vibrational mode of the string corresponds to a particle with distinct spin (J) and mass (M). The spectrum contains a finite number of massless and infinitely many massive excitations with, $M^2 = \frac{n}{2\alpha}$ with $n \in N$.

(iii) The lowest energy state (ground state) is tachyonic, $M^2 < 0$,

(iv) The spectrum of the closed string contains a massless spin-2 (graviton) particle.

The above results were not encouraging enough to sustain interest in string theory as a candidate to describe strong interaction physics. The graviton-like particle was a great embarrassment as the string was first developed as a model of hadrons. However, progress in string theory continued to make it free from above anomalies, particularly the tachyonic states. The major development in string theory at this stage involved inclusion of fermions in string action as any fundamental theory of nature must contain both of these types of particles. In doing this, a new kind of symmetry called 'supersymmetry' was developed.

3.2. Supersymmetry

The supersymmetry is a symmetry between bosons and fermions [58, 59]. It came about in trying to unify gravity with other interactions. According to Einstein's general theory of relativity, gravity is related to the space-time symmetries of Poincare group, a non-compact Lie group, whereas other interactions, namely strong, weak and electromagnetic, are based on local gauge invariance of internal symmetries (or gauge groups) of compact Lie groups namely, $SU(3)$, $SU(2)$ and $U(1)$ respectively. Therefore, any attempt to unify gravity with other interactions requires, the Poincare group and the internal symmetry group to be part of same algebra. However, the answer to such attempts turned out to be no, in the form of the Coleman-Mandula 'no-go theorem', which says that if the Poincare symmetry and internal symmetries were to combine in a Lie algebra, the S-matrices for all processes would be identically zero [60]. However, like all theorems, this theorem also was only as strong as its assumptions, and one of the assumption was that the final algebra is a Lie algebra. It was realized that the notion of Lie algebra can be generalized to a graded Lie algebra (or super algebra) to circumvent the no-go theorem.

A super algebra is an algebra that contains some generators, Q^i, which satisfy anticommutation relations among themselves, and commutation relations with other generators which satisfy commutation relations among themselves. An anticommuting operator Q^i is fermionic, so, these operators generate transformations shown schematically as,

$$Q^i|\text{Boson}\rangle = |\text{Fermion}\rangle\,, \qquad Q^i|\text{Fermion}\rangle = |\text{Boson}\rangle\,.$$

This is called supersymmetry (SUSY). Fermions and bosons are grouped together into supermultiplets which are related under this symmetry. Thus, according to supersymmetry, corresponding to every boson there must be a fermion with the same mass and internal quantum numbers, and vice-versa.

Supersymmetry generators, Q^i, are intrinsically complex objects, so $Q^{i\dagger}$ are also distinct symmetry generators. The index $i = 1, \cdots, \mathcal{N}$ represents different supersymmetric generators in case there are more than one pair.

Supersymmetric partners of currently known particles have not yet been observed experimentally. It is believed that the reason for non observation of these particles is because supersymmetry is a broken symmetry, and as a result the superpartners are much heavier than the known particles. Particle accelerators (LHC) may be on the verge of finding evidence for high energy supersymmetric partners in the near future!

3.3. Superstring theory

Superstring theory attempts to describe both bosons and fermions in terms of the vibrational modes of the fundamental string. In doing that, the field content of the world-sheet is enlarged to include fermionic (anticommuting) variables, Ψ^μ_α, in addition to the bosonic variables, X^μ. For every μ, the two component Ψ^μ_α is a fermionic variable on the world-sheet, quantization of which would give rise to particle states that behave as space-time fermions. The Polyakov string action with world-sheet fermions can be written in conformal gauge as,

$$S_P = -\frac{T}{2}\int d\sigma d\tau \left(\partial_\alpha X^\mu \partial^\alpha X_\mu - i\bar{\Psi}^\mu \rho^\alpha \partial_\alpha \Psi_\mu\right) , \tag{28}$$

where ρ^α are 2-dimensional Dirac matrices,

$$\left\{\rho^\alpha, \rho^\beta\right\} = -2\eta^{\alpha\beta}, \quad \rho^0 = \begin{pmatrix} 0 & -i \\ i & 0 \end{pmatrix}, \quad \rho^1 = \begin{pmatrix} 0 & i \\ i & 0 \end{pmatrix}, \tag{29}$$

and the world-sheet fermions, $\Psi^\mu = \begin{pmatrix} \psi^\mu_- \\ \psi^\mu_+ \end{pmatrix}$, with $\bar{\Psi}^\mu = \Psi^\dagger \rho^0$.

The original action, from which (28) is obtained after gauge fixing, is invariant under: (i) Poincare transformations, (ii) Worldsheet reparametrizations, (iii) Weyl transformations, (iv) super-Weyl transformations (for fermionc variables), and (v) worldsheet supersymmetry. The worldsheet supersymmetry transforms fermions into bosons and vice versa as,

$$\delta X^\mu = \epsilon \Psi^\mu , \quad \delta \Psi^\mu = -i\rho^\alpha \partial_\alpha X^\mu \epsilon , \tag{30}$$

where ϵ is a constant anti-commuting spinor.

The equation of motion for the X^μ fields is the same to that in the bosonic case (Laplace equation), and the equation of motion for the fermionic variables Ψ^μ is the Dirac equation in two dimensions. These are supplemented with boundary conditions. For the closed superstring there are two possibilities for the boundary conditions of fermions, (i) periodic boundary conditions (Ramond (R) sector) $\psi^\mu_{+/-}(\sigma,\tau) = \psi^\mu_{+/-}(\sigma + l, \tau)$ and (ii) anti-periodic boundary conditions (Neveu-Schwarz (NS) sector) $\psi^\mu_{+/-}(\sigma,\tau) = -\psi^\mu_{+/-}(\sigma + l, \tau)$.

In addition there are constraints, coming from the equation of motion of the auxiliary field, to be satisfied by the fermionic variables. These are more involved and called the super-Virasoro constraints.

The quantization of the superstring is achieved, in canonical quantization procedure, by promoting the bosonic variable, X^μ, and the fermionic variable, Ψ^μ, to operators, and imposing commutation and anticommutation relations respectively between them and their canonical conjugates. In doing that the coefficients corresponding to vibrational modes become creation and annihilation operators.

A Fock space of quantum states corresponding to different oscillation modes of superstring are generated by applying these creation operators on the vacuum. The residual symmetry is implemented on quantum states. These vibrational modes of the superstring correspond to an infinite tower of particle states with definite spin (J) and mass (M), satisfying a linear Regge trajectory relation. It also turns out that the spectrum of the superstring can be organized into spacetime supersymmetry multiplets. The spectrum of the fermionic string is truncated by an additional projection, the Gliozzi-Scherk-Olive (GSO) projection [61]. This projection can be chosen in such a way that the tachyon is projected out, and the remaining spectrum is space-time supersymmetric. The main results of superstring theory can be listed as,

(i) The quantized theory of superstring shows that vibrational mode corresponds bosonic and fermionic particles with distinct spin (J) and mass (M),

(ii) The superstring can be quantized consistently only if the dimension of the space-time is 10.

(iii) The ground state is not tachyonic.

(iv) The spectrum of the closed superstring contains a massless spin-2 excitation ($J = 2\,, M = 0$), whereas the spectrum of the open string contains a massless spin-1 excitation ($J = 1\,, M = 0$), along with their space-time supersymmetric partners.

The massless spin-1 excitations of open string were identified with gluons and the massless spin-2 excitation of closed string was identified with graviton. A bold assertion was made that string theory could be a framework for understanding quantum gravity [21, 22].

There are five types of perturbatively consistent superstring theories in 10 space-time dimensions. These are called, Type-I, Type-IIA, Type-IIB, Heterotic $SO(32)$ and Heterotic $E8 \times E8$. All of these are free from tachyon problem, and contain closed strings implying that string theory is inconsistent without closed strings (or gravity).

3.4. String interaction

The simplest string interaction is mediated by joining two open strings at their end points, or splitting an open string into two open strings. For example, the Feynman diagram corresponding to annihilation of two particles is obtained by merging of two open strings into a single open string.

The strength of the interaction is controlled by the value of the dimensionless string coupling constant g_s, which is dynamically determined through the vacuum expectation value, Φ_0, of the dilaton, Φ, as $g_s = e^{\Phi_0}$. The dilaton is a massless excitation present in every string theory.

In quantum field theory, computation of scattering amplitudes for a given process involves summation over all possible Feynman diagrams corresponding to that process. However, in string theory, it is replaced by the summation over world-sheets of different topologies, depending on the choice of the type of the string theory. For example, the worldsheet corresponding to scattering of two closed strings, in the tree level, is topologically equivalent to a sphere with four discs cut out, and in the one-loop it is a torus. These are two dimensional surfaces with holes. Thus, interactions are introduced in string theory through the inclusion of topologically non-trivial surfaces.

Scattering amplitudes, $\mathcal{A}$, of a given process can be computed in perturbation theory, where $\mathcal{A}$ is expanded in a power series in the string coupling g_s as,

$$\mathcal{A} = \sum_{\chi} g_s^{-\chi} A^{(\chi)}\,, \tag{31}$$

where each term $A^{(\chi)}$ is computed separately. The validity of perturbation theory requires $g_s \ll 1$. The strength with which a given world-sheet contributes to the scattering amplitude is governed by χ, the Euler number of the worldsheet, determined by its topology. Though the scattering amplitude is formally given in (31), however, extracting answers from diagrams with more than two loops is very difficult due to the complexity of the mathematics involved in dealing with these surfaces.

The scattering amplitudes in string theory at each order in perturbation theory consist of only one diagram which is not plagued with infinity, in contrast to that in quantum field theory where there are many Feynman diagrams at each order and diagrams containing loops suffer from ultra-violet divergences. Heuristically this can be understood as if the point-like interaction vertices of quantum field theory are now smeared due to finite extent of the string.

The scattering amplitude (31) of string perturbation theory is similar to the scattering amplitude (9) in 't Hooft large N expansion of SU(N) gauge theory with the identification, $g_s \sim \frac{1}{N}$. Feynman diagrams of the gauge theory become surfaces that represent interacting strings. This gives the strongest hint that there could be some connections between string theory and gauge theory.

3.5. Compactification

Superstrings are consistently quantized in 10-dimensional space-time, but observationally space-time of our universe is 4-dimensional. To make the superstrings to describe the universe, the extra 6 dimensions have to be curled up into a small compact space of dimension of the order of the string scale (10^{-33} cm), so that the presence of these extra dimensions will not be detected. Thus, a compact 6-dimensional space needs to be hidden at every point in the 4-dimensional space-time.

It is possible to formulate string theory in non-trivial ten dimensional space-times where only four dimensions are infinitely extended and the remaining six are curled up and compact. One possible realization of compactification can be achieved by making a direct product or factorization ansatz, $\mathcal{M}^{10} = M^4 \times K^6$ for the 10-dimensional space-time, where M^4 is 4-dimensional Minkowski space, and K^6 a 6-dimensional compact manifold. The resulting theory in M^4 depends on the geometry and topology of the compact manifold, K^6. A particular scheme of compactification is known as Kaluza-Klien compactification, in which the extra 6 spatial dimensions may be put on 6 circles, forming a 6-dimensional torus.

The compactification leads to a tower of states, called Kaluza-Klein states, because if one of the spatial dimensions is circular of radius R (periodic), we know from quantum mechanics that the momentum in that dimension is quantized as, $p = \frac{n}{R}(n = 0, 1, 2, 3, ...)$. These states may show up in the mass spectrum, however, choosing the spatial extension, R, of the compactified dimension appropriately the Kaluza-Klien states may be made unobservable in the uncompactified 4-dimensional world.

Another interesting implication of compactification is that strings can wind around a compact dimension. This leads to winding modes in the mass spectrum. A closed string can wind around a periodic (circular) dimension an integral number of times. The energy of the winding mode is equal to the string tension, T, times the total length of the wrapped string, $E_w = w2\pi RT = \frac{wR}{\alpha}$, with $w (= 0, 1, 2, \cdots)$ being winding the number.

Another compactification scheme is by taking the compact manifold K^6 to be Calabi-Yau space. In fact, the string dynamics, in particular the assumption of the factorization ansatz, $\mathcal{M}^{10} = M^4 \times K^6$, and supersymmetry in uncompatified M^4, restricts the compactified space, K^6, to be Calabi-Yau space. It is possible to cleverly compactify [62] the superstring to get realistic gauge groups, $SU(3) \times SU(2) \times U(1)$, of standard model in M^4. The topology of Calabi-Yau space controls the number of families of low mass fermions, quarks and leptons.

3.6. Non-perturbative string theory

The formulation of string theory is of perturbative nature. The Feynman diagrams correspond to surfaces. There is no a priori reason, however, to assume that the string coupling constant,

g_s, which controls the expansion in terms of surfaces, is small. In addition, there are five independent perturbative superstring theories with different basic properties. Thus, for a complete understanding of the string theory we need to include non-perturbative effects. Understanding non-perturbative effects in string theory was possible by the discovery of duality symmetries [63, 64]. Dualities actually are equivalences between two apparently different theories. The idea of duality is very powerful, since mapping to a dual set of variables we can transform a difficult question (strong-coupling behavior) of one theory into a easy one (weak-coupling behavior) of the other. There are two types of duality symmetry in string theory, namely S-duality and T-duality.

The S-duality, which has a correspondence in quantum field theory, relates strong coupling regime of one theory to the weak coupling regime of the other. The T-duality, on the other hand, has no analogue in quantum field theory, it belongs solely to string theory, and originates due to compactification of extra dimensions in string theory. It relates a theory in which extra dimension is compactified on a circle of radius R, to another theory in which extra dimension is compactified on a circle with radius $1/R$.

These dualities play a very important role in unifying five different perturbative superstring theories. Using a combination of S and T dualities it was realized that the five perturbatively defined string theories are actually different phases of a unified theory called M-theory which is defined in 11-dimensions [65, 66]. In this picture, different ground states of the M-theory correspond to various string theories. The perturbation theory only probes the vicinity of each ground state, whereas duality makes non-perturbative correlations across different vacua.

Using T-duality Polchinski [67, 68] showed that string theory necessarily includes, in addition to strings, various higher dimensional objects known as branes [69]. One important class of branes are called Dp-branes (or Dirichlet-p-branes, with $0 \leq p \leq 9$) which are $(p+1)$-dimensional hyper-surfaces in the $(9+1)$-dimensional target space, $\mathcal{M}$. D0-branes are particle-like, D1-branes are string-like, and D2-branes are membrane-like objects, and so on.

There are two ways to think about Dp-branes. On the one hand, in closed string theory these are solitonic solutions of the equations of motion, characterized by a charge and mass per unit volume, called brane tension, which is inversely proportional to the string coupling constant g_s. This dependence on the coupling constant is typical of the solitons in field theory. Because of their masses, Dp-branes can deform the space-time metric around them as manifestation of the gravitational effects.

On the other hand, in open string theory these are (p + 1)-dimensional hyper-surfaces. The end points of open strings, satisfying Neumann boundary conditions in p longitudinal directions and Dirichlet boundary conditions in $9-p$ transverse directions, lie on these surfaces, and move freely along the p-directions of the D-brane, but cannot leave it by moving in the transverse directions. The end points of an open string while moving on a Dp-brane may come in contact and become a closed string which can leave the brane, and closed strings can break touching the Dp-branes and become open strings. The massless excitations of a open string whose end points lie on a Dp-brane are gauge fields and their super-partners constitute a maximally supersymmetric $U(1)$ gauge theory in p+1 dimensions. This gauge theory is localized on the world-volume of the Dp-branes. The gauge coupling constant, g, is related to the string coupling constantg_s as, $g^2 = 4\pi g_s$, for $p = 3$.

4. Gauge-Gravity duality

Strong interaction appears to be too difficult to understand in both the approaches described above, because of QCD becoming strongly coupled at low energies, and string theory becoming more of a theory of quantum gravity. In this situation, it is an interesting idea to look for a duality, that is, to find a theory which is equivalent to QCD, but weakly coupled when QCD is strongly coupled.

There are several examples of duality in physics. A famous example of duality in $(1+1)$-dimensions is between Sine-Gordon model describing massless bosons and Thirring model describing massive fermions. The Sine-Gordon model [70] is a quantum field theory in (1+1)-dimensional space-time of a single scalar field $\phi(x)$ which is self-coupled as described by the lagrangian,

$$\mathcal{L}_{SG} = \frac{1}{2}\partial_\mu\phi\partial^\mu\phi + \frac{\alpha}{\beta^2}\left(\cos\left(\beta\phi\right) - 1\right) , \tag{32}$$

where α is a parameter and β is the coupling constant. The classical equations of motion corresponding to the lagrangian (32) give rise to stable solutions known as solitons and antisolitons. Interestingly, many-soliton solutions obey Pauli's exclusion principle, which can be interpreted as a fermion-like behavior [70]. In turn, the Thirring model [71] is a theory of a self-coupled Dirac field $\psi(x)$ in (1+1)-dimensional space-time described by the lagrangian,

$$\mathcal{L}_T = i\bar{\psi}\gamma^\mu\partial_\mu\psi - m_f\bar{\psi}\psi - \frac{\kappa}{2}\left(\bar{\psi}\gamma^\mu\psi\right)^2 , \tag{33}$$

where m_f is the mass of the fermion field and κ is a dimensionless coupling constant. Coleman [72] compared the n-point Green functions of the Sine-Gordon and Thirring model computed using a perturbative approach in coordinate representation and concluded that they should be equivalent if the coupling constants β and κ are related by,

$$\frac{4\pi}{\beta^2} = 1 + \frac{\kappa}{\pi} . \tag{34}$$

Weak Coupling in the Sine-Gordon model corresponds to strong coupling in the Thirring model. The Thirring model is S-dual to the sine-Gordon model. The fundamental fermions of the Thirring model correspond to the solitons of the Sine-Gordon model.

Given the above example of duality between two completely different theories we may think that a dual to strongly coupled gauge theory like QCD may be found in string theory, because there are several connections between QCD and string theory, namely,

(i) Understanding Regge trajectories (11) in terms of semi-classical spinning relativistic string,
(ii) The confining term in quark-antiquark potential (10) appears to be string-like,
(iii) Feynman diagrams of the $SU(N)$ gauge theory in the large N limit can be identified as surfaces that represent interacting strings.

However, in course of time, strong interaction and string theory got disconnected mainly because of overwhelming success of QCD at high energy, and string theory becoming more of a theory of quantum gravity than strong interaction.

A very important development namely holographic principle, has played the crucial role in finding a dual theory of QCD that brings strong interaction and string theory together again.

4.1. Holographic principle

The holographic principle originated in trying to understand thermodynamics of black holes. A black hole is a region of space from which nothing, not even light, can escape. Thus, a region of space of radius r containing mass M becomes a black-hole, in newtonian description of gravity, if escape speed from that region is greater than the speed of light,

$$\sqrt{\frac{2GM}{r}} > c . \tag{35}$$

Dense enough matter collapses into a black hole [73]. In Einstein's general theory of relativity, which is relativistic description of gravity, black holes appear as the singularities of space-time, where the curvature diverges. The singularities are conjectured [74] to be hidden behind the event horizon, the surface, $r = r_H$,

$$r_H = \frac{2GM}{c^2}\,, \tag{36}$$

through which things can go in but can not come out. Thus, black holes are an essential feature of gravity. Physically black holes may be created in astrophysical processes when stars run out of nuclear fuel and implode under gravitational force. If the mass of the leftover matter after supernova explosion exceeds about 3 solar masses, it will collapse into a black hole.

Using general relativity Hawking [75] proved an useful theorem called the horizon area theorem, according to which the total horizon area, A, in a closed system containing black holes never decreases. It can only increase or stay the same,

$$\Delta A \geq 0\,. \tag{37}$$

Increase in total horizon area comes from the growth of black holes by accretion of "normal" matter or by the coalescence of black holes.

Stationary black holes are unique, in the sense that any arbitrary system having mass M, angular momentum J and charge, Q, will collapse to form a unique black hole. This is known as black hole uniqueness theorem or 'no hair theorem' [76]. Thus, black holes are actually extremely simple macroscopic objects, which can be parametrized by only a few numbers, similar to elementary particles which are characterized by a few quantum numbers.

The black hole uniqueness theorem apparently leads to a paradox of violating second law of thermodynamics, since in the initial state the matter that forms or falls into a black hole has arbitrarily large phase space or entropy, and in the final state the black hole occupies a point in phase space having zero entropy. Resolution of the paradox was given by Bekenstein [77]. Noting striking similarity between second law of thermodynamics, $\Delta S \geq 0$ and Hawking area theorem: $\Delta A \geq 0$, he proposed that black holes must have entropy, $S_{\rm bh}$, which is proportional to the area of their event horizons,

$$S_{\rm bh} \propto A\,. \tag{38}$$

Black holes having entropy gives rise to paradox again, since objects having entropy are hot, so they must radiate, but nothing can come out of black holes. Resolution of this paradox was given by Hawking using quantum mechanics [78]. Virtual pairs of particles and antiparticles are produced and annihilated in vacuum all the time violating energy conservation for short durations allowed by Heisenberg's uncertainty principle. If such an event occurs near the event horizon of a black hole, the virtual pair can be split up by the strong gravity so that it is possible for one with negative energy to fall in the black hole and the other with positive energy to escape. The energy conservation "debt" involved in the vacuum fluctuation is paid by the black hole itself, the negative energy particle makes a negative contribution to the mass of the black hole when it goes in. The black hole's mass decreases by the mass of the escaping particle. The escaping particle is seen by a distant observer as emission by the black hole.

The spectrum of the emitted particles is similar to black body radiation with a temperature. Using semi-classical method Hawking [78] estimated the temperature to be,

$$T_H = \frac{\hbar c^3}{16\pi^2 k_B G M}\,. \tag{39}$$

Using, $dE = c^2 dM$, and first law of thermodynamics, $dE = T_H dS_{\rm bh}$, we get the black hole entropy as

$$S_{BH} = \int \frac{c^2}{T_H} dM = \frac{k_B A}{4 l_P^2}, \tag{40}$$

where $A = 4\pi r_H^2$ is the area of the event horizon. In contrast to statistical physics or local field theory where entropy is a measure of the phase space volume and proportional to volume of the system, black hole entropy, as proposed by Bekenstein, is proportional to the area of the horizon. Proportionality of black hole entropy to the area rather than volume indicates that the degrees of freedom of (quantum) gravity may not be similar to that of local field theories. In the context of string theory using D-branes, the Bekenstein-Hawking formula (40) of black hole entropy has been derived [79] for some special class of black holes.

Application of second law of thermodynamics shows that the entropy of any region is bounded and the upper bound is equal to the entropy of the largest black hole that can be accommodated in the region. Let the entropy, S, of a region of volume V with area A be greater than the entropy, S_{BH}, of the largest black hole which can be accommodated in the region ($S > S_{BH}$). Then, putting in some matter in that region a black hole can be formed, the entropy of the region becomes S_{BH} which is less than original entropy S. It violates second law of thermodynamics. The conundrum of violating second law of thermodynamics can be resolved if the original assumption is wrong. Therefore, in a system with gravity, maximum entropy of any region is proportional to the boundary area of the region.

The perplexing result of quantum gravity that maximum entropy of a given region is proportional to its boundary area, led 't Hooft [80] and Susskind [81] to formulate holographic principle. It states that any theory of gravity in d dimensional space-time should have a description in terms of a quantum field theory defined in a flat (*i.e* without gravitational interactions) $(d-1)$-dimensional space-time. Both the theories have an entropy proportional to the area in d-dimensions, because surface in d-dimension is the same as volume in $(d-1)$-dimensions. In other words, quantum gravity in any space of d-dimension can be formulated in terms of degrees of freedom of a local field theory defined in the boundary space of $(d-1)$ dimension. This situation is analogous to an optical hologram, which stores all information of a 3-dimensional object in a 2-dimensional surface.

The holographic principle suggests that there should exist a relation (equivalence) between gauge theory in $(3+1)$-dimensions with a gravity theory in $(4+1)$-dimensions. This equivalence of a gravity theory in bulk to a gauge theory on surface is called gauge-gravity duality. The remarkable utility of this duality is that, in particular limiting situations when quantum field theory is strongly coupled, the corresponding gravity theory becomes classical. Thus, strongly coupled regime of quantum field theories can be approached doing calculations in classical gravity theory.

An extensively studied particular case of the gauge-gravity duality in the context of string theory is the AdS/CFT correspondence proposed by Maldacena [23]. In the recent time, with the conjecture of Maldacena, string theory has become a powerful analytic tool for studying strongly coupled gauge theories. The most interesting strongly coupled gauge theory is QCD at low energies. Though gravity dual of QCD has not been found yet, but theories close (supersymmetric generalizations) to QCD have been studied using the so-called 'gauge gravity duality.

5. AdS/CFT correspondence

The AdS/CFT correspondence is a special case of gauge-gravity duality, where the gravity theory is defined on a 5-dimensional Anti de-Sitter (AdS_5) space, and the dual gauge theory is a conformal field theory (CFT) defined in the boundary of AdS_5, which is 4-dimensional Minkowski space. Before giving a heuristic derivation of AdS/CFT correspondence, we discuss briefly about the Anti de-Sitter space and conformal field theory.

5.1. *Anti de-Sitter space*

The anti-de Sitter space is a maximally symmetric, homogeneous and isotropic space-time with constant negative curvature (hyperboloid) [82, 83, 84]. Roughly, a negative curvature can be described by the fact that geodesics (shortest line between two points) which are parallel to begin with start moving away from each other while in a spherical or elliptical surface (with positive curvature) they come closer to each other.

Five dimensional anti-de Sitter space, AdS_5, with constant negative curvature, R, can be conveniently defined by the equation,

$$\left[\left(X^0\right)^2-\left(X^1\right)^2-\left(X^2\right)^2-\left(X^3\right)^2\right]-\left(X^4\right)^2+\left(X^5\right)^2=R^2 \tag{41}$$

as an embedding in a 6-dimensional flat space-time with co-ordinates, $(X^0,\cdots,X^5)$, and having $(2,4)$ signature. The defining equation (41) is symmetric under the transformation,

$$X^\mu \to X'^\mu = \Lambda^\mu_\nu\, X^\nu\,, \tag{42}$$

where Λ^μ_ν belongs to $SO(2,4)$, the conformal group in 4-dimension. Thus, $SO(2,4)$ is the isometry group of AdS_5. The line element can be written as,

$$ds^2 = -\left(dX^0\right)^2+\left(dX^1\right)^2+\left(dX^2\right)^2+\left(dX^3\right)^2+\left(dX^4\right)^2-\left(dX^5\right)^2\,, \tag{43}$$

with the constraint (41). Introducing a set of coordinates, called Poincare co-ordinates, given by a four dimensional Lorentz vector, $x^\mu \equiv (x^0,\vec{x})$, and a fifth coordinate, z, as,

$$\begin{aligned} X^i &= \frac{R\,x^i}{z}\,, \quad i=1,\cdots,3; \qquad X^5=\frac{R\,x^0}{z}\,, \\ X^0+X^4 &= z-\frac{\eta_{\mu\nu}x^\mu x^\nu}{z}\,, \qquad X^0-X^4=\frac{R^2}{z}\,, \end{aligned} \tag{44}$$

with $\eta_{\mu\nu}=diag(1,-1,-1,-1)$, the AdS_5 metric in (x^μ,z) co-ordinates can be written as,

$$ds^2=\frac{R^2}{z^2}\left(\eta_{\mu\nu}dx^\mu dx^\nu+dz^2\right)=g_{MN}\,dx^M dx^N\,, \tag{45}$$

where

$$g_{MN}=\frac{R^2}{z^2}\begin{pmatrix}1&0&0&0&0\\0&-1&0&0&0\\0&0&-1&0&0\\0&0&0&-1&0\\0&0&0&0&-1\end{pmatrix}=\frac{R^2}{z^2}\eta_{MN}\,, \tag{46}$$

with η_{MN} as 5-dimensional Minkowaski metric and $\det(g)=\frac{L^{10}}{z^{10}}$. The AdS_5 has slices isomorphic to four-dimensional Minkowski space-time put on z-axis. The Minkowski metric is multiplied by a warp factor $\frac{R^2}{z^2}$, so, an observer living on a Minkowski slice sees all lengths rescaled by a factor of z depending on its position in the fifth dimension.

The Ricci tensor, R_{MN}, calculated from the AdS_5 metric g_{MN} is,

$$R_{MN} = -\frac{4}{R^2} g_{MN}\,, \tag{47}$$

which gives the Ricci scalar, $\mathcal{R}$, of AdS_5 metric as, $\mathcal{R} = -\frac{20}{L^2}$. We see that Ads_5 metric is solution of source free Einstein's equation of general relativity,

$$R_{MN} - \frac{1}{2}\mathcal{R} g_{MN} + \Lambda g_{MN} = 0\,, \tag{48}$$

in five dimension having coordinates $(t, x^1, x^2, x^3, z\ )$ with cosmological constant, $\Lambda = -\frac{12}{R^2}$.

To understand physically the significance of the fifth coordinate, let us consider a 4-dimensional space with coordinates, $[x^\mu \equiv (t, \vec{x})]$, with a variable, l, that corresponds to energy scale,

$$l \equiv E = \text{`}\ \partial_t\ \text{'}\,.$$

Under a scale transformation the coordinates transform as,

$$x^\mu \to x'^\mu = \lambda x^\mu\,, \quad l \to l' = \frac{l}{\lambda}\,. \tag{49}$$

We can construct a 5-dimensional space with co-ordinates, (x^μ, l), that encodes the requirement of scale invariance should have the line element as,

$$ds^2 \;=\; R^2\left(l^2\,\eta_{\mu\nu}dx^\mu dx^\nu + \frac{dl^2}{l^2}\right)\,, \tag{50}$$

where R is the radius of curvature of the AdS_5 space. By changing the coordinate, $l \to \frac{1}{z}$, in (50), we get the line element (45) of AdS_5. Thus, the fifth co-ordinate, z, of AdS_5 is inversely related to the energy scale. Large energy scale, or the ultra violet region, corresponds to small z, and small energy scale, infra red region, corresponds to large z. Thus, physics in the Minkowski space at different energy scale can be explored from AdS_5-space at different values of z.

5.2. Conformal field theories and $\mathcal{N} = 4$ Supersymmetric Yang Mills theory

The conformal transformations include Poincare transformations and scale transformations that preserve the metric up to a scale factor, $g_{\mu\nu}(x) \to g'_{\mu\nu}(x) = \Omega^2(x)\, g_{\mu\nu}(x)$. These transformations form a group called conformal group. The generators are the usual Lorentz generators, $M_{\mu\nu}$, the Poincare translation operators P_μ, and in addition two generators D and K_μ corresponding to scale transformations. The conformal group in 4-dimensions is isomorphic to $SO(2,4)$. A field theory which is invariant under conformal group is called a conformal field theory. If we want to combine the conformal algebra with the supersymmetry algebra, then additional fermionic generators are required to be included in the algebra. The scaling dimension Δ of a field $\phi(x)$ is governed by the transformation rule under scaling of coordinates,

$$D: \quad x^\mu \to \lambda x^\mu\,, \quad \Rightarrow \quad \phi(x) \to \phi'(x) = \lambda^\Delta \phi(\lambda x)\,. \tag{51}$$

Representations of the conformal group are labeled by the scaling dimension Δ and Casimir invariants of Poincare group. The structure of the correlation functions of primary fields, that is n-point functions, in a conformal field theory are completely determined by conformal symmetry.

$\mathcal{N} = 4$ $SU(N_c)$ super Yang-Mills theory in 3 +1 dimensions is an example of superconformal field theory [61, 85, 86], which is invariant under superymmetry as well as conformal

Journal of Physics: Conference Series **374** (2012) 012004 doi:10.1088/1742-6596/374/1/012004

transformations. Because of exact conformal invariance even after quantization, unlike non-supersymmetric 3+1 dimensional pure Yang-Mills theory, it is scale invariant, that is beta function is zero in all orders of perturbation theory and the same is supposed to be true at the non-perturbative level. It means that the couplings do not depend on the energy scale. It contains two types of fermionic operators (generators of supersymmetry), Q^A_α and $\bar{Q}^A_\beta$ $(A = 1, 2, 3, 4)$ which transform under lorentz transformations as four spinors. The theory has the maximal amount of supersymmetry. The supersymmetric algebra is invariant under a $SU(4)$ rotation of the fermionic operators, consequently the lagrangian of $\mathcal{N} = 4$ $SU(N_c)$ SYM is invariant under $SU(4) \sim SO(6)$ global transformations. This is called R-symmetry. The degrees of freedom of this theory are as follows:

(i) A vector field A_μ in the adjoint representation of $SU(N_c)$, which is a singlet under $SO(6)$.

(ii) Six real scalars χ^a in the 6 dimensional vector representation of $SO(6)$, which also transform in the adjoint representation of $SU(N_c)$.

(iii) Four Weyl fermions λ^A_α also transforming in the adjoint representation of $SU(N_c)$ and 4 spinor representation of $SO(6)$, corresponding to the fundamental representation of $SU(4)$.

5.3. The Maldacena Conjecture

The Maldecana conjecture about the duality between gauge theory and gravity, or the AdS/CFT correspondence, arises from two different descriptions of a system of N_c coincident D3 branes.

Let us consider type IIB superstring theory in the presence of N_c D3 branes. On this background, string theory contains perturbative excitations corresponding to both closed string and open strings. Closed strings propagate on the bulk, whereas open strings are attached to the D3 branes and go from one D3 brane to another. The stack of N_c D3 branes are connected by N_c^2 different type of open strings because the strings can begin or end on any of the D3 branes. N_c^2 is the dimension of the adjoint representation of $U(N_c)$. At energies lower than energy of string scale $(1/l_s)$, that is in the limit, $l_s \to 0$, only the massless string states can be excited and the interaction between closed string modes in the bulk and the open string modes in the the stack of D3 branes vanishes. The closed string modes describe free gravity in the bulk, and the N_c^2 massless open string modes and their superpartners constitute maximally supersymmetric $\mathcal{N} = 4$ $U(N_c)$ super Yang-Mills gauge theory on 4 dimensional worldvolume of the stack of branes. However, the rigid motion of the entire system of branes as a whole can be described by the excitations of the $U(1)$ subgroup of $U(N_c)$. Because of the overall translation invariance, this mode decouples from the remaining $SU(N_c)$ modes that describe motion of the branes relative to one another. Thus, there are two decoupled systems, one of free gravity in the bulk and the other a 4-dimensional $\mathcal{N} = 4$ $SU(N_c)$ super Yang-Mills gauge theory on the brane.

The same system can considered from a different point of view. D branes are massive, so a stack of N_c D3 branes is heavy and naturally deform the 10 dimensional space-time. The solution of the equation of motion of low energy string theory effective action for a stack of N_c D3 branes can be written as,

$$ds^2 = \frac{1}{\sqrt{f(r)}}\left(-dt^2 + (dx^1)^2 + (dx^2)^2 + (dx^3)^2\right) + \sqrt{f(r)}\left(dr^2 + r^2 d\Omega_5\right) . \quad (52)$$

The metric inside the parentheses in the first term is the 4-dimensional Minkowski metric of the D3 brane and that in the second term is the flat metric of the 6 transverse coordinates written in spherical coordinates. The function $f(r)$ is given by,

$$f(r) = 1 + \frac{R^4}{r^4} . \quad (53)$$

In the limit, $r \gg R$, $f \simeq 1$, the metric (52) reduces to that of flat space with a small correction corresponding to the gravitational potential due to the massive stack of N_c D3 branes.

On the other hand, for $r \ll R$, $f(r) = \frac{R^4}{r^4}$, in which case the metric (52) reduces to,

$$ds^2 = ds^2_{AdS_5} + R^2 d\Omega_5 \,, \tag{54}$$

where

$$\begin{aligned} ds^2_{AdS_5} &= \frac{r^2}{R^2}\left(-dt^2 + (dx^1)^2 + (dx^2)^2 + (dx^3)^2\right) + \frac{R^2}{r^2}dr^2 \,, \\ &= \frac{R^2}{z^2}\left(\eta_{\mu\nu}dx^\mu dx^\nu + dz^2\right) \,, \qquad \left(\text{with, } r = \frac{R^2}{z}\right) \,, \end{aligned} \tag{55}$$

is the metric (45) of five-dimensional anti-de Sitter space-time. Thus, the ten-dimensional metric factorizes into $AdS_5 \times S^5$ in the strong gravity region. The geometry of 10-dimensional space-time due to the stack of N_c D3-branes is flat at regions far away from the branes, whereas close to the branes a 'throat' geometry of the form $AdS_5 \times S^5$ develops. The size of the throat is set by the curvature, R of AdS_5. There are two distinct sets of degrees of freedom corresponding to gravity, one propagating in the flat region and the other of closed string excitations propagating near the throat. At low energies these two modes decouple from each other.

Thus, in the low-energy limit of both descriptions of a stack of N_c D3-branes there are two decoupled theories, one of which is supergravity in flat space. So, the other theories that appear in both descriptions can be identified. This leads to the Maldacena conjecture: $\mathcal{N} = 4$ $SU(N_c)$ super Yang-Mills theory in 3+1 domensions is equivalent (or dual) to type II B string theory on $AdS_5 \times S^5$.

There are two parameters in the gauge theory side, namely, (1) the gauge coupling constant, g, and (2) number of colors, N_c. The value of 't Hooft coupling constant, $\lambda = g^2 N_c$ decides whether gauge theory is in weakly coupled or strongly coupled regimes. When $\lambda \gg 1$, gauge theory is strongly coupled, and $\lambda \ll 1$ corresponds to weakly coupled regime. On the other hand, string theory on $AdS_5 \times S^5$ has three parameters, namely, (1) string length l_s, (2) string coupling g_s, and (3) radius of curvature of AdS_5 space, R. When $R \gg l_s$, closed string theory reduces to Einstein's classical gravity. The parameters of two sides are related as,

$$g^2 = 4\pi g_s \,, \qquad R^4 = 4\pi g_s l_s^4 N_c \,. \tag{56}$$

The weak coupling regime of gauge theory, $\lambda \ll 1$, corresponds to $g_s N_c \ll 1$, we see from (56) that $R \ll l_s$. It means the radius characterizing the gravitational effect of the D-branes becomes small in string units. Closed strings feel a flat space-time everywhere except very close to the locations of D-branes. In the regime of weak coupling, thus the closed string description is not useful since it would involve complicated sub-string-scale geometry. On the other hand, in the strong coupling regime of gauge theory, $\lambda \gg 1$ corresponds to $g_s N_c \gg 1$, we see from (56) that $R \gg l_s$. It means the radius characterizing the gravitational effect of the D-branes becomes large in string units, so the geometry becomes weakly curved. In this limit the closed string description simplifies and essentially reduces to classical gravity. Thus, the difficult strongly coupled regime of gauge theory is mapped to classical gravity through this remarkable correspondence.

We see that when the gravity side is strongly coupled the gauge theory side is weakly coupled and vice versa. It is a strong-weak coupling duality. This is the reason why this duality is so exciting to study strongly coupled gauge theories. However, finding the dual theory, in general cases is difficult. It has been possible only for those cases which have a D-brane interpretation, and whose low-energy theory is understood.

Though the conjecture has not been proved rigorously, there are many tests that indicate that the conjecture is correct. One of them is symmetry consideration. Symmetries on both sides of the duality are identical. On the gravity side, $AdS_5 \times S^5$ has two symmetry groups, the $SO(6)$ of the 5-sphere, and $SO(4,2)$ of the of the AdS_5 space-time. On the other side, $\mathcal{N}=4$ super Yang-Mills has $SU(4)_R \sim SO(6)$ group as the R-symmetry and $SO(4,2)$ as the conformal symmetry in 4-dimensions.

5.4. Correlation functions

We now wish to understand how the AdS/CFT correspondence can be used to extract information about the strongly coupled gauge theory in 4-dimensions. Let us consider a gauge theory defined by the action, say $S_{4D}[q]$. The correlation functions of suitable gauge invariant operators, $\mathcal{O}(x^\mu) = f(q)$, which are functions of the field variable q encode all information about the gauge theory. The general idea in calculating the correlation functions using AdS/CFT correspondence is that the operators in gauge theory side are related to a field in the AdS_5 space. Corresponding to every operator $\mathcal{O}(x^\mu)$ of the conformal field theory in 4-dimensions, there exists a unique $\phi(x^\mu, z)$ field living in the 5-dimensional AdS_5 space. These fields are called bulk fields, which are related to the boundary field, $\phi_0(x^\mu)$, on the boundary of AdS_5 by the relation,

$$\phi(x^\mu, z) = z^{4-\Delta}\phi_0(x^\mu)\,, \tag{57}$$

where Δ is the conformal dimension of the operator $\mathcal{O}(x^\mu)$. The connected correlation functions can be calculated from the generating functional,

$$Z_{4D}[\phi_0] \equiv e^{W_{4D}[\phi_0]} = \int \mathcal{D}[q]\, \exp\left(iS_{4D}[q] + i\int d^4x\, \mathcal{O}(x)\phi_0(x)\right)\,, \tag{58}$$

by taking the successive functional derivative of the generating functional as,

$$\langle \mathcal{O}(x_1)\mathcal{O}(x_2)\cdots\mathcal{O}(x_n)\rangle_c = (-i)^n \left.\frac{\delta^{(n)} W_{4D}[\phi(x)]}{\delta\phi(x_1)\delta\phi_0(x_2)\cdots\delta\phi_0(x_n)}\right|_{\phi_0=0}\,. \tag{59}$$

These are very difficult to calculate in the strongly coupled regime of gauge theory. The AdS/CFT correspondence provides a procedure to calculate these correlation functions using the gravity theory in AdS_5.

Given the nature of the field, $\phi(x^\mu, z)$, depending on the operator $\mathcal{O}(x^\mu)$, we construct the lagrangian density, $\mathcal{L}(\phi, \partial_M\phi)$, and construct action, $S_{5D}[\phi(x^\mu, z)]$ in 5-dimensional space AdS_5 as,

$$S_{5D}[\phi] = \int d^5x \sqrt{\det(g_{MN})}\, \mathcal{L}(\phi, \partial_M\phi)\,, \tag{60}$$

where g_{MN} is the metric of AdS_5.

The AdS/CFT correspondence [24, 25] identifies the generating functional, $W_{4D}[\phi_0]$, of connected correlation functions of the gauge-invariant operator, $\mathcal{O}(x^\mu)$, in 4-dimensional gauge theory with the 5-dimensional action, $S_{5D}[\phi]$, as,

$$W_{4D}[\phi_0] = i \lim_{\epsilon\to 0} S_{5D}[\phi(x^\mu, \epsilon)]\,, \tag{61}$$

so that the correlation functions can be calculated using $S_{5D}[\phi]$,

$$\langle \mathcal{O}(x_1)\mathcal{O}(x_2)\cdots\mathcal{O}(x_n)\rangle_c = (-i)^{n-1} \lim_{\epsilon\to 0} \left.\frac{\delta^{(n)} S_{5D}[\phi(x,\epsilon)]}{\delta\phi(x_1,\epsilon)\delta\phi(x_2,\epsilon)\cdots\delta\phi(x_n,\epsilon)}\right|_{\phi_0=0}\,. \tag{62}$$

This can be viewed as the definition of gauge theory correlation functions in terms of the action, S_{5D}, in a higher dimensional space. This is an useful formulation of the AdS/CFT correspondence for computing correlation functions of the strongly coupled gauge theory.

6. Holographic QCD

Holographic QCD aims to study strongly coupled regime of QCD using AdS/CFT to get analytical model for hadron structure and their interactions [87, 88]. The AdS/CFT correspondence gives an equivalence between classical gravity theory in 5-dimensions and $\mathcal{N} = 4$ $SU(N_c)$ supersymmetric Yang-Mills theories (SYM) in the large N_c limit. QCD is fundamentally different from SYM theories, it is non-supersymmetric and conformal symmetry of classical QCD is broken by quantum effects. Exact string theory (or gravity) dual of QCD is not known yet. In this situation, to apply the remarkable AdS/CFT correspondence to QCD, we need to modify the AdS geometry to build a realistic gravity dual of QCD. Two Approaches are usually followed to construct holographic models of QCD.

(i) Top down approach: In this approach, one starts from string theory and chooses an appropriate D-brane configuration to get as close as possible to the realistic QCD [89, 90, 91, 92, 93, 94, 95, 96, 97, 98, 99, 100, 101, 102, 103, 104, 105, 106, 107, 108, 109, 110].The most closely related to QCD is of Sakai and Sugimoto [93, 94], where the 'color' Yang-Mills fields are introduced by the massless open string fluctuations of a stack of N_c D4-branes wrapped on a circle, and flavor degrees of freedom through fermionic open string fluctuations between D4-branes and a pair of N_f D8-branes and $\overline{D8}$-branes. The model is nonsupersymmetric and describes $SU(N_c)$ Yang-Mills theory, giving a geometrical picture of chiral symmetry breaking. In the strong coupling limit (large N_c), the stack of N_c color branes has a dual description in terms of a classical gravity theory. we shall not discuss this approach in this article.

(ii) Bottom up approach: In this approach, one looks at the hadron phenomenology of QCD first and then attempts to build its 5D-gravity equivalent model in AdS_5 space-time by introducing appropriate fields and modifying the geometry of the extra dimension to reproduce the phenomenology [111, 112, 113, 114, 115, 116, 117, 118, 119, 120, 121, 122, 123, 124, 125, 126, 127]. The Kaluza-Klien modes of five dimensional fields with quantum numbers of QCD states are identified with the hadronic resonances. The most important aspect of QCD at low energy is confinement. This means that there exists a maximum separation for quarks in the hadrons. The five dimensional model of QCD in AdS_5 must take this into account. Because of confinement, there must exist limiting separation of quarks and therefore a minimum value of energy scale r, or a maximum value of z. Therefore, AdS_5 space should end at a finite value in the fifth axis, $z_0 = \frac{1}{r_0} \equiv \frac{1}{\Lambda R^2}$. Cutoff at z_0 breaks conformal invariance and allows the introduction of the QCD scale. Introduction of cut-off in z-axis (to implement confinement) can be done in two ways,

   (a) Hard wall model: The lagrangian is multiplied by a step function, so that it vanishes beyond $z = z_0$. This is similar to the MIT bag model, but with the boundary condition on the fifth dimension.

   (b) Soft wall model: Introduce an additional field called dilaton field that breaks the scale symmetry. The lagrangian is multiplied by $e^{-D(z)}$, so that it vanishes slowly as z increases.

We shall discuss in the following some elements of how the mass spectrum of scalar and vector particles are generated in the gravity dual models models. We shall also discuss how chiral symmetry appears in holographic QCD.

6.1. Scalar mesons

We start with the action for a scalar (or pseudoscalar) field $\Phi(x^\mu, z)$ in five dimensional AdS_5, in which x^μ represent the four dimensional Minkowski space-time coordinates and z repesents the fifth coordinate. The action is written as,

$$S = \int d^4x\, dz \sqrt{|g|}\, e^{-D(z)} \frac{1}{2} \left(g^{MN} \partial_M \Phi \partial_N \Phi - \mu^2 \Phi^2 \right), \tag{63}$$

where the term $\mu^2\Phi^2$, though looks like a mass term in the five dimensional space, but, as we shall see, the parameter μ is not related to the masses in Minkowski space. A background dilaton field, $D(z)$, has been introduced in the action, which can be taken of the form as,

$$D(z) = \begin{cases} 0, & \text{Hard wall}, \\ \lambda^2 z^2, & \text{Soft wall}. \end{cases}$$

for the hard wall and soft wall scenarios. The metric of AdS_5 as given in (46) can be written, introducing, $A(z) = -\log z + \log R$, for convenience, as,

$$g_{AB} = e^{2A(z)} \eta_{AB}, \qquad g^{AB} = e^{-2A(z)} \eta^{AB}, \qquad \sqrt{|g|} = e^{5A(z)} \tag{64}$$

with η_{AB} as the five dimensional Minkowski metric. We can read off from (63) the lagrangian density as,

$$\mathcal{L} = e^{B(z)} \frac{1}{2} \left(\eta^{MN} \partial_M \Phi \partial_N \Phi - e^{2A(z)} \mu^2 \Phi^2 \right), \tag{65}$$

where $B(z) = 3A(z) - D(z)$. The equation of motion of $\Phi(x^\mu, z)$ can be written as the usual Euler-Lagrange equation as,

$$\partial_A \left(\frac{\partial \mathcal{L}}{\partial \partial_A \Phi} \right) - \frac{\partial \mathcal{L}}{\partial \Phi} = 0, \qquad \Rightarrow \qquad \eta^{\alpha\beta} \partial_\alpha \partial_\beta \Phi - \partial_z^2 \Phi - \partial_z B(z) \partial_z \Phi + e^{2A(z)} \mu^2 \Phi = 0. \tag{66}$$

We decompose, $\Phi(x, z)$, as,

$$\Phi(x, z) = e^{ip_\mu x^\mu} \phi(z), \tag{67}$$

a product of a plane wave part in four dimensional Minkowski space-time and a function $\phi(z)$ depending only on the fifth cordinate z. Substituting $\Phi(x, z)$ in (66) we get the equation for $\phi(z)$ as,

$$-\partial_z^2 \phi(z) - \partial_z B(z) \partial_z \phi(z) + e^{2A(z)} \mu^2 \phi(z) = m^2 \phi(z), \tag{68}$$

where, $m^2 = p^\mu p_\mu$, corresponds to the mass of the scalar(pseudoscalar) particle in four dimensions. Puttting, $\phi(z) = e^{-\frac{B(z)}{2}} \psi(z)$ we can eliminate the term containing first derivative in (68) and transform it to a Schrodinger like equation in $\psi(z)$ as,

$$-\partial_z^2 \psi(z) + V(z) \psi(z) = m^2 \psi(z), \tag{69}$$

with

$$V(z) = \frac{1}{2} \partial_z^2 B(z) + \left(\frac{1}{2} \partial_z B(z) \right)^2 + e^{2A(z)} \mu^2. \tag{70}$$

We now solve the equation for $\psi(z)$, (69), in hard wall and soft wall scenarios.

- Hard wall model: In this model there is no dilation field, so,

$$D(z) = 0\,, \quad \Rightarrow \quad B(z) = 3A(z) = 3\,(\ln L - \ln z)\,, \tag{71}$$

consequently the potential $V(z)$ from (70) becomes,

$$V(z) = \frac{1}{4z^2}\left(4\nu^2 - 1\right)\,, \tag{72}$$

with, $\nu^2 = \mu^2 R^2 + 4$. In addition, the space of AdS_5 is restricted in the fifth direction, $0 \leq z \leq z_0$, with the boundary condition that $\phi(z)$ vanishes at z_0,

$$\phi(z_0) = 0\,. \tag{73}$$

Changing the variable, $y = mz$, and $\psi(y) = \sqrt{y}\,f(y)$ the equation of motion can be transformed to the Bessel equation,

$$\left(-\partial_y^2 - \frac{1}{y}\partial_y + \frac{\nu^2}{y^2}\right) f(y) = f(y) \tag{74}$$

with solutions $f(y) = J_\nu(y)$. The solution of the field equation can be written as,

$$\phi(z) = \left(\frac{z}{R}\right)^{\frac{3}{2}} (mz)^{\frac{1}{2}}\, J_\nu(mz) \quad \sim \quad z^2 J_\nu(mz)\,. \tag{75}$$

The solution is valid for any value of m, thus it does not yield a discrete hadron mass spectrum. To get the discrete mass spectrum, we impose the boundary condition, (73), on the solution, which gives,

$$J_\nu(mz_0) = 0\,. \tag{76}$$

A discrete spectrum of masses of scalar (pseudoscalar) meson is generated through zeros of Bessel function. We get not only one, but a whole series with same index ν. In addition, there are resonances corresponding to the different values of ν. A general feature of string picture is realized. We would get similar pattern of the Regge trajectory, however these excitations are not linear in m^2, but, from the properties of zeros of Bessel function it is seen that these are approximately linear in m.

- Soft wall model: In this model there is a dilation field, $D(z)$, which is taken as,

$$D(z) = \lambda^2 z^2\,, \quad \Rightarrow \quad B(z) = 3A(z) - D(z) = 3\,(\ln L - k \ln z) - \lambda^2 z^2\,, \tag{77}$$

and the five dimensional space-time of AdS_5 is not restricted. The equation for ψ, (69), becomes,

$$-\,\partial_u^2 \psi(u) + \left(\frac{4\nu^2 - 1}{4u^2} + u^2\right)\psi(u) = \left(\frac{m^2}{\lambda^2} - 2\right)\psi(u)\,. \tag{78}$$

with $u = \lambda z$. The solution of this equation are the eigenfunctions,

$$\psi_n(u) = e^{-\frac{u^2}{2}} u^{\nu + \frac{1}{2}} L_n^\nu(u^2) \quad \Rightarrow \quad \phi_n(z) \sim e^{-\frac{\lambda^2 z^2}{2}} z^{2+\nu} L_n^\nu\left(\lambda^2 z^2\right)\,, \tag{79}$$

and eigenvalues,

$$m^2 = 2\lambda^2\,(2(n+1) + \nu)\,, \tag{80}$$

where L_n^ν is the associated Laguerre polynomials. We see that the soft wall model has the desired feature that the excitations are proportional to the squared mass, m^2. The agreement for the experimental values and the theoretical results from the soft wall model is satisfactory.

Journal of Physics: Conference Series **374** (2012) 012004 doi:10.1088/1742-6596/374/1/012004

6.2. Vector particles

The action for the vector field, $A_M(x^\mu, z)$, in five dimensional AdS_5 is written as,

$$S = \int d^4x\, dz \sqrt{|g|}\, e^{-D(z)} \left(g^{MM'} g^{NN'} F_{MN} F_{M'N'} \right), \tag{81}$$

where, F_{MN} is the field tensor given as,

$$F_{MN} = \partial_M A_N - \partial_N A_M , \tag{82}$$

and $D(z)$ is the dilaton field. The lagrangian can be read off from the action as,

$$\mathcal{L} = \frac{1}{4} e^{B(z)} \left(\eta^{MM'} \eta^{NN'} F_{MN} F_{M'N'} \right), \tag{83}$$

with, $B(z) = A(z) - D(z)$. In Minkowski space there are three spin components of a vector particle. We impose two gauge conditions to reduce 5 components of A_M to three components. The gauge conditions are,

$$A_5 = 0 , \qquad \eta^{MK} \partial_K A_M = 0 . \tag{84}$$

The equation of motion are the usual Euler-Lagrange equations wich can be written as,

$$\frac{\partial \mathcal{L}}{\partial A_B} - \partial_A \left(\frac{\partial \mathcal{L}}{\partial_A A_B} \right) = 0 , \quad \Rightarrow \quad -\partial_z B(z) \partial_z A_\alpha - \partial_z^2 A_\alpha + \eta^{\mu\nu} \partial_\mu \partial_\nu A_\alpha = 0 . \tag{85}$$

We decompose A_α into a plane wave part in the Minkowski space and another part depending only on the fifth coordinate z as,

$$A_\alpha(x^\mu, z) = e^{-ip_\mu x^\mu} \epsilon_\alpha(p)\, \phi(z) \tag{86}$$

where $\epsilon_\alpha(p)$ is the polarization vector of a transverse vector field, i.e. $\epsilon.p = 0$. The equation for $\phi(z)$ becomes,

$$\left(-\partial_z^2 - \partial_z B(z)\, \partial_z \right) \phi = m^2 \phi , \tag{87}$$

where, $m^2 = p^\mu p_\mu$ corresponds to the mass of vector particles. To eliminate the term containing first derivative we substitute, $\phi(z) = e^{-\frac{B(z)}{2}} \psi(z)$, so that the equation for $\psi(z)$ becomes a Schrondiger like equation,

$$-\partial_z^2 \psi(z) + V(z) \psi(z) = m^2 \psi(z) , \tag{88}$$

with the potential, $V(z)$, as,

$$V(z) = \frac{1}{4} \left(\partial_z B(z) \right)^2 + \frac{1}{2} \partial_z^2 B(z). \tag{89}$$

We solve (88) in hard wall and soft wall scenarios,

- Hard Wall Model: In this model there is no dilation field, so,

$$D(z) = 0 , \quad \Rightarrow \quad B(z) = A(z) = (\ln L - \ln z) , \tag{90}$$

consequently the potential $V(z)$ from (89) becomes, $V(z) = \frac{3}{4z^2}$, and the space of AdS$_5$ is restricted in the fifth direction, $0 \le z \le z_0$, with the boundary condition that $\phi(z)$ vanishes at z_0, $\phi(z_0) = 0$. The equation for $\psi(z)$ becomes,

$$\partial_z^2 \psi(z) + \frac{3}{4z^2}\psi(z) = m^2 \psi(z)\,, \tag{91}$$

The solution of this equation can be written in terms of the Bessel function as,

$$\psi(z) = \sqrt{mz}\, J_1(mz)\,, \quad \Rightarrow \quad \phi(z) = e^{-\frac{B(z)}{2}}\psi(z) = \sqrt{\frac{m}{R}}\, z\, J_1(mz) \tag{92}$$

The five dimensional field vanishes at the hard wall, $\phi(z_0) = 0$. The masses, m_n, of the vector meson resonances are then determined by the zeros of first order Bessel function. The mass spectrum, as found in the scalar particle case, is not linear in m^2.

- Soft Wall Model: In this model there is a dilaton field, $D(z) = \lambda^2 z^2$, an the AdS_5 is not restricte. The potential, $V(z)$ can be obtained from (89) as,

$$V(z) = \frac{1}{4}\left(\partial_z B(z)\right)^2 + \frac{1}{2}\partial_z^2 B(z) = \frac{3}{4z^2} + \lambda^4 z^2 \tag{93}$$

The equation for ψ becomes,

$$\partial_z^2 \psi(z) + \left(\frac{3}{4z^2} + \lambda^4 z^2\right)\psi(z) = m^2 \psi(z)\,, \tag{94}$$

the solution of which can be written as,

$$\psi_n(z) = e^{-\frac{\lambda^2 z^2}{2}}\,(\lambda z)^{\frac{3}{2}}\, L_n^1(\lambda^2 z^2)\,, \quad \Rightarrow \quad \phi_n(z) = e^{-\frac{B(z)}{2}}\psi(z) = \sqrt{\frac{\lambda^3}{L}}\, z^2\, L_n^1(\lambda^2 z^2)\,, \tag{95}$$

where L_n^1 are the Laguerre polynomials. The mass spectrum is obtained from the eigenvalues,

$$m^2 = 4\lambda^2\,(n+1)\,. \tag{96}$$

The mass spectrum is linear in m^2 in the soft wall scenario for the vector particles also.

6.3. Holographic model of QCD

In the holographic QCD model we consider a 5-dimensional dual of the two flavor, (u, d) sector of QCD. In this case, QCD has a global $SU(2)_L \times SU(2)_R$ chiral symmetry. The conserved currents corresponding to this symmetry are the vector and axial-vector currents which can be expressed in terms of left handed and right handed currents, $j_L^{\mu\,a} = \bar{q}_L \gamma^\mu \tau^a q_L$ and $j_R^{\mu\,a} = \bar{q}_R \gamma^\mu \tau^a q_R$, where τ^a's are $SU(2)$ flavor generators. To describe the chiral dynamics, we shall also need an operator called chiral condensate, $\bar{q}_R^\alpha q_L^\beta$, where α and β are $SU(2)$ flavor indices in the fundamental representation. The holographic QCD does not refer explicitly to quark and gluon degrees of freedom, only deals with the bound states of QCD. According to the AdS/CFT correspondence, there is a one-to-one map between 4-dimensional operators in field theory living on the boundary and 5-dimensional bulk fields in gravity side. The global chiral symmetry $SU(2)_L \times SU(2)_R$ of QCD is made local symmetry in the gravity side. Therefore, we need to introduce local gauge (vector) fields L_M^a and R_M^a of which the values at z = 0 play a role of external sources for $SU(2)_L$ and $SU(2)_R$ currents, $j_L^{\mu\,a}$ and $j_R^{\mu\,a}$ respectively. The vector and axial-vector gauge

	4D: $\mathcal{O}(x)$	5D: $\phi(x^\mu, z)$	p	Δ	m_5^2
Current Operator :	$j^a_{L\mu} = \bar{q}_L\gamma_\mu t^a q_L$	$L^a_M(x^\mu, z)$	1	3	0
	$j^a_{R\mu} = \bar{q}_R\gamma_\mu t^a q_R$	$R^a_M(x^\mu, z)$	1	3	0
Quark bilinear :	$\bar{q}^\alpha_L q^\beta_R$	$\frac{2}{z}X^{\alpha\beta}(x^\mu, z)$	0	3	-3

Table 2. Correspondence of operators in gauge theory side to fields in gravity side.

fields are defined as $V^a_M = (L^a_M + R^a_M)/\sqrt{2}$ and $A^a_M = (L^a_M - R^a_M)/\sqrt{2}$. Since chiral symmetry is known to be broken to $SU(2)_V$ spontaneously, we introduce a field $X^{\alpha\beta}$ corresponding to $\bar{q}^\alpha_R q^\beta_L$ in order to realize the spontaneous chiral symmetry in the AdS_5 side. In table (2) we give the correspondence of operators and fields in QCD side and gravity side respectively. The 5-dimensional mass of the bulk gauge field L, R and X is determined by the relation,

$$m_5^2 R^2 = (\Delta - p)(\Delta + p - 4)\,, \tag{97}$$

[24, 25] where Δ denotes the conformal dimension of the corresponding operator with spin p and R is the AdS_5 curvature. The mass of the gauge field in the bulk of five dimensional space-time turns out to be, $m_5^2 = 0$, which is expected because of gauge symmetry, and the mass of the scalar X is, $m_5^2 = -\frac{3}{R^2}$.

The action in five dimensional space-time for the meson sector in holographic QCD is constructed taking the fields of interest with the appropriate quantum numbers and Lorentz structures as [116, 117],

$$S_{5D} = \int d^4x \int dz\, \sqrt{|g|}\, e^{-D(z)}\, \mathrm{Tr}\left[|DX|^2 + \frac{3}{R^2}X^2 - \frac{1}{4g_5^2}\left(\left|F^L\right|^2 + \left|F^R\right|^2\right)\right], \tag{98}$$

where, $|DX|^2 = (D_M X)^\dagger \left(D^M X\right)$ with the covariant derivative, $D_M X = \partial_M X - iL_M X + iXR_M$, $\left|F^L\right|^2 = F^L_{MN}F^{L\,MN}$ with the field tensors, $F^L_{MN} = \partial_M L_N - \partial_N L_M - i\,[L_M, L_N]$ (and analogous expression for $\left|F^R\right|^2$) and $L_M = \tau^a L^a_M\,, R_M = \tau^a R^a_M$. The gauge coupling, g_5, in five dimensional space-time is fixed by matching the vector correlation function in five dimension to that from the operator product expansion, $g_5 = 12\pi^2/N_c$ [116]. The action (98) contains the dilaton field $D(z)$ for soft wall model, however, by making, $D(z) = 0$, we can consider this lagrangian for hard wall model.

The scalar field, $X(x^\mu, z)$, in five dimensional space is taken as a product,

$$X^{\alpha\beta}(x^\mu, z) = X_0^{\alpha\beta}(z)\, e^{i\tau^a\pi^a(x^\mu, z)}\,, \tag{99}$$

in which the vacuum expectation value, $X_0^{\alpha\beta}(z) = \langle X^{\alpha\beta}(x^\mu, z)\rangle$ depends on z coordinate alone. The pion field $\pi^a(x^\mu, z)$ appearing in (99) is dimensionless and related to the canonically-normalized pion field $\tilde{\pi}^a(x^\mu, z)$ of chiral lagrangians via $\pi^a = \frac{\tilde{\pi}^a}{f_\pi}$ with $f_\pi = 93\,\mathrm{MeV}$. The vacuum expectation value $X_0^{\alpha\beta}(z)$ spontaneously breaks the (approximate) chiral symmetry by forming the quark condensate. The field $X_0^{\alpha\beta}(z)$ is determined solving the equation of motion of this field obtained from the action (98) by turning off all fields except $X_0^{\alpha\beta}(z)$. The equation of motion for $X_0^{\alpha\beta}(z)$ can be derived using usual variational procedure as,

$$\partial_z\left(\frac{1}{z^3}\partial_z X_0^{\alpha\beta}(z)\right) + \frac{3}{z^5}X_0^{\alpha\beta}(z) = 0\,. \tag{100}$$

The general solution to this homogeneous linear second order ordinary differential equation is given by the polynomial,

$$2X_0^{\alpha\beta}(z) = M^{\alpha\beta}z + \Sigma^{\alpha\beta}z^3\,, \tag{101}$$

where two integration constants, $M^{\alpha\beta}$ and $\Sigma^{\alpha\beta}$ are determined by the boundary conditions at $z =$ and $z = z_0$ in the hard wall model as,

$$M^{\alpha\beta} = \lim_{z\to 0}\frac{2X_0^{\alpha\beta}(z)}{z}\,, \qquad \Sigma^{\alpha\beta} = \frac{2X_0^{\alpha\beta}(z_0) - M^{\alpha\beta}z_0}{z_0^3}\,.$$

The physical meanings of $M^{\alpha\beta}$ and $\Sigma^{\alpha\beta}$ are respectively the bare quark mass responsible for explicit breaking of chiral symmetry and the chiral condensate responsible for spontaneous chiral symmetry breaking.

The equations of motion of other fields of interest are obtained varying the action with respect to the corresponding fields. The axial-like gauge, $V_5(x^\mu, z) = 0$, $A_5(x^\mu, z) = 0$ has been found to be convenient to work with. The axial vector field A_μ is written as a sum of a transverse (divergence less) part, $A_{\mu\perp}$, and a longitudinal part ϕ, $A_\mu = A_{\mu\perp} + \phi$ with $\partial_\mu A_{\mu\perp} = 0$. The equations of motion for V_μ^a (vector particles), $A_{\mu\perp}^a$ (transverse component of vector particles), ϕ (longitudinal component of vector particles) and π^a (pseudoscalar particles) become, after taking the Fourier transform with respect to the four dimensional coordinates x^μ, one dimensional equations as,

$$\partial_z\left(\frac{e^{-D(z)}}{2}\partial_z V_\mu^a(q^\mu, z)\right) + \frac{q^2}{2}e^{-D(z)}V_\mu^a(q^\mu, z) = 0\,, \tag{102}$$

$$\left[\partial_z\left(\frac{e^{-D(z)}}{z}\partial_z A_\mu^a(q^\mu, z)\right) + \frac{q^2}{2}e^{-D(z)}A_\mu^a(q^\mu, z) - \frac{g_5^2}{z^3}e^{-D(z)}v(z)^2 A_\mu^a(q^\mu, z)\right]_\perp = 0\,, \tag{103}$$

$$\partial_z\left(\frac{e^{-D(z)}}{z}\partial_z \phi^a(q^\mu, z)\right) + \frac{g_5^2}{z^3}e^{-D(z)}v(z)^2\left(\pi^a(q^\mu, z) - \phi^a(q^\mu, z)\right) = 0\,, \tag{104}$$

$$-q^2\partial_z\phi^a(q^\mu, z) + \frac{g_5^2}{z^2}v(z)^2\partial_z\pi^a(q^\mu, z) = 0\,, \tag{105}$$

with q^μ as the momentum in four dimensions. These equations are solved as eigenvalue equations subject to appropriate boundary conditions of hard wall and soft wall models. In the hard wall model, the masses of the mesons appear from the zeros of the solutions at the infra red cut off z_0. As discussed earlier, the hard wall model does not reproduce the linear relation in square of mass. In soft wall model the mass spectrum appears from the eigenvalues and does reproduce the linear relation in square of mass. The mass spectrum of soft wall model matches well with the experimental values. The solutions can be used to calculate form factors of the mesons and other correlators.

The electromagnetic form factor, for example of a scalar particle, with momentum transfer $q = p - p'$ in holographic QCD can be defined in the same way as the form factor in four dimension,

$$F(Q^2)\,(2\pi)^4\,\delta^4\,(p - p' - q)\,\epsilon.q = \int d^4x dz\sqrt{|g|}g^{LL'}\,\Phi_{p'}^*(x^\mu, z)i\partial_L\Phi_p(x^\mu, z)A_{L'}(x^\mu, z)\,, \tag{106}$$

where $\Phi_p(x^\mu, z)$ are the solutions of equations of motion for scalar particles and $A_{L'}(x^\mu, z)$ is an external electromagnetic field in 5 dimensions with the polarization vector ϵ. The electromagnetic field $A_{L'}$ satisfies Maxwell's equations in 5 dimensions. Using solutions of the equations of motion for appropriate particles we can find the electromagnetic form factors of mesons in hard wall and soft wall scenarios. Pion electromagnetic form factors have been calculated using holographic QCD. The form factor reproduce the experimental data quite well.

The holographic QCD can be extended to baryons [128]. A spin $\frac{1}{2}$ baryon operator in the Minkowski space (boundary of AdS_5) will correspond to a Dirac field in the bulk of AdS_5. The action in five dimensions for the Dirac field can be written with constraints of Lorentz invariance. By solving equation of motion with appropriate boundary conditions tower of hadronic resonance states can be generated. Similar procedure can be adapted as well for higher spin particles.

7. Thermal Gauge Gravity Duality

We want to construct gravity dual of gauge theory at finite temperature to describe, systems like quark gluon plasma. The tool of AdS/CFT correspondence can be applied in these systems to learn about strongly coupled gauge theories at finite temperature by doing calculations in classical gravity [26]. In order to introduce a temperature into the gauge theory there are two approaches which are valid in two different regimes,

(i) Thermal AdS: Temperature can be introduced in the gravity theory in AdS_5 space-time by Wick rotating the Minkowski time of AdS to euclidean time ($t \to i\tau$) in AdS_5 metric,

$$ds^2_{AdS_5} = \frac{R^2}{z^2}\left(d\tau^2 + dx^i dx_i + dz^2\right), \tag{107}$$

and making τ periodic (anti-pereodic) for bosons (fermions), by identifying, $\tau \sim \tau + \beta$, as done in equilibrium thermal field theory. The period (circumference of the 'thermal circle') is related to the temperature by, $\beta = \frac{1}{T}$, so that at high temperature the euclidean time is on a small circle, and at low temperature the circle unwinds and the original theory on Minkowski space is recovered. The temperature T is identified with the temperature of the field theory on the boundary, namely SYM theory since, τ also corresponds to the euclidean time coordinate of the boundary theory. In AdS/CFT correspondence, this description corresponds to the low temperature regime of gauge theory.

(ii) Black hole AdS: In this case, temperature is introduced in the gravity theory in AdS_5 space-time through a black hole metric,

$$ds^2_{AdS_5} = \frac{R^2}{z^2}\left(f(z)d\tau^2 + dx^i dx_i\right) + \frac{R^2}{z^2 f(z)}dz^2, \qquad f(z) = 1 - \frac{z^4}{z_0^4}, \tag{108}$$

since black holes have a temperature due to Hawking process [78]. The event horizon is located at $z = z_0$, which has three flat directions $\vec{x}$. The metric (108) is thus called a black three-brane metric. Hawking temperature, T_H, for the metric (108) is, $T_H = \frac{1}{\pi z_0}$. This description corresponds to high temperature regime of gauge theory.

There should be a temperature where the description of finite temperature SYM by thermal AdS and the black hole AdS coincide. It was shown by Hawking and Page [129] that at some critical temperature, T_{HP}, there is a first order phase transition where thermal AdS metric tunnels to black hole AdS metric. Through gauge gravity correspondence this transition can be interpreted in the gauge theory as a phase transition between confined to de-confined phase. The de-confinement temperature, T_c, obtained using holgraphic QCD in the hard and soft wall models is 122 MeV and 191 MeV respectively [130].

The quark-gluon plasma was expected to be a free gas of quarks and gluons, however, at the experimentally accessible energies the observations seem to indicate that it behaves like a almost perfect-fluid with small shear viscosity [10]. This observation leads to the conclusion that quark-gluon plasma is a strongly coupled system, within the non-perturbative regime of QCD.

The AdS/CFT correspondence can be applied to such situations. It identifies strongly coupled plasma at finite temperature in the gauge theory side with a black brane in gravity side. There exists a one-to-one correspondence between the thermodynamic properties of the black hole and those of the plasma, namely its temperature and entropy. The plasma properties thus should be calculable using black branes. The AdS/CFT correspondence relates the correlators of the energy-momentum tensor, $T_{xy}(t, \vec{x})$, in two space-time points at zero frequency, $\omega = 0$, and the absorption cross section, $\sigma(\omega)$, of a graviton by the static black brane in the bulk,

$$\sigma(\omega) = \frac{8\pi G}{\hbar\omega} \int d^4x\, e^{i\omega t} \left\langle [T_{xy}(t, \vec{x}), T_{xy}(0,0)] \right\rangle . \tag{109}$$

The shear viscosity, η, of plasma is given by a Kubo formula,

$$\eta = \lim_{\omega \to 0} \frac{1}{2\hbar\omega} \int d^4x\, e^{i\omega t} \left[T_{xy}(t, \vec{x}), T_{xy}(0,0)\right] , \tag{110}$$

from which we can relate the shear viscosity and graviton absorption cross section by black brane as,

$$\eta = \frac{\lim_{\omega \to 0} \sigma(\omega)}{16\pi G} . \tag{111}$$

According to a general theorem [132], the graviton absorption cross-section of black hole in the low energy limit ($\omega \to 0$) is, $\sigma(\omega) = A$, where A is the horizon area. The entropy density, s, of the black brane is, $s = \frac{A}{4G}$. We get the viscosity to entropy density ratio of strongly coupled plasma of $\mathcal{N} = 4$ SYM theory as,

$$\frac{\eta}{s} = \frac{\hbar}{4\pi k_B} . \tag{112}$$

The shear viscosity was first computed [27] for the D3 brane corresponding to the $\mathcal{N} = 4$ SYM which is a conformal field theory. This result is extremely robust because it depends only on the universal properties of black hole horizons. All gauge theory dual to gravity should fulfill this relation in spite of their different field contents and type of supersymmetry [28]. This is a kind of universality relation. However, in the absence of gravity dual of QCD it is not clear yet if the result (112) holds for QGP.

8. Summary

In this article we have tried to present in a simplistic way an overview of the developments that has taken place over many years in understanding strong interaction.

In the 1970's QCD was established as the quantum field theoretical description of strong interaction, which by now is extremely well tested experimentally. The degrees of freedom of QCD are quarks and gluons that transform respectively in the fundamental and adjoint representation of the non-abelian local (color) gauge group $SU(3)$. The interaction between quarks is mediated by gluons. QCD has the desired feature of asymptotic freedom. However, QCD remained intractable for the problems of hadron physics at low energies. The mechanism of confinement of quarks into hadron is still not clear.

In the late 1960's phenomenological understanding of strong interaction in terms of S-matrix and Regge trajectories led to the formulation of string theory. It was soon realized that string theory including both bosons and fermions need 10-dimensional space-time to be consistent. It also had inescapably a graviton like massless spin-2 excitation mode. It became a promising candidate for quantum gravity. Energy regime of string theory is in the Planck scale ($\sim 10^{19}$

GeV), making the predictions of string theory almost impossible to verify experimentally. So, string theory was given up as a theory of the strong interactions.

QCD and string theory became two independent successful theory. However, it was always hoped from both sides to find a link between them, from string theory side the link would provide some way for its experimental verification, and from QCD side it may give some new path to understand the low energy phenomena of confinement, chiral symmetry breaking and hadronic interactions. There had been many hints to expect that string theory may be a potential candidate to provide new insights in QCD.

The holographic principle developed in the context of black hole thermodynamics proposed that there must be a duality between a QCD like gauge theory and a gravity theory in higher dimensional space-time. However, it was not possible to find an useful connection between QCD and string theory until the discovery of D-branes in superstring theory.

D branes are special non-perturbative objects in string theory. On the one hand, from the closed string point of view, D branes are soliton like solutions of low energy effective theory of superstring theories. The excitations here are the gravitational modes of closed string vibrations in the space-time curved by by the mass of D branes. On the other hand, from open string point of view, these are hypersurfaces in 10-dimensional space-time, on which open strings end. The excitations in this case are gauge bosons produced due to the motion of charged ends of open strings on the D branes.

The dream of finding an useful connection between QCD and string theory was partially fulfilled with the AdS/CFT correspondence conjectured by Maldacena in 1997, noting two different descriptions of the same object D branes. Tn the original form of the correspondence, the gauge theory is, $\mathcal{N} = 4\,SU(N_c)$ supersymmetric Yang-Mills theory in 4-dimensional Minkowski space-time, and in the gravity theory is type IIB string theory in $AdS_5 \times S^5$ space-time. In the large N_c limit the gravity theory becomes classical.

QCD is very different from $\mathcal{N} = 4\,SU(N_c)$ supersymmetric Yang-Mills theory. So it is necessary to break supersymmetry and conformal invariance by modifying AdS_5 space to apply AdS/CFT correspondence to QCD. Two approaches, namely top-down and bottom-up, have been developed to achieve the above goal. There are considerable success in both the approaches in understanding low energy features of QCD.

Strongly coupled gauge theories at finite temperature have also been studied with the help of AdS?CFT correspondence by introducing black hole horizon in AdS_5 space. These studies give an universal lower limit for the viscosity to entropy ratio for strongly coupled plasma.

Criticisms have been raised against the holographic QCD approach because strictly AdS/CFT works for SYM theories. However, as long as gravity dual to QCD not discovered, we may consider holographic QCD as a effective low-energy model independent of the AdS/CFT correspondence.

We may conclude this article by saying that strong interaction is indeed very difficult to understand, and we have a long way ahead to go to fully understand it.

9. Acknowledgements

ABS is grateful to INFN-LNS, Catania, Italy for hospitality, where part of the work was done.

References

[1] S. Narison, "QCD as a Theory of Hadrons: From Partons to Confinement", Cambridge Univ. Press (2004).
[2] C. E. DeTar, J. F. Donoghue, "Bag Models of Hadrons", Ann. Rev. Nucl. Part. Sc., 33 (1983) 235.
[3] A. W. Thomas, W. Weise, "The structure of the nucleon", Wiley (2001).
[4] A Pich, "Chiral perturbation theory" Rep. Prog. Phys. 58 (1995) 563.
[5] B. Kubis, "An introduction to chiral perturbation theory" in "Physics and Astrophysics of Hadrons and Hadronic Matter", Ed. A. B. Santra, Narosa, New Delhi (2008), arXiv:hep-ph/0703274.
[6] H. J. Rothe, "Lattice Gauge Theories: An Introduction", World Scientific (2005).

[7] C. Gattringer, C. B. Lang, "Quantum chromodynamics on the lattice", Springer (2010).
[8] J. C. Collins, M. J. Perry, "Superdense Matter: Neutrons Or Asymptotically Free Quarks? ," Phys. Rev. Lett. 34 (1975) 1353
[9] N. Cabibbo, G. Parisi, "Exponential Hadronic Spectrum and Quark Liberation," Phys.Lett. B59 (1975) 67.
[10] E. V. Shuryak, "What RHIC experiments and theory tell us about properties of quark-gluon plasma?", Nucl. Phys. A 750 (2005) 64, arXiv:hep-ph/0405066.
[11] J. Casalderrey-Solana, H. Liu, D. Mateos, K. Rajagopal, U. A. Wiedemann, "Gauge/String Duality, Hot QCD and Heavy Ion Collisions," arXiv:hep-ph/1101.0618.
[12] D. Mateos, "Gauge/string duality applied to heavy-ion collisions: limitations, insights and prospects", J. Phys. G: Nucl. Part. Phys. 38 (2011)124030, arXiv:hep-th/1106.3295.
[13] R. Dolen, D. Horn, C. Schmid, "Finite-Energy Sum Rules and Their Application to N Charge Exchange", Phys. Rev. 166 (1968) 1768.
[14] C. Schmid, "Direct-Channel Resonances from Regge-Pole Exchange", Phys. Rev. Letters 20, 689 (1968).
[15] G. Veneziano, "Construction of a crossing-symmetric, Regge-behaved amplitude for linearly rising trajectories", Nuovo Cimento A 57 (1968) 190.
[16] Y. Nambu, "Quark model and the factorization of the Veneziano amplitude," in Symmetries and Quark Models, ed. R. Chand, Gordon and Breach (1970).
[17] H. B. Nielsen, "An almost physical interpretation of the integrand of the n-point Veneziano amplitude," submitted to the 15th International Conference on High Energy Physics, Kiev (1970).
[18] L. Susskind, "Dual-symmetric theory of hadrons," Nuovo Cim. 69A (1970) 457.
[19] V. A. Matveev, R.M. Muradian, A. hep-th/ 9805156.N. Tavkhelidze, "Automodellism in the large - angle elastic scattering and structure of hadrons," Lett. Nuovo Cim. 7 (1973) 719.
[20] S. J. Brodsky, G. R. Farrar, "Scaling Laws at Large Transverse Momentum.," Phys. Rev. Lett 31 (1973) 1153.
[21] J. Scherk and J. Schwarz, Dual models for non-hadrons, Nucl. Phys. B81 (1974) 118.
[22] T. Yoneya,"Connection of Dual Models to Electrodynamics and Gravidynamics," Prog. Theor. Phys. 51 (1974) 1907.
[23] J. M. Maldacena, "The Large N limit of superconformal field theories and supergravity," Adv. Theor. Math. Phys. 2 (1998) 231, arXiv:hep-th/9711200.
[24] S. S. Gubser, I. R. Klebanov, "Gauge theory correlators from non-critical string theory," A. M. Polyakov, Phys. Lett. B428 (1998) 105, arXiv:hep-th/9802109.
[25] E. Witten, "Anti-de Sitter space and holography,"Adv. Theor. Math. Phys. 2 (1998) 253, arXiv:hep-th/9802150.
[26] E. Witten,"Anti-de Sitter space, thermal phase transition, and confinement in gauge theories" Adv. Theor. Math. Phys. 2 (1998) 505, arXiv:hep-th/9803131.
[27] G. Policastro, D. T. Son and A. O. Starinets, "The shear viscosity of strongly coupled $\mathcal{N} = 4$ supersymmetric Yang-Mills plasma, Phys. Rev. Lett. 87 (2001) 081601, arXiv:hep-th/0104066.
[28] P. Kovtun, D. T. Son and A. O. Starinets, "Holography and hydrodynamics: Diffusion on stretched horizons," JHEP 0310 (2003) 064, arXiv:hep-th/0309213.
[29] H, Yukawa, "On the Interaction of Elementary Particles," Proc. Phys. Math. Soc. Jap. 17 (1935) 48.
[30] G. P. S. Occhialini, C. F. Powell, "Processes involving charged mesons," Nature 159 (1947) 186.
[31] M. Gellmann, Y. Neeman, "The Eightfold Way," W. A. Benjamin, 1964, New York.
[32] M. Gell-Mann, "A Schematic Model of Baryons and Mesons," Phys.Lett. 8 (1964) 214.
[33] G. Zweig, "An SU(3) model for strong interaction symmetry and its breaking," in Developments In The Quark Theory Of Hadrons, Vol. 1, CERN (1964)
[34] O. W. Greenberg, "Spin and Unitary Spin Independence in a Paraquark Model of Baryons and Mesons," Phys. Rev. Lett. 13 (1964) 598.
[35] M. Y. Han and Y. Nambu, "Three-Triplet Model with Double SU(3) Symmetry," Phys. Rev. 139 B1006.
[36] J. I Friedman and H. W. Kendall, "Deep Inelastic Electron Scattering," Ann. Rev. Nucl. Sci., 22 (1972) 203.
[37] H. Fritzsch, M. Gell-Mann and H. Leutwyler, "Advantages of the Color Octet Gluon Picture," Phys. Lett. B 47 (1973) 365.
[38] C. N. Yang, R. L. Mills, "Conservation of Isotopic Spin and Isotopic Gauge Invariance," Phys. Rev. 96 (1954) 191.
[39] D. J. Gross and F. Wilczek, "Ultraviolet Behavior of Non-Abelien Gauge Theories," Phys. Rev. Lett. 30 (1973) 1343.
[40] H. D. Politzer, "Reliable Perturbative Results for Strong Interactions?," Phys. Rev. Lett. 30 (1973) 1346.
[41] Particle Data Group, http://pdg.lbl.gov.
[42] G. 't Hooft, "A Planar Diagram Theory for Strong Interactions," Nucl. Phys. B72 (1974) 461.

[43] E. Witten, "Baryons in the 1/N Expansion," Nucl. Phys. B160 (1979) 57.
[44] J. Gasser and H. Leutwyler," Chiral Perturbation Theory to One Loop," Annals Phys. 158 (1984) 142.
[45] G. S. Bali, "QCD forces and heavy quark bound states,"Phys. Rept. 343, 1 (2001), arXiv:hep-ph/0001312.
[46] T.T. Takahashi, H. Matsufuru, Y. Nemoto and H. Suganuma, "Three-Quark Potential in SU(3) Lattice QCD," Phys. Rev. Lett. 86, 18 (2001), arXiv:hep-lat/0006005.
[47] T. Kawanai, S. Sasaki, "Interquark potential with finite quark mass from lattice QCD," Phys. Rev. Lett. 107 (2011) 091601, arXiv:hep-lat/1102.3246.
[48] G.F. Chew, S. C. Frautschi, "Regge trajectories and the principle of maximum strength for strong interactions," Phys. Rev. Lett. 8 (1962) 41.
[49] R. Hagedorn, "Statistical thermodynamics of strong interactions at high-energies," Nuovo Cim. Suppl. 3 (1965) 147.
[50] M.B. Green, J.H. Schwarz, E. Witten, "Superstring theory, Vols. I & II," Cambridge University Press (1987).
[51] D. Lust, S. Theisen, "Lectures on string theory," Lect.Notes Phys. 346 (1989).
[52] J. Polchinski, "String theory, Vols. I & II," Cambridge Univ. Press (1998)
[53] C. V. Johnson, "D-branes," Cambridge Univ. Press (2003).
[54] R. J. Szabo, "An Introduction to String Theory and D-Brane Dynamics," Imperial College Press, (2004).
[55] B. Zwiebach, "A First Course in String Theory," Cambridge University Press (2004).
[56] K. Becker, M. Becker, and J. Schwarz, "String Theory and M-Theory: A Modern Introduction," Cambridge Univ. Press (2007).
[57] E. Kiritsis, "String theory in a nutshell," Princeton Univ. Press (2007).
[58] J. Wess, J. Bagger, "Supersymmetry and supergravity," Princeton Univ. Press (1992).
[59] S. Weinberg, "The Quantum Theory of Fields, Volume III: Supersymmetry," Cambridge Univ. Press (2000).
[60] S. Coleman and J. Mandula, "All Possible Symmetries of the S Matrix," Phys. Rev. 159 (1967) 1251.
[61] F. Gliozzi, D.I. Olive and J. Scherk,"Supersymmetry, Supergravity Theories and the Dual Spinor Model," Nucl. Phys. B122 (1977) 253.
[62] P. Candelas, G. T. Horowitz, A. Strominger, E. Witten,"Vacuum Configurations for Superstrings," Nucl. Phys. B258 (1985) 46.
[63] S. E. Hjelmeland and U. Lindstrom, "Duality for the non-specialist," arXiv:hep-th/9705122.
[64] A. Sen, "An introduction to non-perturbative string theory," in "Cambridge 1997, Duality and supersymmetric theories", arXiv:hep-th/9802051.
[65] C. Hull and P. K. Townsend, "Unity of superstring dualities," Nucl. Phys. B438 (1995) 109, arXiv:hep-th/9410167.
[66] E. Witten, "String theory dynamics in various dimensions," Nucl. Phys. B443 (1995) 85, arXiv:hep-th/9503124.
[67] J.Polchinski, "Dirichlet Branes and Ramond-Ramond charges," Phys. Rev. Lett. 75 (1995) 4724, arXiv:hep-th/9510017.
[68] J. Polchinski, "TASI lectures on D-branes," arXiv:hep-th/9611050.
[69] A. Giveon and D. Kutasov, "Brane Dynamics and Gauge Theory," Rev. Mod. Phys. 71 (1999) 983, arXiv:hep-th/9802067.
[70] T. H. R. Skyrme, "A Nonlinear field theory," Proc. Roy. Soc. A 260 (1961) 127.
[71] W. Thirring, "A Soluble relativistic field theory?," Ann. Phys. (N.Y.) 3 (1958) 91.
[72] S. Coleman, "Quantum sine-Gordon equation as the massive Thirring model," Phys. Rev. D 11, (1975) 2088.
[73] J. R. Oppenheimer, H. Snyder, "On continued gravitational contraction," Phys. Rev. 56 (1939) 455.
[74] R. Penrose, "Gravitational collapse: the role of general relativity," Riv. del Nuovo Cimento 1, (1969) 252.
[75] S. W. Hawking, "Black holes in general relativity," Commun. Math. Phys. 25 (1972) 152.
[76] W. Israel, "Event Horizons in Static Vacuum Space-Times," Phys. Rev., 164 (1967) 1776.
[77] J. D. Bekenstein, "Black holes and entropy," Phys. Rev. D7 (1973) 2333.
[78] S. W. Hawking, "Particle Creation by Black Holes," Commun. Math. Phys. 43 (1975) 199.
[79] A. Strominger, C. Vafa, "Microscopic origin of the Bekenstein-Hawking entropy," Phys. Lett. B379 (1996) 99, hep-th/9601029.
[80] G.t Hooft, "Dimensional Reduction in Quantum Gravity," in Salam Festschrit, eds. A. Aly, J. Ellis and S. Randjbar-Daemi, World Scientic, Singapore, 1993, arXiv:gr-qc/9310026.
[81] L.Susskind, "The World as a Hologram," J. Math. Phys. 36 (1995) 6377, arXiv:hep-th/940989.
[82] O. Aharony, S. S. Gubser, J. M. Maldacena, H. Ooguri, and Y. Oz, "Large N field theories, string theory and gravity," Phys. Rept. 323 (2000) 183, arXiv:hep-th/9905111.
[83] J. L. Petersen, "Introduction to the Maldacena conjecture on AdS/CFT," Int. J. Mod. Phys. A14 (1999) 3597, arXiv:hep-th/9902131.

[84] I. R. Klebanov, "TASI Lectures: Introduction to the AdS/CFT Correspondence," arXiv:hep-th/0009139.
[85] E. D'Hoker, D. Z. Freedman, "Supersymmetric gauge theories and the AdS / CFT correspondence," arXiv:hep-th/0201253.
[86] L. Brink, J. Scherk and J.H. Schwarz, "Supersymmetric Yang-Mills Theories.," Nucl. Phys. B121 (1977) 77.
[87] I. R. Klebanov, "QCD and String Theory," Int. J. Mod. Phys. A21 (2006) 1831, arXiv:hep-ph/0509087.
[88] S. S. Gubser, A. Karch, "From gauge-string duality to strong interactions: a Pedestrian's Guide," arXiv:hep-th/0901.0935.
[89] I. R. Klebanov, M. J. Strassler, "Supergravity and a confining gauge theory: Duality cascades and chi SB resolution of naked singularities," JHEP 0008 (2000) 052, arXiv:hep-th/0007191.
[90] Karch, A., Katz, E. "Adding avor to AdS/CFT," JHEP 0206 (2002) 043, arXiv:hep-th/0205236.
[91] M. Kruczenski, D. Mateos, R. C. Myers, D. J. Winters, " Towards a holographic dual of large N(c) QCD," JHEP 0405 (2004) 041, arXiv:hep-th/0311270.
[92] J. Babington, J. Erdmenger, N. J. Evans, Z. Guralnik, I. Kirsch, "Chiral symmetry breaking and pions in nonsupersymmetric gauge / gravity duals.," Phys. Rev. D 69 (2004) 066007, arXiv:hep-th/0306018.
[93] T. Sakai and S. Sugimoto, "Low energy hadron physics in holographic QCD," Prog. Theor. Phys. 113 (2005) 843, arXiv:hep-th/0412141.
[94] T. Sakai and S. Sugimoto, "More on a holographic dual of QCD," Prog. Theor. Phys. 114 (2006) 1083, arXiv:hep-th/0507073.
[95] J. Hirn, N. Rius and V. Sanz, "Geometric approach to condensates in holographic QCD," Phys. Rev. D 73 (2006) 085005, arXiv:hep-ph/0512240.
[96] K. Ghoroku, N. Maru, M. Tachibana and M. Yahiro, "Holographic model for hadrons in deformed AdS(5) background," Phys. Lett. B 633 (2006) 602, arXiv: hep-ph/0510334.
[97] E. Antonyan, J. A. Harvey, D. Kutasov, "Chiral symmetry breaking from intersecting D-branes," Nucl.Phys. B784 (2007) 1, arXiv:hep-th/0608177.
[98] N. Evans, A. Tedder and T. Waterson, "Improving the infra-red of holographic descriptions of QCD," JHEP 0701 (2007) 058, arXiv:hep-ph/0603249.
[99] C. Csaki and M. Reece, "Toward a systematic holographic QCD: A Braneless approach," JHEP 0705 (2007) 062, arXiv:hep-ph/0608266.
[100] R. Casero, E. Kiritsis and A. Paredes, "Chiral symmetry breaking as open string tachyon condensation," Nucl.Phys. B787 (2007) 98, arXiv:hep-th/0702155.
[101] J. Erdmenger, K. Ghoroku and I. Kirsch, "Holographic heavy-light mesons from non-Abelian DBI," JHEP 0709 (2007) 111, arXiv:hep-th/0706.3978.
[102] O. Bergman, S. Seki and J. Sonnenschein, "Quark mass and condensate in HQCD," JHEP 0712 (2007) 037, arXiv: hep-th/0708.2839.
[103] O. Bergman, G. Lifschytz, M. Lippert, "Holographic Nuclear Physics," JHEP 0711 (2007) 056, arXiv: hep-th/0708.0326.
[104] A. Dhar and P. Nag, "Sakai-Sugimoto model, Tachyon Condensation and Chiral symmetry Breaking," JHEP 0801 (2008) 055, arXiv:hep-th/0708.3233.
[105] U. Gursoy and E. Kiritsis, "Exploring improved holographic theories for QCD: Part I," JHEP 0802 (2008) 032, arXiv:hep-th/0707.1324.
[106] U. Gursoy, E. Kiritsis and F. Nitti, "Exploring improved holographic theories for QCD: Part II," JHEP 0802 (2008) 019, arXiv:hep-th/0707.1349.
[107] J. Erdmenger, N. Evans, I. Kirsch, "Mesons in Gauge/Gravity Duals - A Review," Eur. Phys. J. A35 (2008) 81, arXiv:hep-th/0711.4467.
[108] B. A. Burrington, J. Sonnenschein, "Holographic Dual of QCD from Black D5 Branes," JHEP 0908 (2009) 057 arXiv:hep-th/0903.0628.
[109] D. K. Hong, H. Yee, "Holographic aspects of three dimensional QCD from string theory," JHEP 1005 (2010) 036 arXiv:hep-th/1003.1306.
[110] T. Imoto, T. Sakai, S. Sugimoto, "Mesons as Open Strings in a Holographic Dual of QCD," Prog. Theor. Phys. 124 (2010) 263, arXiv:hep-th/1005.0655.
[111] J. Polchinski and M. J. Strassler, "Hard scattering and gauge/string duality," Phys. Rev. Lett. 88 (2002) 031601, arXiv:hep-th/0109174.
[112] J. Polchinski, M. J. Strassler, "Deep inelastic scattering and gauge/string duality," JHEP 0305 (2003) 012, arXiv:hep-th/0209211.
[113] H. Boschi-Filho, N. R.F. Braga, "Gauge/string duality and scalar glueball mass ratios," JHEP 0305 (2003)009, arXiv:hep-th/0212207.
[114] H. Boschi-Filho, N. R. F. Braga, "QCD/string holographic mapping and glueball mass spectrum," Eur.Phys.J.C32 (2004)529, arXiv:hep-th/0209080.
[115] S. J. Brodsky, G. F. de Teramond, "Light front hadron dynamics and AdS/CFT correspondence," Phys.

Lett. B582 (2004) 211-221, arXiv:hep-th/0310227.
[116] J. Erlich, E. Katz, D. T. Son and M. A. Stephanov, "QCD and a holographic model of hadrons," Phys. Rev. Lett. 95 (2005) 261602, arXiv:hep-ph/0501128.
[117] L. Da Rold and A. Pomarol, "Chiral symmetry breaking from five dimensional spaces," Nucl. Phys. B721 (2005) 79, arXiv:hep-ph/0501218.
[118] G. F. de Teramond, S. J. Brodsky, "Hadronic spectrum of a holographic dual of QCD,"Phys. Rev. Lett. 94 (2005) 201601, arXiv:hep-th/0501022.
[119] S. J. Brodsky and G. F. de Teramond, "Hadronic spectra and light-front wavefunctions in holographic QCD," Phys. Rev. Lett. 96 (2006) 201601, arXiv: hep-ph/0602252.
[120] J. Erlich, G. D. Kribs and I. Low, "Emerging holography," Phys. Rev. D 73 (2006) 096001, arXiv:hep-ph/0509002.
[121] L. Da Rold and A. Pomarol, "The Scalar and pseudoscalar sector in a five-dimensional approach to chiral symmetry breaking," JHEP 0601 (2006) 157, arXiv:hep-ph/0510268.
[122] A. Karch, E. Katz, D. T. Son and M. A. Stephanov, "Linear confinement and AdS/QCD," Phys. Rev. D 74 (2006) 015005, arXiv:hep-ph/0602229.
[123] A. Pomarol, A. Wulzer, "Baryon Physics in Holographic QCD," Nucl.Phys. B809 (2009) 347, arXiv:hep-ph/0807.0316.
[124] H. Kim, Y. Kim, U. Yakhshiev, "Mesons and nucleons from holographic QCD in a unified approach," JHEP 0911 (2009) 034, arXiv:hep-ph/0908.3406.
[125] W. de Paula, T. Frederico, "Scalar mesons within a dynamical holographic QCD model.," Phys.Lett. B693 (2010) 287, arXiv:hep-ph/0908.4282.
[126] K. Jo, B. Lee, C. Park, S. Sin, "Holographic QCD in medium: A Bottom up approach," JHEP 1006 (2010) 022 , arXiv:hep-ph/0909.3914.
[127] S.S. Afonin, "Low-energy holographic models for QCD," Phys. Rev. C83 (2011) 048202, arXiv:hep-ph/1102.0156.
[128] D. K. Hong, T. Inami, H. Yee, "Baryons in AdS/QCD," Phys. Lett. B646 (2007) 165, arXiv:hep-ph/0609270.
[129] S. W. Hawking and D. N. Page, 'Thermodynamics of Black Holes in anti-De Sitter Space', Commun. Math. Phys. 87 (1983) 577.
[130] C. P. Herzog, "A Holographic Prediction of the Deconfinement Temperature," Phys. Rev. Lett. 98 (2007) 091601, arXiv:hep-th/0608151.
[131] D. T. Son, A. O. Starinets, "Viscosity, Black Holes, and Quantum Field Theory," Ann. Rev. Nucl. Part. Sci. 57 (2007) 95, arXiv:hep-th/0704.0240.
[132] S. R. Das, G. W. Gibbons and S. D. Mathur, "Universality of low energy absorption cross sections for black holes," Phys. Rev. Lett. 78 (1997) 417, arXiv:hep-th/9609052.

5th DAE-BRNS Workshop on Hadron Physics (Hadron 2011) IOP Publishing
Journal of Physics: Conference Series **374** (2012) 012005 doi:10.1088/1742-6596/374/1/012005

On nuclear states of $\overline{K}$ mesons

S.Wycech

National Centre for Nuclear Studies, Warsaw, Poland

E-mail: wycech@fuw.edu.pl

Abstract. The search for deeply bound nuclear states of $\bar{\mathrm{K}}$ mesons is presented and the main experimental and theoretical uncertainties are briefly discussed. The need for new experiments is stressed and some atomic experiments are suggested. The involved physics is illustrated on a simple fixed nucleon model which becomes a basis for variational calculations.

1. Introduction

The interest in the search for nuclear states of anti-K mesons stems from several sources:

- A new branch of nuclear physics emerges. It opens the way for interesting laboratory discoveries with some astrophysical consequences.
- The $\overline{K}$ meson may be deeply bound and create high density clusters inside nuclei.
- A new mechanism of nuclear attraction is met. It is generated by subsequent excitations of nucleons into strange baryons $\Lambda(1405)$ and $\Sigma(1385)$. It differs from the standard nuclear force mechanism that generates attraction by the meson exchanges.

This text presents the main problems. As yet, there are no definite solutions, experiments have not been fully confirmed and their interpretation is frequently questioned. However, this field is very active and it involves several laboratories in Japan, Europe and US. A recent review [1] is recommended.

It has been known since discoveries of K-mesic atoms that the $\overline{K}$ is attracted by nuclei. It was also realized that absorption from atomic states was so strong that if nuclear states can be formed their lifetimes would be very short and discovery rather improbable and uninteresting. Already at that time it was well understood that the basic $\overline{K}$N interaction is strongly energy dependent and it may be different in the nuclear matter.

Later on the experimental possibilities were reduced as most of the low energy kaonic beam facilities were dismantled. As a consequence, the interest in the theories almost disappeared.

New life was given to this field with the calculations of Yamazaki and Akaishi predicting deeply bound states in light systems $\overline{K}$NN and $\overline{K}$NNN. New experiments performed in Japan indicated positive results. These turned out to be over-optimistic but the impetus was given to further studies.

At this moment there are two experimental indications of a $\overline{K}$NN state shown in Table 1. The structure of this state might be quite complicated but in the kaonic channel it is a mixture of K^-pp and K^0np commonly named K^-pp which apparently constitutes the dominant component.

Journal of Physics: Conference Series **374** (2012) 012005 doi:10.1088/1742-6596/374/1/012005

Table 1. Experimental measurements of the K^-pp binding energies and half widths $E_B - i\Gamma/2$ [MeV]. FINUDA - [2] and DISTO - [3].

FINUDA	DISTO
115(7) -i 67(14)/2	103(8)-i118(18)/2

Discussion of these results and related questions are given below in three sections :
- Section (2) presents basic models of $\overline{K}$ N interactions
- Section (3) characterizes K-mesic atomic and nuclear states
- Section (4) presents a variational model used to calculate the $\overline{K}$-few nucleon levels.

2. $\overline{K}$ N multiple channel system

The scenario for low energy $\overline{K}$N interaction is determined by a system of coupled channels. These channels and relevant thresholds are given in Table 2. Experiments provide elastic $K^-p \rightarrow K^-p$ and inelastic $K^-p \rightarrow, K^0n$, $K^-p \rightarrow \Lambda, \pi$, Σ, π scattering cross sections from the threshold up to some 250 MeV/c. These are dominated by S waves amplitudes. For isospin 0 there are two relevant channels $\overline{K}N, \Sigma\pi$ and for isospin 1 three channels $\overline{K}N$, $\Lambda\pi$, Σ, π. These data, which may be found in ref. [4], are about 40 years old and more precise results are still unavailable. The important contribution of this century has been the measurement of the $1S$ level shift in kaonic hydrogen atom which produces precise K^-p scattering length [5] consistent with the scattering result.

Table 2. Channels coupled to K^-p and their threshold energies. The $\Sigma\pi$ thresholds are averaged over isospin multiplets.

Λ, π	Σ, π	$\Sigma(1385)$	$\Lambda(1405)$	K^-p
1255	1331	1385	$1405 - 1428$	1432

The scattering data are parameterized, for each isospin I, in terms of reaction matrix $\widehat{K^I}$ related to the scattering matrix $\widehat{T}$ by

$$(\widehat{T^I})^{-1} = (\widehat{K^I})^{-1} - i\widehat{Q} \quad (1)$$

where $\widehat{Q}$ is a diagonal matrix of channel momenta. Unitarity requires $\widehat{K^I}$ to be real and symmetric which allows overall nine free parameters, three for I=0 and six for I = 1. With the best fit values one finds the elastic amplitudes T^I_{KK} in terms of scattering lengths A^I

$$T^I_{KK} = A^I \;/\; (1 - iQ_{KK}A^I) \quad (2)$$

It turns out that the isospin 0 amplitude extrapolated below $\overline{K}$N threshold and into complex energy plane displays a pole. It apparently describes $\Lambda(1405)$ which in this way may be interpreted as $\widehat{K^I}$ quasi-bound state decaying into Σ, π channel. The existence of $\Lambda(1405)$ is apparently the result of short ranged $\widehat{K^I}$ interactions. This is not the only possibility. If $\Lambda(1405)$ contains a genuine three quark component, it may be introduced into $\widehat{K}$ matrix formalism as an additional pole coupled to the channel states. The best fit to the data does not indicate any large component of this type [4]. As far as the the nuclear states are concerned the actual origin of $\Lambda(1405)$ is not crucial. What matters is the actual spectral profile of this state.

Many technical questions arise at this stage. First, the determination of $\widehat{K}$ matrices is not complete, the data allows to fix six parameters out of nine allowed. Second, there is some energy dependence of these parameters. These uncertainties affect particularly the subthreshold region rendering sizable uncertainties of the position and widths of $\Lambda(1405)$. Several methods have been used to fix the $\Lambda(1405)$ parameters. A natural one is to observe its shape in reactions like $pp \to K\Lambda(1405)$ with subsequent decay $\Lambda(1405) \to \Sigma, \pi$. Such an analysis is complicated by several allowed isospin states of Σ, π, uncertain formation mechanism and final state interactions. Indeed, the very recent progress in this search involving electro and photo-production of $\Lambda(1405)$ indicates six different locations of the resonant maxima in the Σ, π spectra which depend on the formation process and Σ, π charge states. No understanding has so far not been attempted, a brief presentation and references may be found in [6].

For nuclear interactions of $\overline{K}$ mesons it is the elastic amplitude T_{KK} that is of prime importance as it determines the optical potential for this meson. So far the best method to check the subthreshold extrapolation is a mixed use of the reaction matrix and dispersion relations. Lucid presentation of this approach is given in ref.[4]. This method - " the phenomenological approach " - locates the pole in the complex energy plane of $T_{KK}(E)$ at $E_{real} \approx 1410$ MeV.

Another method recently studied - "the chiral approach" - assumes additional SU(3) symmetry in the reaction matrix introduced within the chiral perturbation theory [7],[8],[9]. The number of free parameters is comparable to that used in the phenomenological approach but the physics is different. One difference is the existence of another pole in the complex energy plane of $T_{KK}(E)$ due to an additional resonance in the Σ, π channel. However, the basic difference is the position of the main pole related to $\Lambda(1405)$ arising at $E_{real} \approx 1420$ MeV. As discussed in next sections the change of the $\overline{K}$N quasi-bound state position has tremendous impact on the levels of nuclear states of $\overline{K}$.

3. Nuclear states of $\overline{K}$

The dominant attraction comes from the S-wave $\Lambda(1405)$ and an additional contribution is due to $\Sigma(1385)$ which couples to $\overline{K}$N in P wave. These states are additional to the scattering states and by the "level repulsion rule" one should expect that at energies above resonances the $\overline{K}$ interaction in nuclear systems is repulsive while below resonances it is attractive. This is seen clearly in K-mesic atoms. The level shifts and widths in such atoms are shifted from electromagnetic values by the effects of strong interactions. The latter may be described by optical potential which in its simplest form is given by

$$V_K^{optical}(r) = -\frac{2\pi}{\mu_{KN}}\, T_{KK}(-E_B - E_{\text{recoil}})\, \rho(r), \tag{3}$$

where ρ is the nuclear density and μ is the reduced mass. Atomic levels which can be tested involve interactions at a distant nuclear surface where an average binding energies $E_B \simeq 15$ and average KN recoil energies $E_{\text{recoil}} \sim 10-15$ MeV are calculable. Average KN scattering lengths A are repulsive but for bound nucleons T_{KK} refers to sub-resonant region and becomes attractive. That allows for a semi-quantitative understanding of the level shifts [10], [11], [12]. A number of calculations performed in remote past also indicated sizable dependence of T_{KK} on the density of nuclear matter. In particular, the Pauli principle and self-consistency of the nuclear propagator results in a weaker binding of $\Lambda(1405)$. In consequence, the optical potential tested in kaonic atoms is not a linear function of the density as the effective "in medium " T_{KK} becomes weaker or even repulsive in the nuclear center. Unfortunately these calculations depend on the details of the self-consistency procedure and have not been tested experimentally. Nevertheless, recent precise studies of the optical potential indicate such an effect [20].

There is a long way from the atomic optical potential tested at nuclear surface for a zero energy meson to an optical potential for a meson deeply bound in a dense nuclear matter. Inspection of Table 2 shows that binding of a hundred MeV closes the dominant Σ, π decay mode and leaves small phase space for the other Λ, π mode. Is such scenario conceivable ? If one looks for a solution of the Dyson equation in nuclear matter

$$T(E) = T_0(E) + T_0(E)(G[E, V(E)] - G_0[E])T(E) \tag{4}$$

where G_0 is the free propagator for $\overline{K}$ N pair and G is a propagator in nuclear matter there exists a self-consistent solution at $E \simeq 90$ MeV with a small Im $T(E)$. This led to conjecture that narrow (about 10-20 MeV) states may exist in large nuclei [13].

On the other hand, Yamazaki and Akaishi have shown that quasi-bound states may exists also in a few nucleon, in particular in the $\overline{K}$NN, systems [14]. A number of calculations followed. The contemporary status is given in Table 3. The dominant configuration of the indicated state is $I_{NN} = 1$ and the total isospin $I_{tot} = 1/2$, but some models predict also other structures. The calculations differ in terms of technical methods used and in terms of the $\overline{K}$N input. The first four references use fairly similar phenomenological interactions. The calculations apply molecular technique [14], Faddeev equations with separable forces [15] [16], and a variational technique with realistic NN forces [17]. The latter is discussed in the next section. The first calculation operates with single $\overline{K}N$ channel the other three introduce the multiple scattering also in the decay states. This accounts for the largest part of difference in the binding energy. On the other hand, the last two calculations [18], [19] use chiral input. The different $\Lambda(1405)$ position is reflected in much weaker binding energy. Two of the listed works include also P wave interactions mediated by $\Sigma(1385)$. In the case of strong binding the energy of $\overline{K}N$ subsystem is located mostly below the resonance. The binding is enhanced by about 10 MeV and the widths of the state is reduced by the related reduction of the phase space, [17]. In the case of weak binding, the main effect of $\Sigma(1385)$ is a sizable enhancement of the width [18].

Comparison of Tables 1 and 3 creates an impression that the chiral approach is refuted by the data. This is likely, but not necessarily true. First, the experiments need to be confirmed. Second, the FINUDA measurement has been performed as a capture of K^- on Li from an atomic state and a subsequent observation of the Λ and proton. Final state interactions are important [23], [22] and may deform the spectrum. On the other hand, the DISTO result obtained in $pp \to K^+\Lambda, p$ reaction indicates surprisingly broad state which rises suspicions that the process is not well understood. Proximity of the Σ, π, N threshold may also have some effect. On the theory side, one obvious conclusion from Table 3 is the necessity for further studies of the $\Lambda(1405)$ structure.

Table 3. The $\overline{K}$ NN system. The second line - calculated binding energies of the lowest state. The third line gives related widths, all in [MeV]. These calculations assume only S-wave interactions.

[14]	[15]	[16]	[17]	[18]	[19]
48	55-70	$\simeq 80$	$\simeq 70$	20	30-40
61	95-110	$\simeq 73$	≈ 85	20-40	≈ 50

3.1. Variational calculations

This section presents a method [17] which may be used to calculate the $\bar{\text{K}}$-few-nucleon levels. It is inspired by Brueckner's [21] calculation of the scattering length of a meson on two nucleons.

Here, it is turned into an eigenvalue problem. An advantage over the Faddeev approach is a simple way to use realistic NN forces. The advantage over the optical potential method is the inclusion of correlations induced by the meson. Realistic calculation requires several steps of complication : (1) The $\bar{K}$-few-N levels are found within the fixed nucleon approximation with a simple S wave $\bar{K}$ N interaction. (2) The nucleon degrees of freedom and NN interactions are introduced and a related Schrödinger equation is solved as a variational problem (3) The method is extended to multiple channel situations. (4) Both S and P wave $\bar{K}$N interactions are allowed.

With up to four nucleons this method is simple to implement. Here it is presented in two nucleon case, with a simplified notation and several shortcuts. Consider scattering of a meson on two identical nucleons fixed at coordinates $\mathbf{x}_i (i = 1, 2)$, $r = |\mathbf{x}_1 - \mathbf{x}_2|$. The wave function is assumed to be

$$\Psi(\mathbf{x},\ \mathbf{x}_1, \mathbf{x}_2) = \chi_K(\mathbf{x};\ \mathbf{x}_1, \mathbf{x}_2)\ \chi_{NN}(\mathbf{x}_1, \mathbf{x}_2), \tag{5}$$

where $\mathbf{x}$ is the meson coordinate. The meson wave function χ_K is given by the solution of the multiple scattering equation

$$\chi_K(\mathbf{x}, \mathbf{x}_1, \mathbf{x}_2) = \chi_K(\mathbf{x})^o - \Sigma_i \int \mathbf{dy}\ \frac{\exp[ip \mid \mathbf{x}\text{-}\mathbf{y} \mid]}{4\pi \mid \mathbf{x}\text{-}\mathbf{y} \mid}\ U_{KN}(\mathbf{y}, \mathbf{x}_i)\ \chi_K(\mathbf{y}, \mathbf{x}_1, \mathbf{x}_2). \tag{6}$$

One looks for solutions of Eq. (6) with no incident wave $\chi_K(\mathbf{x})^o$. The momentum p becomes a complex eigenvalue $p(x_i)$ which determines the energy and width of the system for given nucleon positions x_i. The potential $U_{KN} = 2\mu_{KN} V_{KN}$ is chosen in a separable form

$$V_{KN}(\mathbf{x} - \mathbf{x_i}, \mathbf{x}' - \mathbf{x_i}) = \lambda\ v(\mathbf{x} - \mathbf{x_i})\ v(\mathbf{x}' - \mathbf{x_i}), \tag{7}$$

where v is a form-factor and λ is a strength parameter. The eigenvalue equation (6) may be now reduced to matrix equation for wave amplitudes ψ_i defined at each scatterer i by

$$\psi_i = \lambda \int \mathbf{dx}\ v(\mathbf{x} - \mathbf{x_i})\ \chi_{\mathbf{K}}(\mathbf{x}, \mathbf{x_1}, \mathbf{x_2}). \tag{8}$$

To find equations for ψ_i one introduces the off-shell $\bar{K}$ N scattering matrices T and matrix elements of the propagator

$$G_{1,2}(\mathbf{x_1}, \mathbf{x_2}) = \int \mathbf{dydx}\ v(\mathbf{x} - \mathbf{x_1})\ \frac{\exp(\mathbf{iP} \mid \mathbf{x} - \mathbf{y} \mid)}{4\pi \mid \mathbf{x} - \mathbf{y} \mid}\ v(\mathbf{y} - \mathbf{x_2}). \tag{9}$$

The diagonal value, $G_{i,i} \equiv G$, determines the meson nucleon scattering matrix t by the well known relation

$$t(E) = (1 + \lambda\ G)^{-1}\ \lambda \tag{10}$$

and this yields the full off-shell scattering amplitude T

$$T(k, E, k') = -v(k)\ t(E)\ v(k'). \tag{11}$$

Here, k, k' are the initial and final momenta while the form-factor $v(k)$ is given by the Fourier transform of $v(r)$.

Simple manipulations allow to obtain the standard multiple scattering equations

$$\psi_1 + t\ G_{1,2}\ \psi_2 = 0, \qquad \psi_2 + t\ G_{21}\ \psi_1 = 0, \tag{12}$$

which describe the meson bouncing off the two nucleons. When the determinant $D = 1 - (t\ G_{12})^2$ is put to zero, the binding "momenta" $p(r)$ may be obtained numerically. Two different solutions corresponding to $1 + tG = 0$ or $1 - tG = 0$ may exist. The first solution is symmetric $\psi_2 = \psi_1$

and describes the meson in an S wave state with respect to the NN center of mass. The second solution is antisymmetric $\psi_2 = -\psi_1$ and describes a P wave solution. With the rank one separable interaction this latter solution does not exist in the full range of r. However, it may with more complicated S wave or P wave interactions.

Eigenvalues corresponding to unstable quasi-bound states are obtained in the second quadrant of complex $p(r) = p_R + ip_I$ plane. In this quadrant the kernel has asymptotic behavior

$$G \sim \exp(-p_I r)\exp(ip_R r)/r \tag{13}$$

as required by the asymptotic form of the bound state wave function χ_K. At short distances G is regularized by the form-factor. The $G(r)$ describes meson nucleon correlations which differ from the correlation inherent to $\Lambda(1405)$. The eigen-value $p(r)$ determines the energy of the meson bound to the fixed NN pair

$$E = p(r)^2/(2\mu_{KN}). \tag{14}$$

The motivation for this definition of E follows from the example discussed below. If the $\bar{\mathrm{K}}$N interaction is dominated by a quasi-bound state, such as $\Lambda(1405)$, then the related pole dominates the scattering amplitude and in some energy region $t = \gamma^2/(E - E^*)$, where γ is a coupling constant and $E^* = E_r - i\Gamma_r/2$ is the $\Lambda(1405)$ complex binding energy. The $\bar{\mathrm{K}}$NN eigenvalue $p(r)$ is given by the equation $1 + tG = 0$, which now takes the form

$$E = E^* - \gamma^2 G(r,p). \tag{15}$$

The solution $E(r) \equiv E_B(r) - i\Gamma(r)/2$ depends on the N-N separation r. Since Re $G(r,p)$ close to the resonance is positive, the binding of $\bar{\mathrm{K}}$ to fixed NN is stronger than its binding to a nucleon, $\mid E_B(r) \mid > \mid E_r \mid$. Increasing the separation $r \to \infty$ one obtains $G \to 0$ and $E(r) \to E^*$, i.e. the $\bar{\mathrm{K}}$ meson becomes bound to one of the nucleons. In the same limit the lifetime of $\bar{\mathrm{K}}$ becomes equal to the lifetime of $\Lambda(1405)$. Hence, the separation energy is understood here as the energy needed to split the $\bar{\mathrm{K}}$-N-N system into the $\Lambda(1405)$-N system. The condition required is $p(\infty) = q_r$ where q_r is the resonant momentum in the $\bar{\mathrm{K}}$N center of mass system i.e. $q_r^2/2\mu_{KN} = E^*$. Definition (14) fulfils this condition.

The difference between the binding at a given separation r and its asymptotic value generates a potential $V_K(r)$, which contracts the nucleons to a smaller radius. It is defined as

$$ReV_K(r) = E_B(r) - E_B(\infty) \tag{16}$$

while the corresponding imaginary part is

$$ImV_K(r) = -\Gamma(r)/2. \tag{17}$$

Typical profiles of $E_B(r)$ are plotted in Figure 1.

In the next step of this calculation the nucleon degrees of freedom are restored. The solution of the full $\bar{\mathrm{K}}$NN bound state problem is given by equation

$$(-\frac{\Delta_x}{2m} - \frac{\Delta_1}{2M} - \frac{\Delta_2}{2M} + V_{KN1} + V_{KN2} + V_{NN})\Psi = E\Psi. \tag{18}$$

which with the wave function from Eq. (5) and given V_{NN} may be solved by variational methods.

The explicit description of the decay channels requires more components in the amplitudes ψ_i. We refer to paper [17] which includes decay channels and finds a number of KNNN and KNNNN states. It may be interesting to note that a loosely bound state in the K^-nn system is also predicted to exist as a consequence of $\Sigma(1385)$.

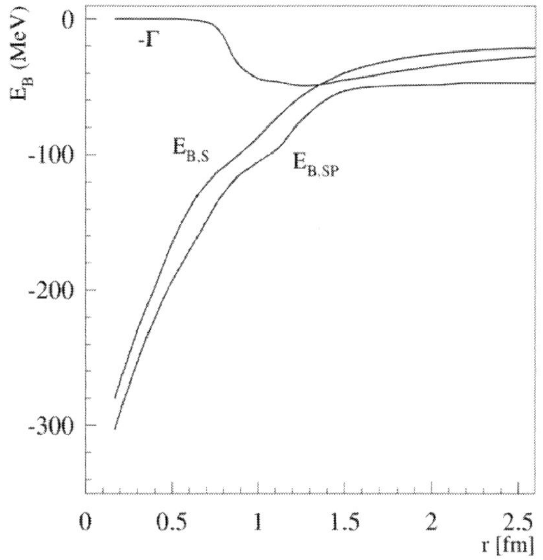

Figure 1. The kaon binding in the $\bar{K}$-pp configuration for fixed N-N separation r. $E_{B,S}$ is obtained with the S-wave and $E_{B,SP}$ with $S+P$ wave $\bar{K}$-p interactions. Γ is the width.

4. Conclusions

The search for nuclear states of anti-kaons is a fascinating story of successes and failures. One still has to answer some basic questions :

(a) What is the binding mechanism? How many mesons can be bound in one nucleus ?

(b) Are the technical questions under control?

(c) Can the widths be narrow?

Ad (a), it seems clear that the profile of $\Lambda(1405)$ in the basic $\bar{K}$N channel is of prime importance. One way to study it would be a measurement of radiative transitions $(K^-p)_{atom} \rightarrow \gamma+\Lambda(1405)$. Another possibility exists in studies of the highest available levels in K-mesic atoms. In such states the optical potentials of equation (3) involves predominantly valence nucleons and the scattering amplitude may be traced as a function of the separation energy. Figure 2 shows the absorptive amplitude extracted from atomic data. Unfortunately, the data describing the essential region of $\sim$ 25 MeV below the threshold are uncertain. New experiments are needed.

Ad (b), it is not clear whether the optical potential or other average field approximations are a good description in view of strong N- $\bar{K}$-N and $\bar{K}$-N correlations.

Ad (c), if experimental results are correct the widths on nuclear states might be determined by two nucleon capture of the kaon. There is no reliable knowledge in this question.

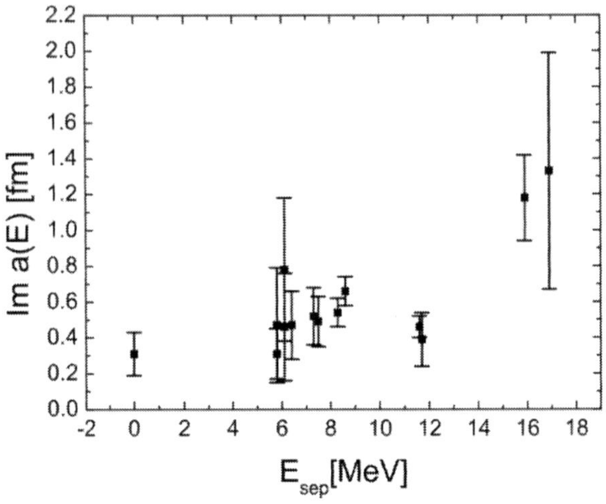

Figure 2. The isospin averaged, $N\bar{K}$ amplitudes $a =< T^I_{KK} >$ in the sub-threshold region extracted from upper levels of kaonic atoms. To trace the position of resonance one needs to add about 10 MeV of the recoil energy to the values of E_{sep}.

Acknowledgements
Financial support from European Hadron 3 - LEANNIS project is acknowledged.

References
[1] J. Zmeskal, Progr.Part.Nucl.Phys.**61**,512(2008).
[2] M. Agnello *et al.* (FINUDA Collaboration), Phys. Rev. Lett. **94**(2005)212303.
[3] M. Maggiora *et al.* (DISTO Collaboration), Nucl. Phys. **A835**(2010)43.
[4] A.D. Martin Nucl.Phys. **B179**(1981)33, in *Low and intermediate energy kaon-nucleon physics*, Ed. E.Ferrari and G.Violini, Reidel Publ.1981.
[5] M. Iwasaki *et al.* Phys. Rev.Lett. **78**(1997)3067 ; G. Beer *et al.* Phys. Rev. Lett **94**(2005) 212302.
[6] S. Wycech, Proceedings of EXA 11, Hyperfine Interactions (2011).
[7] E. Oset, A. Ramos and C. Bennhold, Phys. Lett. **B527** (2002)99.
[8] N. Kaiser, P.B. Siegel and W. Weise, Nucl. Phys. A **594**,(1995)325.
[9] B. Borasoy, U.-G. Meißner and R. Nißler,Phys. Rev. **C74**(2006) 055201 (2006).
[10] M. Alberg, E.M. Henley and L. Wilets, Ann. Phys. NY **96** (1976)43.
[11] S.Wycech, Nucl.Phys. B **28**(1971)541.
[12] W.Bardeen and E.W.Torigoe, Phys. Rev. C **3**(1971)1785.
[13] S. Wycech, Nucl. Phys. **A450** (1986) 399c.
[14] Y. Akaishi and T. Yamazaki, Phys. Rev. C **65**(2002) 044005.
[15] N. V. Shevchenko, A. Gal, J. Mareš and J. Révai, Phys. Rev. C **76**(2007)044004
[16] Y. Ikeda and T. Sato, Phys. Rev. C **76**(2007) 035203.
[17] S. Wycech and A.M. Green Phys. Rev. **C79**(2009)014001.
[18] T. Hyodo and W. Weise, Phys. Rev. **C77**(2008) 035204.
[19] M.Bayar, J. Yamagata-Sekihara and E.Oset, Phys. Rev. C **84**(2011)015209.
[20] A.Cieply, E. Friedman, A. Gal, D. Gazda and J. Mareš, Phys. Lett. B **702**(2011)402.
[21] K. A. Brueckner, Phys. Rev. **89**(1953)834.
[22] Grishma Pandejee , N.J. Upadhyay and B.K. Jain, Phys. Rev. C **82**(2010)034608.
[23] V. K. Magas, E. Oset, A. Ramos and H. Toki, Phys. Rev. C **74**(2006)025206.

5th DAE-BRNS Workshop on Hadron Physics (Hadron 2011) IOP Publishing
Journal of Physics: Conference Series **374** (2012) 012006 doi:10.1088/1742-6596/374/1/012006

Colour Confinement and Deformed Baryons in Quantum Chromodynamics

Syed Afsar Abbas
Department of Physics, Aligarh Muslim University, Aligarh - 202 002, India
E-mail: drafsarabbas@yahoo.in

Abstract. The confinement of coloured entities in Quantum Chromodynamics (QCD) is traced to colour singletness of the observed entities. This is believed to arise from colour singlet state of quark-antiquark for mesons and a fully colour antisymmetric state for baryons. This demands a spherically symmetric baryon in the ground state. However it is pointed out that a deformed baryon in the ground state has been found to be extremely successful phenomenology. There are convincing experimental supports for a deformed nucleon as well. This means that something has been missed in the fundamental theory. In this paper this problem is traced to a new colour singlet state for baryons which has been missed hitherto and incorporation of which provides a consistent justification of a deformed baryon in the ground state. Interestingly this new colour singlet state is global in nature.

1. Introduction

Quantum Chromodynamics (QCD) has been an extremely successful theory of the strong interaction. However colour confinement still remains an unsolved problem of QCD. In spite of intense efforts one has not been able to understand it fully within QCD. Because of this it has even been compared to Fermat's Last Problem in mathematics[1].

If not understood fundamentally, phenomenologically one has attained a reasonable understanding of confinement in QCD[2]. As per our best understanding, colour singletness is the basis for confinement. Mesons are confined due to colour singlet states for the quark-antiquark. For three quarks one obtains saturation in colour space by having the colours of the quarks constituting a totally antisymmetric wave function. Hence for baryons, the wave function in the rest of the degrees of freedom is fully symmetric in the exchange of any two quarks.

As per this framework of confinement of three quarks in a baryon, these should sit in a spherically symmetric s-state[3]. This model does pretty well, say, fitting the magnetic moments of nucleon quite well etc. This model forms the basis of the quark model [3] and may be called the canonical phenomenological model of the baryons. Note that spherical symmetricity of baryons is the backbone of this model.

Though pretty successful, it has had failures too. This has led to it being labeled as Naive Quark Model (NQM). This has prompted people to go beyond NQM and propose various other phenomenological models.

However, the "Naive" tag on quark model is quite unjustified if we are able to rectify its flaws within the quark model itself. For example, it was noted by Glashow[4] that the quark model does pretty badly for the axial vector coupling constant g_A where, in quark model its value is

5/3[3] while the experimental number is 1.24. Similar problem exist for D/F parameter of the weak decay of baryons. To rectify the problem Glashow[4] suggested an appropriate amount of mixture of D-state in the nucleonic ground state.

$$|N\rangle \;=\; \sqrt{1-P_D}\;|N_S\rangle \;+\; \sqrt{P_D}\;|N_D\rangle \tag{1}$$

where $|N_S\rangle$ and $|N_D\rangle$ give the spherically symmetric part $(56, 0^+)$ and the D-state admixture of $(70, 2^+)$ to the nucleon respectively (see Appendix). The first one is the spherically symmetric part and the second one the deformed part. P_D is the D-state probability. This clearly suggests that the nucleon is deformed in the ground state itself. Glashow found that a P_D of ~ 0.25 yielded good fits to the experimentally measured g_A and D/F ratios. Later Vento et al.[5] found that the ratio of $\pi N\Delta$ to πNN coupling constant is fitted very well with the same $P_D \sim 0.25$.

The present author has found that the same $P_D \sim 0.25$ does very well for a large number of physically relevant quantities in the quark model:
(a) A comprehensive fit to allowed semileptonic decays of all the $1/2^+$ baryon octet[6]. (b) Pion-Nucleon-Delta coupling constant, $E2/M1$ ratio, double delta coupling constant, constituent quark masses etc.[7]. (c) A consistent understanding of the contentious spin structure of the nucleon[8]. (d) Consistent understanding of the magnetic dipole strength and the Gamow-Teller strengths in nuclei[9].

Such comprehensive fits to a large number of physically measured quantities is very impressive. Most remarkable is the fact that it is just one parameter $P_D \sim 0.25$ which does the job for all. We therefore feel that a nucleon deformed in the ground state should be taken seriously. Experimentally too there are strong evidences of deformed baryons in the ground state. For example recent improved experimental determination of the $p \to \Delta^+$ transition quadrapole moment[10] gives convincing evidence of an intrinsically deformed nucleon. This is in conflict with our current understanding of confinement of baryons, it should be spherical. So it means that we are actually missing some fundamental aspect of the reality of confinement. Which aspect of confinement could we be missing which could account for such large (25%) deformation of the nucleon?

One knows that one gluon exchange between quarks has a tensor term. The tensor term would induce a D-state admixture to the ground state of nucleon[5]-[9]. But this D-state admixture can not be more than a few percent at best. It definitely can not induce a D-state admixture of 25%. So where does this 25% deformation of the nucleon in the ground state come from? As per our current understanding of confinement[1]-[3] the nucleon should be spherical. How come, it becomes so largely deformed? Seeking a mechanism for this within a consistent franework of QCD is an aim of this paper.

However if a colour singlet state exists it would mediate long range forces. From experiments we know that there are no such long range forces existing in nature. Hence this colour singlet state cannot be a representation of a gauge particle. However one may ask - could it manifest itself for physical reality in some other manner?

This problem may be resolved by assuming that this singlet state **cannot act as a force propagator** but it may manifest itself in some manner for describing say a static situation - such as for example in a bound system. In fact, as such it may provide a basis for confinement. And indeed it does - as colour confinement for quark-antiquatrk mesonic states. This in fact is well known and well accepted. **Hence this colour singlet state finds application as a quantum mechanical state providing confinemet to a bound system (like meson) rather than being a representation of a mediator of a long range force.**

Let us write this in full

$$\sum_{i,j} \phi^a(i)\phi_a(j) = \sum_{i \neq j} \phi^a(i)\phi_a(j) + \sum_{i=j} \phi^a(i)\phi_a(j) \tag{2}$$

where 'a' stands for colour and i,j stand for location of the colour fields in space. The first term provides the well known colour state for meson confinement. We shall show here as to how the second term provides the mising confinement term for baryons.

Let us look at colour singlet state for mesons

$$\phi^a(4)\phi_a(3) \Rightarrow \frac{1}{\sqrt{3}}\left(\bar{R}(4)R(3) + \bar{B}(4)B(3) + \bar{G}(4)G(3)\right) \tag{3}$$

Here we are explicitly exhibiting the locations of colour and anticolour. Here anticolour is situated at position 4 and colour at position 3. Reason for this somewhat odd numbering of positions will become clear below. Using the identity

$$\phi^a(4) = \epsilon^{abc}\phi_b(1)\phi_c(2) \tag{4}$$

where the anticolour at position 4 creates colours at positions 1 and 2 which are in antisymmetric state. As we already know from the studies of diquark[11] that position 4 should be the Centre of Mass of the positions 1 and 2 of identical quarks. Thus

$$\phi^a(4)\phi_a(3) = \epsilon^{abc}\phi_b(1)\phi_c(2)\phi_a(3) \tag{5}$$

For baryons the positions 1, 2 ,3 at which the three colours are sitting there and which are all independent, give a totally antisymmetric state on the exchange of any two colours at positions 1, 2, and 3. These three independent positions 1, 2 and 3 are shown in Fig.1(a). All this is well understood for meson and baryon colour singletness.

Some further comments to explain what has been done so far. Mesons consisting of quark-antiquark are colour singlet as per eqn (3) above. Baryons consisting of 3 quarks are colour singlet due to the completely antisymmetric state given in eqn (5) above. Is there any connection between the colour singlet states of mesons (eqn (3)) and the colour singlet states of the baryons (eqn (5))? As $3 \times 3 = \bar{3} + 6$ there indeed is! This connection is expressed by the unique SU(3) transformation expessed in eqn (4) (note that the same is not true in SU(2)). This connection is well recognized by the picture of baryons as made up of a quark and a diquark. Diquark is nothing but what is given in eqn (4). This diquark picture is well studied and well rewieved in Ref [11] and for example provides an explanation of the similarity of the Regge trajectories for mesons and baryons.

Next, we look at the colour singlet state given by the second term in eqn (2). Let us write this as (labelling '3' for the arbitrary location):

$$\frac{1}{\sqrt{3}}\left(\bar{R}(3)R(3) + \bar{B}(3)B(3) + \bar{G}(3)G(3)\right) \tag{6}$$

Now using Eq.(4) above

$$\frac{1}{\sqrt{6}}\left[\left(B(1)G(2) - G(1)B(2)\right)R(3) + \left(G(1)R(2) - R(1)G(2)\right)B(3) + \left(R(1)B(2) - B(1)R(2)\right)G(3)\right] \tag{7}$$

where the anticolour at position 3 creates a two colour antisymmetric state at positions 1 and 2 respectively such that position 3 is the Centre of Mass of these. This is the point at which the third colour sits. This situation is depicted in Fig.1(b). Position 3 is rigid at the Center of Mass while position 1 and 2 can be exchanged. However in colour space, because of the Levi-Civita tensor ϵ^{abc} in Eq.(4) there is still antisymmetry between the exchange of any two colours. I will discuss this point below. Note that this state is colour singlet and has baryon number one as well. As per colour singlet hypothesis this colour singlet state for $B = 1$ should contribute to confinement. Here we claim that this is the colour singlet confined state which has been missed so far for the confinement of baryons.

If r is the distance between particle 3 and particle 1 and 2, respectively (Fig.1(b)) then clearly orbital terms like r^2, r^4, r^6..... or even powers of r will occur. These corresponds to angular momentum states $L = 2, 4, 6.....$ respectively. We note that for both $J = 1/2$ and $J = 3/2$ baryons these will provide states of Regge trajectory and thereby forming a natural explanation of the same in this model.

Figure 1. Location of quarks in positions 1, 2 and 3 for (a) as per Eq.(4) and (b) as for Eq.(5) and Eq.(6).

Next note that $L = 2$ with proper combination with the spin degrees of the three quarks will form the sought after D-state admixture as indicated above (see Appendix). Hence this gives a consistent mechanism for the D-state admixture of the nucleons in the ground state and shows that as per colour singlet confinement hypothesis, there are two states, one provides the S-state (already known) plus the new one giving the D-state admixture.

So we see that one colour singlet state has been missed for baryons. It is the new state and as discussed above, it is this which is making nucleon and delta and all the other baryons deformed. This new state also could explain the Regge trajectory states as well.

In Eq.(5) i.e. $\epsilon^{abc}\phi_b(1)\phi_c(2)\phi_c(3)$ is the well known colour antisymmetric state, the antisymmetry is imposed by the Levi-Civita term ϵ^{abc} where a, b and c are colour modes of $SU(3)_c$. However, the same antisymmetry is obtained if we do not change the colour states but exchange the position labels 1, 2 and 3. The duality is due to the fact that these positions label quarks with all the properties i.e. flavour, spin, orbital state and colour state also. So colour sits on quarks and any exchange of one lead to the other.

Next, the new colour singlet state is depicted in Eq.(6) and Eq.(7) can be seen in terms of Eq.(4) again provided the positions under 1 and 2 only being exchanged while 3 sits at the Center of Mass (Fig.1(b)). Full antisymmetry in the colour space arises, independent of the position at 3, due to the Levi-Civita term ϵ^{abc}. Hence here notice that colour degrees of freedom break up from the quarks other degrees of freedom so as to provide antisymmetry.

This means that the new symmetry as imposed upon the system is antisymmetric $SU(3)_C$ space and symmetric in the rest of the space. This symmetry is produced by having $1 \leftrightarrow 2$ exchange only i.e. the mixed symmetric states like ρ- and λ- (see Appendix). Note that in Eq.(A.8) the $\rho - \rho$ and $\lambda - \lambda$ terms each are separately symmetric only in $1 \leftrightarrow 2$ exchange. Only after being added do they provide symmetry with respect to '3'. And it is this that gives the D-state admixture to the baryon in the ground state.

The subtleties of emerging colour (antisymmetry), spin and flavour (symmetry) symmetries can be understood in the following way. First in $SU(3)_C$ space write mixed colour state $[\phi_a(1)\phi_b(2) - \phi_b(1)\phi_a(2)]\,\phi_c(3)$. This is antisymmetric in exchange $a \leftrightarrow b$ and no specific symmetry with respect to colour 'C'. The same situation with respect to locations $1 \leftrightarrow 2$ antisymmetry and none with respect to the exchange of 3. Now let the situation be that of Fig.1(b) when 3 is rigid and non movable Center of Mass of positions 1 and 2 which in t urn can be exchanged.

Next insert the Levi-Civita antisymmetrization tensor $\epsilon^{abc}\,[\phi_a(1)\phi_b(2) - \phi_b(1)\phi_a(2)]\,\phi_c(3)$. Now this induces antisymmetry with respect to the exchange of colour 'C' also. As position 3 is rigid, the position '3' does not change but this exchange is in colour space.

Next in the quark flavour, spin and orbital space in Eq.(A-8)

$$\chi^{\lambda}\left[\chi^{S}\psi^{2}_{2m\lambda}\right]^{J=1/2} = \frac{1}{\sqrt{2}}\left(\Psi_S + \Psi_\lambda\right) \tag{8}$$

$$\chi^{\rho}\left[\chi^{S}\psi^{2}_{2m\rho}\right]^{J=1/2} = \frac{1}{\sqrt{2}}\left(\Psi_S - \Psi_\lambda\right) \tag{9}$$

which have only symmetry with respect to $1 \leftrightarrow 2$ and none with respect to flavour-spin at position 3. Clearly these are spurious excitations with respect to the Center of Mass in the Harmonic Oscillator Model. These spurious excitations are gotten rid of, and pure intrinsic excitations arise only on summing the above two as in Eq.(A-8). This induces new symmetry with respect to flavour-spin at the location 3. Hence what we show geometrically in Fig.1(b) is **exactly** what is obtained when only intrinsic excitations with respect to the Center of Mass are calculated as in Eq. (A-8).

Note that for confinement we are getting two colour singlet states - one colour anti-symmetric in $SU(3)_C$ (giving spherically symmetric configuration of three quarks) and the other also colour anti-symmetric in $SU(3)_C$ (giving deformed D-state admixed state of three quarks). The requirement of colour singletness is unable to distinguish between the two - both being required at fundamental level. Therefore, just being singlet in colour can not be the only condition for confinement of quarks. Note that in the second case quarks are antisymmetric in colour space inspite of the fact that one of them is always rigidly located at the Center of Mass position with respect to the other two. The system still manages to give the corresponding totally symmetric state in the other degrees of freedom of the three quarks. The only thing common between the two colour singlet states is the requirement of colour antisymmetry in $SU(3)_C$. Just the colour singletness would suffice for the confinement of quark-antiquark in a meson. However, it appears that the more basic property for confinement of quarks in baryons is that of antisymmetry in the colour space. We need not demand of colour singletness as an additional requirement. Antisymmetry in colour space would itself ensure colour singletness for three quarks as done above for the two singlet states.

So it may be summarized that one part of the colour singlet state was missed out when one only concentrates upon the canonical antisymmetric colour states for baryons. That gave a spherically symmetric nucleon. The missed color singlet state for baryon provides a D-state admixture to the baryons wave function and which makes the baryon deformed in the ground state itself. What we found on fundamental grounds here, matches and is consistent with the phenomenological quark model calculations of deformed baryons done earlier.

Appendix A.

The nucleonic ground state as given in Eq.(1) is

$$|56, 0^+\rangle \equiv |N_S\rangle \tag{A.1}$$

where

$$|N_S\rangle = \frac{1}{\sqrt{2}}\left(\chi^{\rho}\phi^{\rho} + \chi^{\lambda}\phi^{\lambda}\right)\psi^{0}_{00S} \tag{A.2}$$

where for example

$$\phi^{\lambda}_{p} = -\frac{1}{\sqrt{6}}\left(udu + duu - 2uud\right) \tag{A.3}$$

$$\phi^{\rho}_{p} = \frac{1}{\sqrt{2}}\left(udu - duu\right) \tag{A.4}$$

and

$$\chi^{\lambda}_{\frac{1}{2}} = -\frac{1}{\sqrt{6}}(\uparrow\downarrow\uparrow + \downarrow\uparrow\uparrow -2\uparrow\uparrow\downarrow) \tag{A.5}$$

$$\chi^{\rho}_{\frac{1}{2}} = \frac{1}{\sqrt{2}}(\uparrow\downarrow\uparrow - \downarrow\uparrow\uparrow) \tag{A.6}$$

And

$$|70, 2^{+}\rangle \equiv |N_D\rangle \tag{A.7}$$

where

$$|N_D\rangle = \frac{1}{\sqrt{2}}\left\{\phi^{\rho}\left[\chi^{S}\psi^{2}_{2m\rho}\right]^{J=\frac{1}{2}} + \phi^{\lambda}\left[\chi^{S}\psi^{2}_{2m\lambda}\right]^{J=\frac{1}{2}}\right\} \tag{A.8}$$

where for example ($X = \rho \mathrm{or} \lambda$)

$$\left[\chi^{S}\psi^{2}_{2mX}\right]^{J=\frac{1}{2}}_{J_Z=\frac{1}{2}} = -\frac{1}{\sqrt{10}}\chi^{S}_{\frac{3}{2}}\psi^{2}_{2-1X} + \frac{1}{\sqrt{5}}\chi^{S}_{\frac{1}{2}}\psi^{2}_{20X} - \frac{3}{\sqrt{10}}\chi^{S}_{-\frac{1}{2}}\psi^{2}_{21X} + \frac{2}{\sqrt{5}}\chi^{S}_{-\frac{3}{2}}\psi^{2}_{22X} \tag{A.9}$$

For full details see Ref.[3] and Ref.[4]-[9].

References

[1] Smilga A 2001 *Lectures on Quantum Chromodynamics* (Singapore, World Scientific).
[2] Alkofer R and Greensite J 2007, *J. Phys G* **34**, S3.
[3] Close F E 1979 *An Introduction to Quarks and Partons* (London, Academic Press).
[4] Glashow S L 1979, *Physica A* **96**, 27.
[5] Vento V, Baym G and Jackson A D 1981, *Phys. Lett. B* **10**, 97.
[6] Abbas A 1988, *Europhys. Lett.* **5**, 287.
[7] Abbas A 1989, *Phys. Rev. Lett.* **63**, 334; Abbas A 1991, *J. Phys. G* **17**, L181; Abbas A 1992, *J. Phys. G* **18**, L89.
[8] Abbas A 1989 *J. Phys. G* **15**, L73; Abbas A 1992 *J. Phys. G* **15**, L195.
[9] Abbas A 1987 *Phys. Rev. C* **36**, R1663; Abbas A 1987 *J. Phys. G* **13**, 1337.
[10] Bernstein A M 2003 *Eur. Phy. J. A* **17**, 349; Tiator L et al., 2003 *Eur. Phys. J. A* **17**, 357; Blanpied G et al. 2001 *Phys. Rev. C* **64**, 025203.
[11] Anselmino M, Predezzi E, Ekelin S, Fredriksson S and Lichtenbeg D B 1993 *Rev. Mod. Phys.* **65**, 199; Szczekowski M 1989 *Int. J. Mod. Phys. A* **4**, 3985; Paria L and Abbas A 1996 *Pramana* **46**, 417.

Vector- and Pseudoscalar-baryon coupled channel systems

K. P. Khemchandani[1†], A. Martínez Torres[2], H. Kaneko[1], H. Nagahiro[3], A. Hosaka[1]
[1]Research Center for Nuclear Physics, Osaka University, Ibaraki 567-0047, Japan.
[2] Yukawa Institute for Theoretical Physics, Kyoto University, Kyoto 606-8502, Japan.
[3]Department of Physics, Nara Women's University, Nara 630-8506, Japan.

E-mail: †kanchan@rcnp.osaka-u.ac.jp

Abstract. In this manuscript, I will report the details of our recent work on the vector meson-baryon (VB) interaction, which we studied with the motivation of finding dynamical generation of resonances in the corresponding systems. We started our study by building a formalism based on the hidden local symmetry and calculating the leading order contributions to the scattering equations by summing the diagrams with: (a) a vector meson exchange in the t-channel (b) an octet baryon exchange in the s-, u-channels and (c) a contact interaction arising from the part of the vector meson-baryon Lagrangian which is related to the anomalous magnetic moment of the baryons. We find the contribution from all these sources, except the s-channel, to be important. The amplitudes obtained by solving the coupled channel Bethe-Salpeter equations for the systems with total strangeness zero, show generation of one isospin 3/2, spin 1/2 resonance and three isospin 1/2 resonances: two with spin 3/2 and one with spin 1/2. We identify these resonances with Δ (1900) S_{31}, $N^*(2080)$ D_{13}, $N^*(1700)$ D_{13}, and $N^*(2090)$ S_{11}, respectively.

We have further extended our study by including pseudoscalar meson-baryon (PB) as the coupled channels of VB systems. For this, we obtain the PB $\rightarrow$ VB amplitudes by using the Kroll-Ruddermann term where, considering the vector meson dominance phenomena, the photon is replaced by a vector meson. The calculations done within this formalism reveal a very strong coupling of the VB channels to the low-lying resonances like $\Lambda(1405)$ and $\Lambda(1670)$, which can have important implications on certain reactions producing them. In addition to this, we find that the effect of coupling the higher mass states to the lighter channels is not restricted to increasing the width of those states, it can be far more strong.

1. Introduction

The vector mesons, which are understood as $(q\bar{q})_{l=1}$ composite systems, have attracted a lot of attention historically, beginning from the observation of the universality of the ρ-meson couplings to the light hadrons, which can be understood in terms of the fluctuation of a photon to a vector meson that subsequently interact with the hadrons. Further findings, like the dominance of the the vector mesons in explaining the electromagnetic form factors of the hadrons [1] stimulated the formulation of a theory where the vector mesons are treated as the dynamical gauge bosons of the hidden local symmetry of the non-linear sigma model [2, 3, 4]. The resulting theory has been widely used in hadron physics involving the vector mesons (see Refs. [5, 6, 7, 8, 9, 10, 11] as examples).

Using this effective field theory based on the hidden local symmetry, we have focussed on understanding the low-energy interaction of vector mesons with baryons, with the idea of finding dynamical generation of baryon resonances, motivated by the large branching ratios of some of the excited baryons to the vector meson-baryon decay channels. Generation of resonances in vector meson-baryon systems has earlier been reported in several phenomenological studies also [12, 13, 14, 15, 16, 17]. We proceeded with our study by reminding ourselves that in case of the vector meson-baryon interaction, we cannot rely on the low energy theorems which have worked well in case of the pseudoscalar mesons (see, for example, Refs. [18, 19, 20, 21, 22, 23, 24] where several baryon resonances have been found to get generated in the pseudoscalar-baryon or even in two/three pseudoscalar-baryon systems). Keeping this in mind, we built a formalism [25] on the basis of the hidden local symmetry and calculated the four diagrams shown in Fig. 1, which correspond to: (a) a vector meson exchange in the t-channel, (b) a contact interaction demanded by the gauge invariance (corresponding to the hidden local symmetry (HLS)) of the vector meson-baryon interaction Lagrangian, which can be written for the SU(2) case as:

$$\mathcal{L}_{\rho N} = -g\bar{N}\left\{F_1\gamma_\mu\rho^\mu + \frac{F_2}{4M}\sigma_{\mu\nu}\rho^{\mu\nu}\right\}N, \tag{1}$$

and (c, d) the s- and u-channel octet baryon exchange.

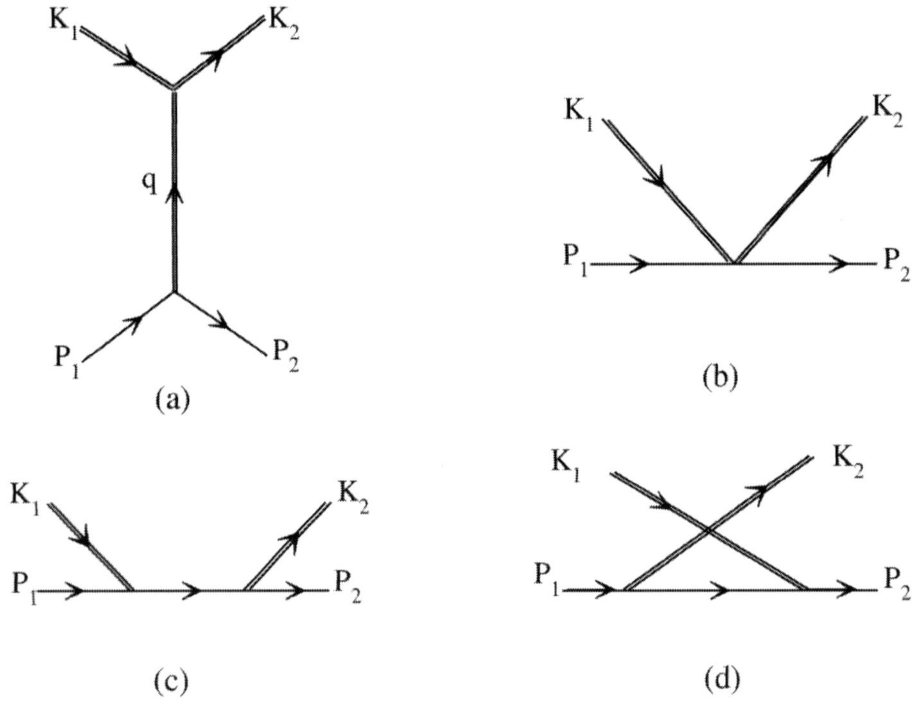

Figure 1. Different diagrams contributing to the vector meson-baryon interactions (see details in the text).

Writing the HLS gauge invariant Lagrangian given by Eq. (1), explicitly, we get [25]

$$\mathcal{L} = -g\bar{N}\left(\gamma^\mu\rho_\mu + \frac{\kappa_\rho}{4M}\left[\partial_\mu\rho_\nu - \partial_\nu\rho_\mu + ig\left(\rho_\mu\rho_\nu - \rho_\nu\rho_\mu\right)\right]\sigma^{\mu\nu}\right)N, \tag{2}$$

where the nucleon field transforms as $N \to h(x)N$ with $h(x)$ being an element of the hidden local symmetry, $F_1(q^2=0)=1$, $F_2(q^2=0)=\kappa_\rho \sim 3.22$ is the anomalous magnetic coupling for the ρNN vertex, M is the mass of the nucleon and

$$\rho_\mu = \frac{1}{2}\vec{\tau}\cdot\vec{\rho_\mu} = \frac{1}{2}\begin{pmatrix} \rho^0_\mu & \sqrt{2}\rho^+_\mu \\ \sqrt{2}\rho^-_\mu & -\rho^0_\mu \end{pmatrix}. \quad (3)$$

Using Eq. (2), we can write Lagrangians for the vertices involving one and two meson fields, respectively, as

$$\mathcal{L}_{\rho NN} = -g\bar{N}\left(\gamma^\mu \rho_\mu + \frac{\kappa_\rho}{4M}\left(\partial_\mu \rho_\nu - \partial_\nu \rho_\mu\right)\sigma^{\mu\nu}\right)N \quad (4)$$

$$\mathcal{L}_{\rho\rho NN} = -ig^2\bar{N}\left(\rho_\mu\rho_\nu - \rho_\nu\rho_\mu\right)\sigma^{\mu\nu}N, \quad (5)$$

where the former is used to obtain the amplitudes for the s- and u-channel diagrams [25]

$$V^s_{\rho N} = \frac{g^2}{4}\left[1-\frac{\kappa_\rho m}{2M}\right]^2 \frac{1}{m+2M}\left(1-2\vec{t_\rho}\cdot\vec{t_N}\right)\left(1-2\vec{s_\rho}\cdot\vec{s_N}\right), \quad (6)$$

$$V^u_{\rho N} = \frac{g^2}{4}\left[1+\frac{\kappa_\rho m}{2M}\right]^2 \frac{1}{m-2M}\left(1+2\vec{t_\rho}\cdot\vec{t_N}\right)\left(1+2\vec{s_\rho}\cdot\vec{s_N}\right), \quad (7)$$

and latter one gives a contact interaction with the amplitude given by

$$V^{CT}_{\rho N} = \frac{g^2\kappa_\rho}{M}\vec{t_\rho}\cdot\vec{t_N}\vec{s_\rho}\cdot\vec{s_N}. \quad (8)$$

It should be emphasized here that this contact interaction arises from the $[\rho_\mu, \rho_\nu]$ term of the vector tensor, which is essentially required to obtain the HLS gauge invariance of the Lagrangian given by Eq. (2).

To calculate the t-channel diagrams, together with Eq. (4), we also need

$$\mathcal{L}_{3\rho} = -2igTr\left(\partial_\mu V_\nu V^\mu V^\nu - \partial_\mu V_\nu V^\nu V^\mu\right), \quad (9)$$

which comes from the non-Abelian kinetic term of the ρ-meson. The resulting t-channel amplitude has the form [25]

$$V^t_{\rho N} = -\frac{1}{2f_\pi^2}\left(K^0_1 + K^0_2\right)\vec{\epsilon_1}\cdot\vec{\epsilon_2} \qquad \text{for } I = 1/2 \quad (10)$$

$$= \frac{1}{4f_\pi^2}\left(K^0_1 + K^0_2\right)\vec{\epsilon_1}\cdot\vec{\epsilon_2} \qquad \text{for } I = 3/2, \quad (11)$$

which is spin independent in nature. However the contact, s- and u-channel interactions lead to spin dependent amplitudes. A comparison of Eqs. (6), (7), (8), (10) and (11) shows that the contribution from all the four diagrams is equally important (it can be easily seen by using the KSRF relation: $g = m_\rho/\sqrt{2}f_\pi$). We have calculated the amplitudes for other VB channels using a more general Lagrangian [25], which we will give in the next section. In the subsequent section, we will review the results of the full coupled channel scattering equations for the total strangeness zero VB systems.

Having built a reliable formalism for the vector-meson baryon interactions, we extended our study by including pseudoscalar-baryon systems as coupled channels [26]. Since several resonances decay with large branching ratios to both PB and VB channels, it is possible, then, that a coupled PB-VB dynamics can play important role in explanation of the properties of

Journal of Physics: Conference Series **374** (2012) 012007 doi:10.1088/1742-6596/374/1/012007

some of such resonances. However, since most PB and VB channels lie well separated on the energy scale, we did not know if we were going to get important information by coupling them. Hence, we decided to concentrate on studying the low-lying resonances alone since those are better understood. This also allowed us to rely on the t-channel calculations of PB $\to$ PB and VB $\to$ VB vertices, which simplified the formalism. However, as we will discuss in Section 3, it turned out that the coupling between the PB and VB channels can result in very useful and significant findings, implying the requirement of a more detailed calculations.

2. Formalism

2.1. Vector-baryon interaction

To calculate the VB transition amplitudes, we generalized the SU(2) vector meson-baryon Lagrangian of Eq. (2) to SU(3) to get

$$\begin{aligned}\mathcal{L}_{VBB} &= -g\Big\{\langle\bar{B}\gamma_\mu\left[V_8^\mu, B\right]\rangle + \frac{1}{4M}\Big(F\langle\bar{B}\sigma_{\mu\nu}\left[\partial^\mu V_8^\nu - \partial^\nu V_8^\mu, B\right]\rangle \\ &+ D\langle\bar{B}\sigma_{\mu\nu}\left\{\partial^\mu V_8^\nu - \partial^\nu V_8^\mu, B\right\}\rangle\Big) + \langle\bar{B}\gamma_\mu B\rangle\langle V_0^\mu\rangle \\ &+ \frac{C_0}{4M}\langle\bar{B}\sigma_{\mu\nu}V_0^{\mu\nu}B\rangle\Big\}, \end{aligned} \tag{12}$$

where the constants $D = 2.4$ and $F = 0.82$. These values were found to well reproduce the magnetic moments of the baryons in Ref. [27]. Further, in our normalization scheme,

$$V = \frac{1}{2}\begin{pmatrix} \rho^0 + \omega & \sqrt{2}\rho^+ & \sqrt{2}K^{*^+} \\ \sqrt{2}\rho^- & -\rho^0 + \omega & \sqrt{2}K^{*^0} \\ \sqrt{2}K^{*^-} & \sqrt{2}\bar{K}^{*^0} & \sqrt{2}\phi \end{pmatrix} \tag{13}$$

and

$$B = \begin{pmatrix} \frac{1}{\sqrt{6}}\Lambda + \frac{1}{\sqrt{2}}\Sigma^0 & \Sigma^+ & p \\ \Sigma^- & \frac{1}{\sqrt{6}}\Lambda - \frac{1}{\sqrt{2}}\Sigma^0 & n \\ \Xi^- & \Xi^0 & -\sqrt{\frac{2}{3}}\Lambda \end{pmatrix}, \tag{14}$$

which give the octet (V_8), and the singlet (V_0) fields of the vector mesons by considering the $\omega - \phi$ mixing into account. For this, we write the wave functions of the ω- and the ϕ-meson, under the ideal mixing assumption, as

$$\begin{aligned}\omega &= \sqrt{\frac{1}{3}}\omega_8 + \sqrt{\frac{2}{3}}\omega_0 \\ \phi &= -\sqrt{\frac{2}{3}}\phi_8 + \sqrt{\frac{1}{3}}\phi_0.\end{aligned} \tag{15}$$

With these inputs, we obtained the amplitudes for different processes involving all the VB channels. All these amplitudes are listed in all details in Ref. [25].

2.2. Coupling pseudoscalar- and vector meson-baryon systems

The new information required to carry out a coupled channel study of the PB-VB systems is to write the PB $\leftrightarrow$ VB vertex consistently. This is done by using the Kroll-Ruderman (KR) theorem to write the Lagrangian for the $\gamma N \to \pi N$ process and by replacing the γ by a vector meson via the notion of the vector meson dominance. More concretely, we write the πN Lagrangian from the Gell-Mann-Levi's linear sigma model,

$$\mathcal{L}_{\pi N} = \bar{\psi}\left[i\gamma^\mu \partial_\mu - g_{\pi NN}\left(\sigma + i\vec{\tau}.\vec{\pi}\gamma_5\right)\right]\psi, \tag{16}$$

and introduce a vector meson field as a gauge boson of the hidden local symmetry

$$i\partial\!\!\!/ \longrightarrow i\partial\!\!\!/ - g\phi\!\!\!/, \tag{17}$$

to obtain the Lagrangian for the $\rho N \to \pi N$ vertex (see Ref [26] for more details)

$$\begin{aligned} \mathcal{L}_{\pi N \rho N} &= -i\frac{g}{2f_\pi}\bar{N}\left[\pi, \rho^\mu\right]\gamma_\mu\gamma_5 N \\ &\to -i\frac{gg_A}{2f_\pi}\bar{N}\left[\pi, \rho^\mu\right]\gamma_\mu\gamma_5 N, \end{aligned} \tag{18}$$

where $\pi = \vec{\tau}\cdot\vec{\pi}$, $\rho = \vec{\tau}\cdot\dfrac{\vec{\rho}}{2}$ and $f_\pi = 93$ MeV is the pion decay constant.

The SU(3) Lagrangian corresponding to Eq. (18) can be written as

$$\mathcal{L}_{PBVB} = \frac{-ig}{2f_\pi}\left(\tilde{F}\langle\bar{B}\gamma_\mu\gamma_5\left[\left[P, V_\mu\right], B\right]\rangle + \tilde{D}\langle\bar{B}\gamma_\mu\gamma_5\left\{\left[P, V_\mu\right], B\right\}\rangle\right), \tag{19}$$

where the trace $\langle ... \rangle$ has to be calculated in the flavor space and $\tilde{F} = 0.46$, $\tilde{D} = 0.8$ such that $\tilde{F} + \tilde{D} \simeq g_A = 1.26$. The ratio $\tilde{D}/(\tilde{F} + \tilde{D}) \sim 0.63$ here is close to the quark model value of 0.6, and the empirical values of $\tilde{F}$ and $\tilde{D}$ can be found, for example, in Ref. [28]. Using Eq. (19), amplitudes for all the processes within SU(3) have been obtained and are given in Ref. [26].

In the next section, we will discuss some of the most important results of our work on this topic.

3. Results and discussions

3.1. Strangeness 0 resonances found in the VB systems

Having calculated the amplitudes for different VB coupled channels with total strangeness 0, which are ρN, ωN, ϕN, $K^*\Sigma$ and $K^*\Lambda$, we solve the Bethe-Salpeter equation

$$T = V + VGT, \tag{20}$$

where the kernels $V(= V^t + V^{CT} + V^u + V^s)$ are effectively treated as contact interactions, which, in turn, can be factorized out of the loop integral involved in Eq. (20). Such a treatment, which is suitable for studies involving low-energy interactions, reduces solving of integral Eq. (20) to an algebraic one.

Since we are looking for dynamical generation of resonances, which implies working with low-energy dynamics, we calculate all the amplitudes in the s-wave. This means that the states found in our work can have $J^\pi = 1/2^-, 3/2^-$. Further, the total isospin of the VB system can be 1/2 or 3/2. Let us first discuss the results obtained for the total isospin 1/2 case. In Fig. 2, we show the squared amplitude for the $\rho N \to \rho N$ process. We find it very illustrative to show the results obtained by considering the t-channel exchange only for the VB interaction and compare them with those obtained by considering all the four diagrams of Fig. 1.

From Fig. 2, one can clearly see that the ρN amplitudes obtained from the t-channel are identical for spin 1/2 and 3/2 (upper-left and right panel), as expected from a spin independent interaction. However, a drastic change is seen when the amplitudes are obtained by solving Eq. (20) with the sum of the s-, t-, u-channels and the contact interaction as the kernel. In this case no peak is seen for the spin 1/2 configuration (see the lower left panel) while a more pronounced peak is found for spin 3/2 (shown in the lower right panel). Thus the spin degeneracy of the t-channel amplitude gets lifted by adding the spin dependent contact interaction and the s-, u-channel diagrams. To identify the peak seen in the spin 3/2 amplitude with a known resonance, we go to the complex plane and find a pole at $1637 - i35$ MeV. We find it to couple strongly to the ρN and $K^*\Lambda$ channel. These properties are very similar to those of the $N^*(1700)$ D_{13} resonance listed by the particle data group (PDG) [29]. Our findings are also in good agreement with other independent phenomenological studies [15, 30, 31].

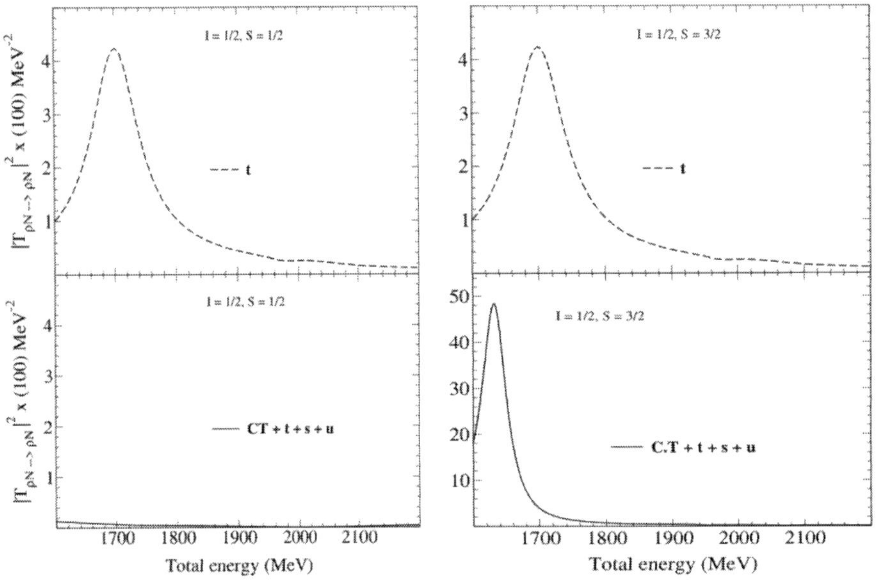

Figure 2. Squared amplitudes for the $\rho N \to \rho N$ process in the isospin 1/2 configuration. The upper panels show the results obtained by considering the t-channel diagram for VB interactions whereas the lower panels show the effect of taking all the four diagrams of Fig. 1 into account.

We find similar changes in all the coupled channels. In Fig. 3, we show the square amplitude for another channel: $K^*\Sigma$. In this case, once again, the upper-left and -right panels of Fig. 3 depict the same amplitudes for both spin 1/2 and 3/2, which show a peak near 2 GeV. We found that these peaks corresponded to a double pole structure in the complex plane [25]. The lower panels show the results obtained by considering all four diagrams of Fig. 1. In this case, both the states in spin 1/2 and 3/2 survive, although the peak positions differ by about 100 MeV. We find that the peak in the spin 3/2 amplitude corresponds to a pole in the complex plane at $2071 - i70$ MeV. This pole is found to couple strongly to the ωN, ϕN and $K^*\Lambda$ channels and can be related to the $N^*(2080)$ D_{13}. In the spin 1/2 case, we find a pole at $1977 - i22$ MeV. There is no known $J^\pi = 1/2^-$ state in Ref. [29] with a mass very close to this pole, however, an enhancement of the cross section is seen near 2 GeV region in the photo-production of the ϕ-meson on a nucleon studied by the LEPS group [32]. This enhancement could possibly be explained in terms of the spin $1/2^-$ resonance found in our work. In fact, a phenomenological analysis of the $\gamma N \to \phi N$ reaction showed that a better fit to the LEPS data [32] was found when a $J^\pi = 1/2^-$ resonance was included in their study as compared to the one obtained by including a $1/2^+$ resonance [33].

We have not discussed the results for total isospin $3/2$ so far, where we also find very important contribution from all the diagrams shown in Fig. 1. The t-channel interaction in isospin $3/2$ is repulsive (for spin $1/2$ as well as $3/2$) and, thus, does not give rise to formation of any states. However, the vector meson-baryon contact interaction is strongly attractive for isospin $3/2$ and

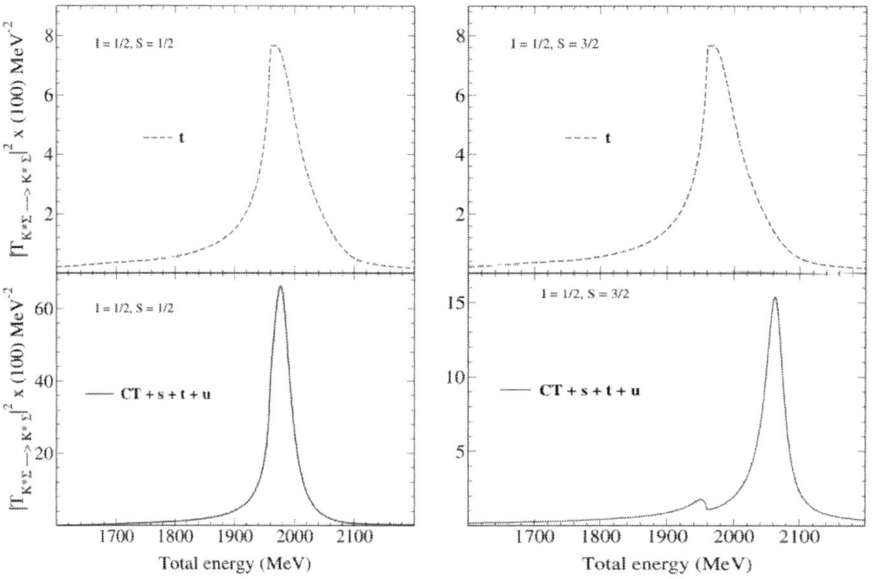

Figure 3. Same as Fig. 2 but for the $K^*\Sigma \to K^*\Sigma$ process.

spin $1/2$, which leads to the formation of a resonance near 2 GeV. In Fig. 4 we show the squared amplitude for the isospin $3/2$ VB channels, to which only ρN and $K^*\Sigma$ contribute. The left panel of this figure shows the amplitudes obtained by considering the t-channel interaction only while the right panel shows the results obtained by summing the s-, t-, u- channel exchanges, and the contact interaction. A pole was found at $2006 - i112$ MeV corresponding to the peak shown in the right panel. This state can be related with $\Delta(1900)\ S_{31}$.

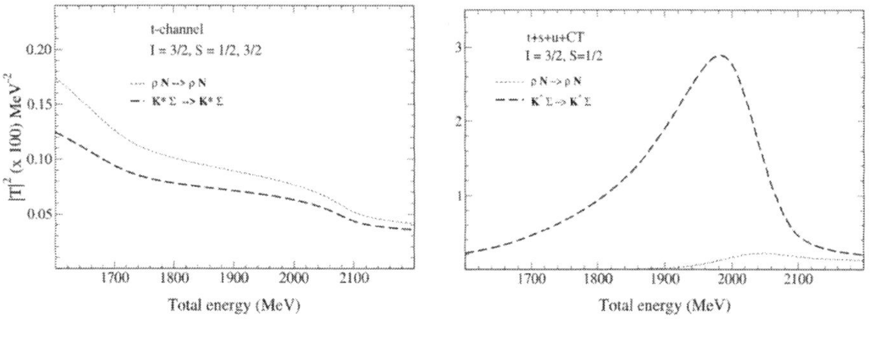

Figure 4. Squared amplitude for VB channels with total strangeness zero in the isospin $3/2$ configuration.

3.2. Strangeness −1 PB-VB coupled systems

Strangeness −1 is probably the most explored sector when it comes to discussing the dynamically generated resonances. Starting from the work of Dalitz [34] to more recent studies reported in Refs. [19, 20, 21, 35, 36, 37], strangeness −1 systems have been the center of attention when looking for resonances. There are at least two low-lying resonances which are good candidates for qualifying as the dynamically generated resonances in PB systems. However, these resonances have not been coupled to the vector meson-baryon systems. We made such a study within the formalism explained in the previous section (see Ref. [26] for more details). We found that the known properties of these resonances, i.e., the corresponding pole positions, their coupling to different PB channels and the PB amplitudes on the real axis, did not change much by coupling the VB channels. However, we found that the low-lying resonances couple very strongly to the heavy, closed VB channels. It is important to mention here that the large couplings of the low-lying Λ's to the VB channels found in our work do not imply the presence of a large fraction of the VB component in the wave function of these resonances since the large mass difference between the two would suppress it. Therefore, the interpretation of the low-lying Λ's as PB molecular states does not change. However, our findings could have some implications on, for example, the photoproduction of the Λ resonances where the production mechanism proceeding through the exchange of a vector meson could become important [38]. This should be verified in future.

We show the pole positions for the low-lying resonances and their couplings to the different PB and VB channels for both uncoupled as well as coupled case (which corresponds to the KR coupling 6 between the PB and VB channels [26]) in Table 1. As can be seen from this table, the couplings of the poles to the VB channels is null when the latter ones are not coupled to PB, while a very large coupling gets developed on coupling both systems. The larger couplings of the poles to various VB channels have been highlighted by boldfacing them in Table 1.

Table 1. g^i couplings of low-lying resonances to the PB and VB channels for different strengths of the coupling between PB-VB systems: 0 indicates no coupling between the two and 6 indicates fully coupled PB-VB systems (see [26] for more details).

	Λ(1405)				Λ(1670)	
	Pole1		Pole2			
PB-VB coupling	0	6	0	6	0	6
$M_R - i\Gamma/2 \rightarrow$ (MeV)	$1377 - i63$	$1357 - i53$	$1430 - i15$	$1412 - i11$	$1767 - i25$	$1744 - i28$
Channels ↓	Couplings (g^i) of the poles to the different channels					
$\bar{K}N$	$1.4 - i1.6$	$1.1 - i1.4$	$2.4 + i1.1$	$2.8 + i0.5$	$0.2 - i0.5$	$0.3 - i0.6$
$\pi\Sigma$	$-2.3 + i1.4$	$-2.2 + i1.4$	$-0.2 - i1.4$	$-0.2 - i1.1$	$0.1 + i0.2$	$0.1 + i0.3$
$\eta\Lambda$	$0.2 - i0.7$	$0.1 - i0.6$	$1.3 + i0.3$	$1.5 + i0.1$	$-1.0 + i0.3$	$-1.0 + i0.3$
$K\Xi$	$-0.4 + i0.4$	$-0.6 + i0.4$	$0.0 - i0.3$	$0.0 - i0.3$	$3.2 + i0.3$	$3.4 + i0.2$
$\bar{K}^*N$	$0.0 + i0.0$	$-1.7 + i0.7$	$0.0 + i0.0$	$\mathbf{-0.1 - i5.3}$	$0.0 + i0.0$	$-0.3 + i1.1$
$\omega\Lambda$	$0.0 + i0.0$	$-0.7 - i0.3$	$0.0 + i0.0$	$0.2 - i1.8$	$0.0 + i0.0$	$0.1 - i0.1$
$\rho\Sigma$	$0.0 + i0.0$	$\mathbf{1.3 + i6.8}$	$0.0 + i0.0$	$\mathbf{-2.4 - i1.6}$	$0.0 + i0.0$	$\mathbf{0.3 - i3.5}$
$\phi\Lambda$	$0.0 + i0.0$	$1.0 + i0.5$	$0.0 + i0.0$	$-0.3 + i2.6$	$0.0 + i0.0$	$-0.2 + i0.1$
$K^*\Xi$	$0.0 + i0.0$	$\mathbf{1.3 + i5.7}$	$0.0 + i0.0$	$-2.0 - i0.5$	$0.0 + i0.0$	$0.5 - i1.2$

Further, although our purpose was to study low-lying resonances first, for which we relied on the t-channel exchange to calculate PB → PB and VB → VB amplitudes, to draw any concrete conclusion about higher mass resonances, we should take into account a more complete VB →

VB interactions as shown in Ref. [25]. Still, within the present formalism, we could test if the widths of the resonances in the 1800-2100 MeV region increased a lot by coupling the lower mass (PB) open channels. We found that the higher mass resonance poles did not always attain larger widths when coupling them to PB channels, sometimes poles disappeared and other times new poles appeared by coupling VB to the light PB channels [26]. For instance, we found that a new pole gets generated near 1440 MeV in total isospin 1 case when PB and VB are coupled. This pole can be related to $\Sigma(1480)$ [29, 39, 40]. However, to make stronger claims on the properties of the resonances found in 1800-2100 MeV, we must carry out a more detailed calculations by obtaining the VB amplitudes from the different diagrams shown to be important in Ref. [25]. Such a study is being carried out at this moment and we expect it to finalize soon.

4. Summary and outlook

We can summarize our study on the baryon resonances coupled to vector meson-baryon systems as follows:

- A gauge invariant Lagrangian under the hidden local symmetry gives rise to significantly important tree-level contributions from the s-, t- and u- channel exchange diagrams and a contact term. Our findings show that none of these contributions should be assumed to be negligible *a priori*.
- The resulting vector meson-baryon interaction is very spin-isospin dependent. This is something which should be expected when two particles with similar mass and non-zero spin interact.
- Many low-lying resonances like $\Lambda(1405)$ couple strongly to VB channels, which is a very useful information, for instance, to study photoproduction of $\Lambda(1405)$.
- When heavier VB resonance poles are allowed to couple to PB channels, they do not always get wider. Sometimes they can disappear. Also new poles can appear sometimes. For example, a new isospin 1 resonance is found to arise due to the PB-VB coupled dynamics near 1430 MeV

  Finally, as an outlook we would like to say that a PB-VB coupled channel calculation should be made with a more detailed VB interaction. Further, it is important to use these amplitudes to study relevant reactions, which have been studied experimentally.

5. Acknowledgments

This work is partly supported by the Grant-in-Aid for Scientific Research on Priority Areas titled Elucidation of New Hadrons with a Variety of Flavors" (E01: 21105006 for K.P.K and A.H) and (22105510 for H. N) and the authors acknowledge the same. A. M. T is thankful to the support from the Grant-in-Aid for the Global COE Program The Next Generation of Physics, Spun from Universality and Emergence from the Ministry of Education, Culture, Sports, Science and Technology (MEXT) of Japan.

References

[1] J. J. Sakurai, *Currents and Mesons* (Univ. Chicago Press, Chicago, 1969).
[2] M. Bando, T. Kugo and K. Yamawaki, Phys. Rept. **164**, 217 (1988).
[3] M. Bando, T. Kugo, S. Uehara, K. Yamawaki and T. Yanagida, Phys. Rev. Lett. **54**, 1215 (1985).
[4] M. Bando, T. Kugo and K. Yamawaki, Nucl. Phys. B **259**, 493 (1985).
[5] H. Kaneko, A. Hosaka and O. Scholten, arXiv:1112.4776 [hep-ph].
[6] Y. Yamaguchi, S. Ohkoda, S. Yasui and A. Hosaka, Phys. Rev. D **84**, 014032 (2011) [arXiv:1105.0734 [hep-ph]].
[7] R. Molina, H. Nagahiro, A. Hosaka and E. Oset, Phys. Rev. D **83**, 094030 (2011) [arXiv:1009.4881 [hep-ph]].
[8] J. Yamagata-Sekihara and E. Oset, Phys. Lett. B **690**, 376 (2010) [arXiv:1001.1816 [hep-ph]].

[9] A. Martinez Torres, K. P. Khemchandani, D. Gamermann and E. Oset, Phys. Rev. D **80**, 094012 (2009) [arXiv:0906.5333 [nucl-th]].
[10] R. Molina, H. Nagahiro, A. Hosaka and E. Oset, Phys. Rev. D **80**, 014025 (2009) [arXiv:0903.3823 [hep-ph]].
[11] K. Nawa, A. Hosaka and H. Suganuma, Phys. Rev. D **79**, 126005 (2009) [arXiv:0901.3080 [hep-th]].
[12] O. Krehl, C. Hanhart, S. Krewald and J. Speth, Phys. Rev. C **62**, 025207 (2000) [arXiv:nucl-th/9911080].
[13] M. F. M. Lutz, G. Wolf and B. Friman, Nucl. Phys. A **706**, 431 (2002) [Erratum-ibid. A **765**, 431 (2006)] [arXiv:nucl-th/0112052].
[14] P. Gonzalez, E. Oset, J. Vijande, Phys. Rev. C **79**, 025209 (2009).
[15] M. Doring, C. Hanhart, F. Huang, S. Krewald and U. G. Meissner, Nucl. Phys. A **829**, 170 (2009) [arXiv:0903.4337 [nucl-th]].
[16] S. Sarkar, B. -X. Sun, E. Oset, M. J. Vicente Vacas, Eur. Phys. J. A **44**, 431–443 (2010).
[17] E. Oset and A. Ramos, Eur. Phys. J. A **44**, 445 (2010) [arXiv:0905.0973 [hep-ph]].
[18] P. C. Bruns, M. Mai, U. -G. Meissner, Phys. Lett. **B697**, 254-259 (2011). [arXiv:1012.2233 [nucl-th]].
[19] N. Kaiser, P. B. Siegel, W. Weise, Nucl. Phys. **A594**, 325-345 (1995). [nucl-th/9505043].
[20] E. Oset and A. Ramos, Nucl. Phys. **A635**, 99–120 (1998).
[21] D. Jido, T. Sekihara, Y. Ikeda, T. Hyodo, Y. Kanada-En'yo, E. Oset, Nucl. Phys. **A835**, 59-66 (2010) [arXiv:1003.4560 [nucl-th].
[22] A. Martinez Torres, D. Jido, Phys. Rev. **C82**, 038202 (2010) [arXiv:1008.0457 [nucl-th]]; A. Martinez Torres, K. P. Khemchandani, E. Oset, Phys. Rev. **C79**, 065207 (2009) [arXiv:0812.2235 [nucl-th]].
[23] A. Martinez Torres, K. P. Khemchandani and E. Oset, Phys. Rev. C **77**, 042203 (2008) [arXiv:0706.2330 [nucl-th]]; K. P. Khemchandani, A. Martinez Torres, E. Oset, Eur. Phys. J. **A37**, 233-243 (2008) [arXiv:0804.4670 [nucl-th]]; K. P. Khemchandani, A. Martinez Torres, E. Oset, AIP Conf. Proc. **1322**, 379-383 (2010).
[24] A. Martinez Torres, K. P. Khemchandani, D. Jido and A. Hosaka, Phys. Rev. D **84**, 074027 (2011) [arXiv:1106.6101 [nucl-th]].
[25] K. P. Khemchandani, H. Kaneko, H. Nagahiro and A. Hosaka, Phys. Rev. D **83**, 114041 (2011) [arXiv:1104.0307 [hep-ph]].
[26] K. P. Khemchandani, A. Martinez Torres, H. Kaneko, H. Nagahiro and A. Hosaka, Phys. Rev. D **84**, 094018 (2011) [arXiv:1107.0574 [nucl-th]].
[27] D. Jido, A. Hosaka, J. C. Nacher, E. Oset, A. Ramos, Phys. Rev. **C66**, 025203 (2002). [hep-ph/0203248].
[28] T. Yamanishi, Phys. Rev. D **76**, 014006 (2007).
[29] K. Nakamura *et al.* (Particle Data Group), J. Phys. **G37**, 075021 (2010).
[30] Yang, S. N., Kiswandhi, A., Xie, J. -J., [arXiv:1108.1677 [nucl-th]]
[31] Kiswandhi, A., Yang, S. N., [arXiv:1108.1657 [nucl-th]]
[32] T. Mibe *et al.* [LEPS Collaboration], Phys. Rev. Lett. **95**, 182001 (2005). [nucl-ex/0506015].
[33] S. Ozaki, A. Hosaka, H. Nagahiro, O. Scholten, Phys. Rev. **C80**, 035201 (2009). [arXiv:0905.3028 [hep-ph]].
[34] R. H. Dalitz and S. F. Tuan, Phys. Rev. Lett. **2**, 425 (1959).
[35] D. Jido, J. A. Oller, E. Oset, A. Ramos and U. G. Meissner, Nucl. Phys. A **725**, 181 (2003) [nucl-th/0303062].
[36] Y. Ikeda, T. Hyodo and W. Weise, Phys. Lett. B **706**, 63 (2011) [arXiv:1109.3005 [nucl-th]]
[37] T. Hyodo and D. Jido, Prog. Part. Nucl. Phys. **67**, 55 (2012) [arXiv:1104.4474 [nucl-th]].
[38] S. -i. Nam, J. -H. Park, A. Hosaka, H. -C. Kim, arXiv:0806.4029 [hep-ph].
[39] I. Zychor, V. Koptev, M. Buscher, A. Dzyuba, I. Keshelashvili, V. Kleber, H. R. Koch, S. Krewald *et al.*, Phys. Rev. Lett. **96**, 012002 (2006).
[40] J. -J. Wu, S. Dulat, B. S. Zou, Phys. Rev. **C81**, 045210 (2010); J. -J. Wu, S. Dulat, B. S. Zou, Phys. Rev. **D80**, 017503 (2009).

Unitary chiral dynamics of two hadrons in a finite volume: the KD, ηD_s system and the $D_{s^*0}(2317)$ resonance

A. Martínez Torres[1], L. R. Dai[1,2], C. Koren[1,3], D. Jido[1] and E. Oset[1,3]
[1]Yukawa Institute for Theoretical Physics, Kyoto University, Kyoto 606-8502, Japan.
[2]Department of Physics, Liaoning Normal University, Dalian, 116029, China.
[3]Departamento de Física Teórica and IFIC, Centro Mixto Universidad de Valencia-CSIC, Institutos de Investigación de Paterna, Aptdo. 22085, 46071 Valencia, Spain

Abstract. We investigate the KD and ηD_s system in a finite volume and study the properties of the $D_{s^*0}(2317)$ resonance, which is generated in this coupled channel system. We calculate the energy levels in a cubic box and considering them as synthetic lattice data we solve the inverse problem of determining the bound states and phase shifts in the infinite volume. We observe that it is possible to obtain accurate KD phase shifts and the position of the $D_{s^*0}(2317)$ state from the synthetic lattice data considered and that a careful analysis of the finite volume data can shed some light on the nature of the $D_{s^*0}(2317)$ resonance as a KD molecule or otherwise.

1. Introduction

Quantum Cromodynamics (QCD) is considered nowadays as the theory for the strong interactions, with the quarks being the degrees of freedom. However, while at high energies, because of the asymptotic freedom of the quarks, the theory has been successfully tested by the experiment, the situation is completely different at low or intermediate energies due to the confinement regime of QCD, which makes perturbative methods non applicable.

It is interesting to notice that in the low energy region the QCD spectrum exhibits a very peculiar fact: the presence of an isospin triplet (pions) with a mass much smaller than the rest of the QCD states. An extension to SU(3) can be done by considering the lowest octet of pseudoscalar states, i.e., π, K, and η. This fact can be understood by the presence in the QCD Lagrangian of a chiral symmetry for the light quark sector (u, d, s) which spontaneously breaks down and generates, through the Goldstone theorem, the aforementioned states (see, for example, Refs. [1, 2, 3, 4]).

A theory which considering these facts describes successfully the interaction between hadrons is the so called chiral perturbation theory or χPT. This theory consist in a series of Lagrangians in a power momentum expansion which is valid up to some high energy scale, $\Lambda_{\chi PT}$, with $\Lambda_{\chi PT} \sim 1$ GeV. Hence, the χPT expansion is valid for momenta $p << \Lambda_{\chi PT} \sim 1$GeV. Although χPT is a powerful tool at low energies, its convergence is limited to a narrow interval (for example, in the study of the meson-meson scattering it is possible to reach energies around 0.5 GeV, while in the meson-baryon scattering one can goes up to the threshold region). This means that one of the most interesting facts of the strong interacting phenomena, which is the

Published under licence by IOP Publishing Ltd

formation of resonances, cannot be studied with χPT alone. One could think in increasing the energy region of applicability of the theory just by including higher orders terms in the Lagrangian. However, this has the disadvantage of losing the predictive power of the theory, since the number of free parameters increases tremendously with the order: the leading order χPT Lagrangian has basically no free parameters, while in the next-to-leading order there are 12 parameters and if one goes to the next-next-to-leading order then there are more than 100. But, even then a resonance cannot be obtained by summing a series perturbatively. As a consequence, the development of non-perturbative techniques is needed in order to extend the energy region of applicability of the theory, but without losing its predictive power. A simple way of accomplish this issue is considering the lowest order chiral Lagrangians and implement the theory with unitarity, which has given rise to the so mentioned unitary chiral perturbation theory ($U\chi PT$).

In the last decade, two-body meson-baryon and meson-meson systems, as for example, $\bar{K}N$, $\pi\Sigma$, $\pi\Lambda$, πN, $\pi\pi$, $K\bar{K}$, $\pi\eta$, etc., have been studied extensively with $U\chi PT$ and dynamical generation of many baryon and meson resonances in these systems has been found, for instance, the $\Lambda(1405)$, $\Lambda(1520)$, $N^*(1535)$, etc., in the baryon sector and the $f_0(980)$, $a_0(980)$, $\sigma(600)$, etc., in the meson sector [5, 6, 7, 8, 9, 10, 11, 12, 13]. The $U\chi PT$ theory has also been used to study three body hadron systems, like two pseudoscalars and a baryon, two pseudoscalars and a vector meson, three pseudoscalars, or clusters of a hadron and a resonance formed from the interaction of two hadrons and generation of many baryon and mesons states have been found [14, 15, 16, 17, 18, 19, 20, 21, 22, 23, 24].

Recently, the $U\chi PT$ has also been extended to investigate the interaction of two hadrons in a finite volume [25, 26, 27, 28, 29]. As it is well known, one of the challenges of lattice QCD calculations is to determine the spectra of mesons and baryons. However, in the case of resonances the main difficulty is that they do not correspond to isolated energy levels in the spectrum of the QCD Hamiltonian on the lattice. Thus, an additional effort has to be done. In the case of one or two channels, the problem is formally solved under the framework of Lüscher [30, 31, 32, 33], which relates the measured discrete value of the energy in a finite volume to the scattering phase shift at the same energy, for the same system in the infinite volume.

In the present case we study the KD, ηD_s system, in which the $D_{s^*0}(2317)$ state is formed [34, 35, 36, 37], in a finite volume following the model developed in Ref. [28]. We start with the "phenomenological" potential of Ref. [37] and use the KD, ηD_s channels to evaluate the energy levels in a box using periodic boundary conditions. Using this energy levels to generate synthetic lattice data, we face the inverse problem of extracting from them the KD phase shifts in the infinite volume and the position of the $D_{s^*0}(2317)$. We show that with a few data from the lattice (about ten) corresponding to two levels of the box with various values of L (the box size), we can reproduce the results in the infinite volume, both for the energy of the bound state and the KD phase shifts, with a good accuracy.

2. Formalism

The scattering matrix in the $U\chi PT$ is obtained by solving the Bethe-Salpeter equation

$$T(E) = [1 - V(E)G(E)]^{-1}V(E), \tag{1}$$

with E the energy of the system in the center of mass frame, V represents the matrix for the transition potentials between the channels and G, in our case, the loop function of two meson propagators, which is defined as

$$G = i\int \frac{d^4q}{(2\pi)^4}\frac{1}{q^2 - m^2 + i\epsilon}\frac{1}{(P-q)^2 - M^2 + i\epsilon} \tag{2}$$

with m and M the masses of the two mesons and P the four momentum of the external meson-meson system. This loop function is divergent and needs to be regularized. In the dimensional regularization scheme, this integral is evaluated and for a meson-meson system gives the result [38, 39]

$$\begin{aligned} G_i^D(E) = \frac{1}{(4\pi)^2}\Bigg\{ & a_i(\mu) + \log\frac{m_i^2}{\mu^2} + \frac{M_i^2 - m_i^2 + s}{2s}\log\frac{M_i^2}{m_i^2} \\ & + \frac{Q_i(\sqrt{s})}{\sqrt{s}}\Big[\log\left(s-(M_i^2-m_i^2)+2\sqrt{s}Q_i(\sqrt{s})\right)+\log\left(s+(M_i^2-m_i^2)+2\sqrt{s}Q_i(\sqrt{s})\right) \\ & - \log\left(-s+(M_i^2-m_i^2)+2\sqrt{s}Q_i(\sqrt{s})\right)-\log\left(-s-(M_i^2-m_i^2)+2\sqrt{s}Q_i(\sqrt{s})\right)\Big]\Bigg\}, \end{aligned} \tag{3}$$

where $s=E^2$, Q_i is the on-shell momentum of the particles in the ith channel, μ a regularization scale and $a_i(\mu)$ a subtraction constant (note that there is only one independent parameter, because a change in μ can be absorbed into a_i).

In a finite box of side length L, instead of integrating over the energy states of the infinite volume as done in Eq. (2), we have to perform a sum over the discrete momenta given by the periodic boundary conditions, i.e., $\vec{q}_L = (2\pi/L)\vec{n}$ with $\vec{n}\in\mathbb{Z}^3$. In this case, the loop function associated with the one of Eq. (3) in a finite volume is given by (see Ref. [28] for more details)

$$\tilde{G}_i(E) = G_i^D(E) + \lim_{q_{\max}\to\infty}\left[\frac{1}{L^3}\sum_{\vec{q}_L}^{q_{\max}} I_i(\vec{q}_L) - \int\limits_{q<q_{\max}}\frac{d^3q}{(2\pi)^3}I_i(\vec{q})\right] \tag{4}$$

with

$$\begin{aligned} I_i(x) &= \frac{1}{2\omega_i(x)\,\omega_i'(x)}\frac{\omega_i(x)+\omega_i'(x)}{E^2-(\omega_i(x)+\omega_i'(x))^2+i\epsilon}, \\ \omega_i(x) &= \sqrt{m_i^2+x^2}, \quad \omega_i'(x) = \sqrt{M_i^2+x^2} \\ x &= \vec{q}_L, \vec{q}. \end{aligned} \tag{5}$$

In this way, to determine the T matrix in the box we just need to replace the G function in Eq. (1) by Eq. (4). The eigenenergies of the box correspond to energies that produce poles in the T matrix. Thus we search for these energies by looking for zeros of the determinant

$$\det(1 - V\tilde{G}) = 0. \tag{6}$$

In Fig. 1 we show the first five energy levels obtained for a cubic box and for different values of L. For large values of L the lowest energy level in the box should converge to the value of the energy of the bound state of the infinite volume case. This is a standard feature for the ground level bound state in the lattice. As it can be seen in Fig. 1, we find that with values of $L = 3\ m_\pi^{-1}$ there is already a good convergence to the value in the infinite volume.

3. The inverse problem

Let us face now the problem of determining the pole position of the $D_{s^*0}(2317)$ state and the KD phase shift in the infinite volume from the energy levels shown in Fig. 1. To do this, we take the first two energy levels and consider 10 points to which we associate an error of ± 10

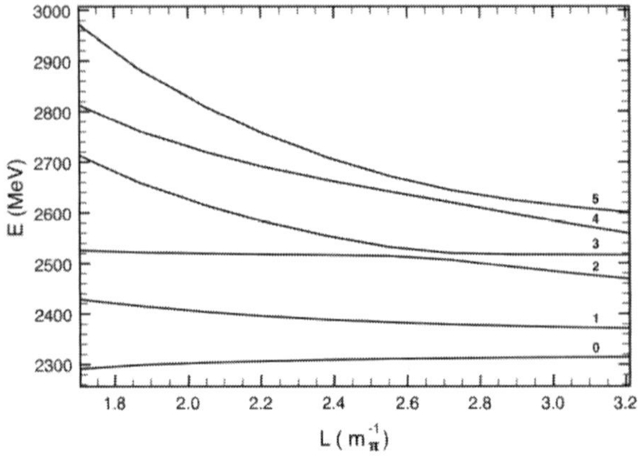

Figure 1. Energy levels for the KD, ηD_s system as functions of the side length of the box, L.

MeV. We call the data generated in this form as synthetic lattice data. Then we perform a fit to these synthetic data by solving Eq. (6) with KD and ηD_s as coupled channels and under the assumption that we have a potential with the same energy dependence than the one obtained in Ref. [37], i.e.,

$$V_{ij} = a_{ij} + b_{ij}[s - (m_K + M_D)^2], \quad (7)$$

with m_K and M_D the mass of the Kaon and D meson, respectively. We look for the minimum value of χ^2, χ^2_{min}, and obtain a set of parameters for a_{ij}, b_{ij}. Then we generate random sets of the parameters close to those of the minimum, such that χ^2 is only increased below $\chi^2_{min} + 1$. In Fig. 2 we show the results of the levels reconstructed from the best fits to the data, with a band corresponding to the random choices of parameters satisfying the condition $\chi^2 < \chi^2_{min} + 1$.

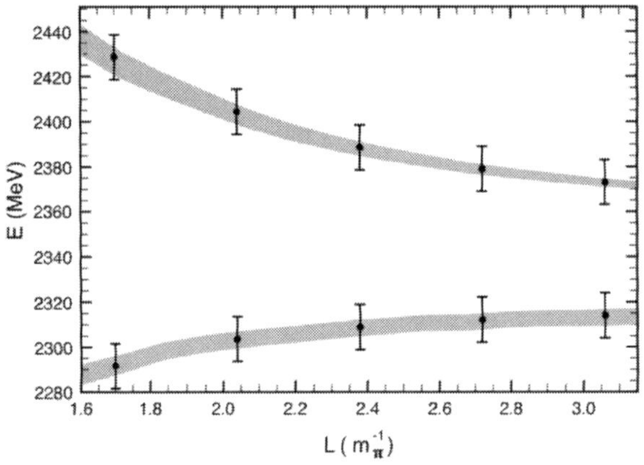

Figure 2. Energy levels as functions of the side length of the box L, reconstructed from fits to the "data" of Fig. 1 using the potential of Eq. (7). The band corresponds to different choices of parameters within errors.

With the different potentials obtained from the fits we can solve Eq. (1) and determine the T matrix in infinite volume and, thus, calculate the phase shift for the KD channel. The results for the KD phase shift for total isospin 0 and s-wave, δ_0^0, are shown in Fig. 3 and, as it can be seen, the agreement with the exact result (solid line) is quite good.

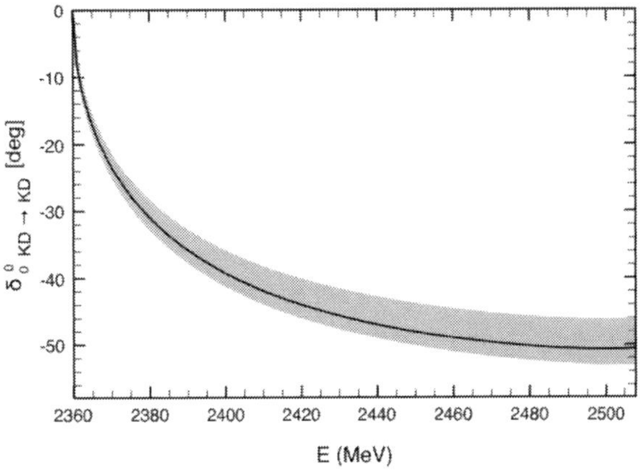

Figure 3. Phase shifts for KD scattering derived from the coupled channels unitary approach of Ref. [37] (solid line). The band corresponds to the results obtained from the fits to the "data" of Fig. 1 using the potential of Eq. (7) with two channels.

Since the ηD_s channel is far away from the region at which the resonance is generated, we may consider the possibility of explaining the synthetic lattice data considered by taking into account only the KD channel. In Fig. 4 we can see the results obtained for the KD phase shifts from the fits to the synthetic lattice data of Fig. 1 with one channel potential . We observe that for low energies the results are very similar to those obtained with two channels. However, as the energy increases, the results with two channels are closer to the exact solution.

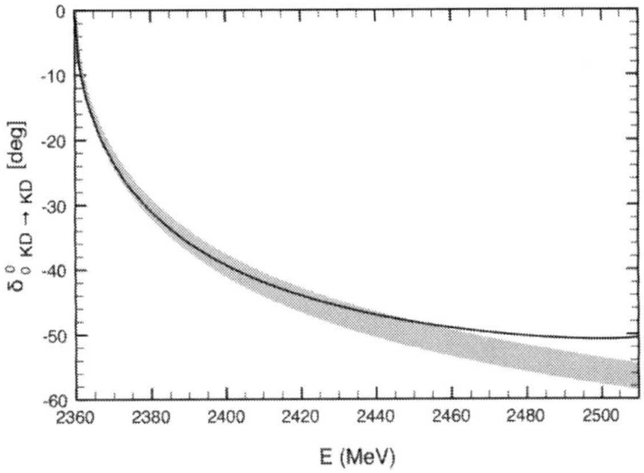

Figure 4. Phase shifts for KD scattering derived from the coupled channels unitary approach of Ref. [37] (solid line). The band corresponds to using the fits to the "data" of Fig. 1 using the potential of Eq. (7) with only the KD channel.

Apart from the phase shift, we can also determine from the fits the energy of the bound state in infinite volume, getting good results with one or with two channels: $E = 2317 \pm 5$ MeV in both cases.

Journal of Physics: Conference Series **374** (2012) 012008 doi:10.1088/1742-6596/374/1/012008

4. The bound state and its nature

A very interesting issue is to investigate the possibility of determining the nature of the $D_{s^*0}(2317)$ from the lattice data. For this purpose we can use a sum rule whose origin stems from normalization to unity of the wave function of the bound state and which states [40, 41]

$$\sum_i g_i^2 \left.\frac{dG_{ii}}{ds}\right|_{E=E_\alpha} = -1 \tag{8}$$

where

$$g_i^2 = \lim_{s \to s_R} (s - s_R) T_{ii} \tag{9}$$

are the residues of the T_{ii} scattering matrices (coupling squared) at the pole of the bound state, $\sqrt{s_R}$, and G_{ii} are the propagator of the two particles of the corresponding channels (the loop function G^D that we use here). Note, however, that there is a caveat in our formulation, since in the derivation of Eq. (8) an energy independent potential is assumed, while we are now producing potentials with a moderate energy dependence. Taking into account the energy dependence of Eq. (7) we find that approximately one should have now (see Ref. [28] for more details)

$$\sum_i g_i^2 \left\{ \frac{b_{ii}}{[a_{ii} + b_{ii}(s - (m_D + m_K)^2)]^2} + \frac{dG_i}{ds} \right\} = -1. \tag{10}$$

One can see from the derivation in Ref. [40] that each term in Eq. (8) accounts (with reversed sign) for the probability of the bound state to be a bound state of the pair of particles of the channel considered. In the case that we had the coupling of the bound state to another hypothetical elementary particle outside the space of pair of particles considered, there would be an extra term in the sum of Eq. (8), -Z, where Z would account for the overlap of the bound state with this hypothetical genuine particle. The diversion of the sum of Eq. (8) with respect to -1 would indicate the amount of the bound state which cannot be considered a bound state of the two particles.

Obviously with the synthetic lattice data which we have produced the answer is trivial since the state was dynamically generated from the interaction of the KD, ηD_s channels and Eq. (10) is fulfilled up to the level of 3%. But if the real lattice results were different than those obtained here the answer might be different. For example, there is the possibility that the data might be such that one could find a gross deviation from having KD as the large component of the $D_{s^*0}(2317)$ wave function, reflecting the fact that the $D_{s^*0}(2317)$ would be a genuine state, not generated by the KD interaction. For this purpose we have made a test introducing a different potential where a CDD pole (Castillejo, Dalitz, Dyson) [42] is introduced by hand. The potential in just one channel would now be

$$V = V_M + \frac{g_{CDD}^2}{s - s_{CDD}}, \tag{11}$$

where V_M is assumed to be energy independent and g_{CDD}^2, s_{CDD} are the parameters of the CDD pole. In this case, the sum rule for this particular energy dependence read as (see Ref. [28] for

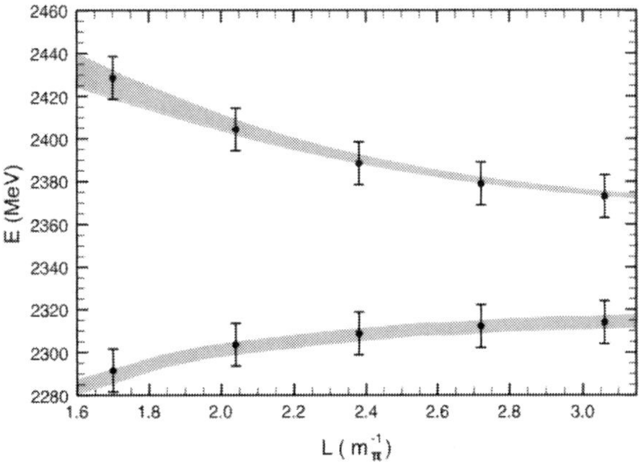

Figure 5. Fits to the "data" of Fig. 1 using a potential of the type of Eq. (11) that contains a CDD pole.

more details)

$$-g^2\frac{dG}{ds} = \frac{1}{-\frac{dV^{-1}}{ds}\left(\frac{dG}{ds}\right)^{-1}+1} = 1-Z$$
$$= \frac{1}{-\frac{g_{CDD}^2 G^2}{(s-s_{CDD})^2\frac{dG}{ds}}+1}. \qquad (12)$$

Using the potential of Eq. (11) we try to fit the synthetic lattice data of Fig. 1 (obtained with the coupled channels approach of Ref. [37]). The results are shown in Fig. 5 and, as we can see, the fit to the data is quite good. In Fig. 6 we show the results for the phase shifts associated with the CDD potential, which coincide with those obtained with the two channel potential up to energies around 2400 MeV. After this, there is a large dispersion of the results and they divert significantly from those obtained with the two channel approach. Certainly, the use of an additional "lattice" level would put big constraints in the phase shifts in that region helping us decide between the two options.

However, it is interesting to analyze the results obtained: the CDD pole is found at the energy of about 2500 MeV, far away from the energy at which the state $D_{s^*0}(2317)$ is formed. If one restricts oneself to low energies, the fit with the CDD would be acceptable. Yet, the fact that the CDD pole has appeared so far away from the energy of the $D_{s^*0}(2317)$ state indicates that the data do not want a CDD pole being responsible for this state. The CDD pole far away in this case simply generates a smooth energy dependent potential in the region of the low energies. The interesting thing is that if we calculate now Z from Eq. (12) we find that $Z \sim 0.15$, indicating that the bound state is basically a KD bound state, with the precision that the limited low energy data provide. A more precise determination would require the consideration of an additional level of the box.

5. Conclusions

We have studied the KD, ηD_s system in a finite volume and generate the energy levels of the system in a cubic box of side length L. By assuming that the results obtained for these energy

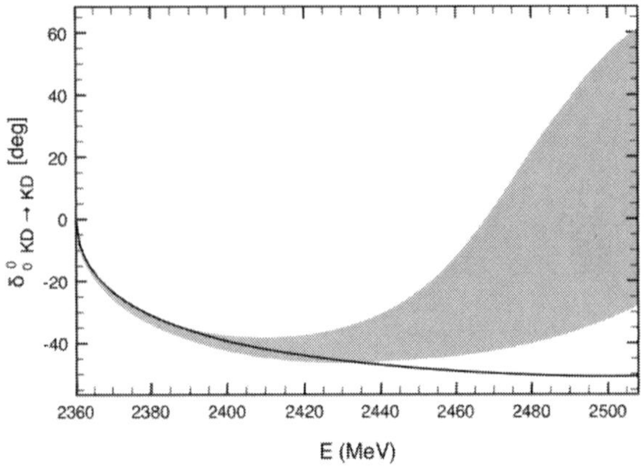

Figure 6. Phase shifts from the fits with a CDD pole to the first two levels obtained with the coupled channels approach of Ref. [37] shown in Fig. 5. The solid line corresponds to the results with the two channel analysis of Fig. 3.

levels would correspond to results given by lattice calculations, we have faced the problem of determining the KD phase shift in infinite volume as well as the pole position and nature of the $D_{s^*0}(2317)$, a state generated in the KD, ηD_s system. We found that the two channel method proposed by us works well, but if we restrict ourself to low energies of the KD system we also see that the one channel analysis is quite good, but as we approach the threshold of ηD_s, the two channel analysis provides better phase shifts. We have also investigated whether we could say something from the lattice results concerning the nature of the bound state obtained. We found that indeed it was possible to induce some information on the nature of this state as basically a molecule of KD, by inspecting the spectra in finite volume. Alternatively we tried to analyze the lattice data in terms of a CDD potential that could accommodate a non molecular state and we could find a good fit to the data but with an artificial pole at an energy very far away, such that the potential was essentially constant in the region of interest and the Z value for overlap with a genuine state was small, compatible with zero within the precision that the data allow.

All these exercises show that the data provided by QCD lattice calculations on energy levels in a box contain the information to decide on the nature of the bound state $D_{s^*0}(2317)$.

References

[1] U. G. Meissner, Rep. Prog. Phys. **56**, 903 (1993)
[2] V. Bernard, N. Kaiser and U. G. Meisnner, Int. J. Mod. Phys. E **4**, 193 (1995).
[3] A. Pich, Rep. Prog. Phys. **58**, 563 (1995).
[4] J. A. Oller, E. Oset, A. Ramos, Prog. Part. Nucl. Phys. **45**, 157-242 (2000).
[5] E. Oset and A. Ramos, Nucl. Phys. A **635**, 99 (1998)
[6] T. Inoue, E. Oset and M. J. Vicente Vacas, Phys. Rev. C **65**, 035204 (2002)
[7] J. A. Oller and E. Oset, Nucl. Phys. A **620**, 438 (1997) [Erratum-ibid. A **652**, 407 (1999)]
[8] J. Nieves and E. Ruiz Arriola, Nucl. Phys. A **679**, 57 (2000)
[9] B. Borasoy, R. Nissler and W. Weise, Eur. Phys. J. A **25**, 79 (2005)
[10] D. Jido, J. A. Oller, E. Oset, A. Ramos and U. G. Meissner, Nucl. Phys. A **725**, 181 (2003)
[11] J. A. Oller, E. Oset and J. R. Pelaez, Phys. Rev. Lett. **80**, 3452 (1998).
[12] V. Bernard, N. Kaiser and U. G. Meissner, Phys. Rev. C **52**, 2185 (1995).
[13] L. Roca, S. Sarkar, V. K. Magas and E. Oset, Phys. Rev. C **73**, 045208 (2006).
[14] A. Martínez Torres, K. P. Khemchandani and E. Oset, Phys. Rev. C **77** 042203 (2008); Eur. Phys. J. A **35** (2008) 295.
[15] K. P. Khemchandani, A. Martínez Torres and E. Oset, Eur. Phys. J. A **37**, (2008) 233.

[16] A. Martínez Torres et al., Phys. Rev. D **78** (2008) 074031.
[17] A. Martínez Torres, D. Jido, Phys. Rev. C **82**, 038202 (2010).
[18] A. Martínez Torres, D. Jido, Y. Kanada-En'yo, Phys. Rev. C **83**, 065205 (2011) .
[19] L. Alvarez-Ruso, J. A. Oller, and J. M. Alarcon, Phys. Rev. D**80**, 054011 (2009).
[20] M. Albaladejo, J. A. Oller, and L. Roca, Phys. Rev. D **82**, 094019 (2010).
[21] L. Roca and E. Oset, Phys. Rev. D **82**, 054013 (2010).
[22] J. J. Xie, A. Martinez Torres and E. Oset, Phys. Rev. C **83**, 065207 (2011).
[23] M. Bayar, J. Yamagata-Sekihara and E. Oset, Phys. Rev. C **84**, 015209 (2011).
[24] A. Martínez Torres, K. P. Khemchandani, D. Jido and A. Hosaka, Phys. Rev. D **84**, 074027 (2011).
[25] M. Doring, U. -G. Meissner, E. Oset and A. Rusetsky, Eur. Phys. J. A **47**, 139 (2011).
[26] M. Doring, J. Haidenbauer, U. -G. Meissner and A. Rusetsky, Eur. Phys. J. A **47**, 163 (2011).
[27] M. Doring and U. G. Meissner, JHEP **1201**, 009 (2012).
[28] A. Martínez Torres, L. R. Dai, C. Koren, D. Jido and E. Oset, arXiv:1109.0396 [hep-lat], Phys. Rev. D in print.
[29] L. Roca and E. Oset, arXiv:1201.0438 [hep-lat].
[30] M. Lüscher, Commun. Math. Phys. **105** (1986) 153 (1986).
[31] M. Lüscher, Nucl. Phys. B **354** (1991) 531.
[32] M. Lage, U.-G. Meiğner and A. Rusetsky, Phys. Lett. B **681**, 439 (2009).
[33] V. Bernard, M. Lage, U.-G. Meiğner and A. Rusetsky, JHEP **1101**, 019 (2011).
[34] E. E. Kolomeitsev, M. F. M. Lutz, Phys. Lett. B **582**, 39-48 (2004).
[35] J. Hofmann, M. F. M. Lutz, Nucl. Phys. A **733**, 142-152 (2004).
[36] F. -K. Guo, P. -N. Shen, H. -C. Chiang, R. -G. Ping, B. -S. Zou, Phys. Lett. B **641**, 278-285 (2006).
[37] D. Gamermann, E. Oset, D. Strottman, M. J. Vicente Vacas, Phys. Rev. D **76**, 074016 (2007).
[38] J. A. Oller, U. G. Meissner, Phys. Lett. B**500**, 263-272 (2001).
[39] E. Oset, A. Ramos, C. Bennhold, Phys. Lett. B **527**, 99-105 (2002).
[40] D. Gamermann, J. Nieves, E. Oset, E. Ruiz Arriola, Phys. Rev. D **81**, 014029 (2010).
[41] T. Sekihara, T. Hyodo, D. Jido, Phys. Rev. C **83**, 055202 (2011).
[42] L. Castillejo, R. H. Dalitz, F. J. Dyson, Phys. Rev. **101**, 453-458 (1956).

5th DAE-BRNS Workshop on Hadron Physics (Hadron 2011) IOP Publishing
Journal of Physics: Conference Series **374** (2012) 012009 doi:10.1088/1742-6596/374/1/012009

Nucleon-Nucleon t-matrix Effective Interaction for $(p, 2p)$ Reactions.

Arun K. Jain, B. N. Joshi and Sudhir R. Jain

Nuclear Physics Division, Bhabha Atomic Research Center,
Mumbai-400 085, India.

E-mail: arunjain@barc.gov.in, bnjoshi@barc.gov.in, srjain@barc.gov.in

Abstract. In $(p, 2p)$ reactions the p-p interaction is the main knockout interaction which acts in its full glory. Recent finite range $(p, 2p)$ calculations used a parametrized spin-isospin dependent finite range $t_{NN}(r)$ - effective interaction operator which is a function of the distance between the nucleons alone. In the present work we have evaluated the nucleon-nucleon (N-N) t-matrix effective interaction using the Reid soft core N-N interaction potential. In this we find that the proper t- matrix effective interaction is not only a function of spin and isospin of the interacting nucleons but is a strong function of the relative orbital angular momentum, L also. The radial behaviour of the $t_L(r)$ is also found to be very unconventional in that it is found to go to zero at $r = 0$ while in the conventional usage the $t(r)$ is taken to peak at $r = 0$ because of its representation in terms of Yukawa functions. This very different behaviour of our calculated N-N t-matrix effective interaction operator is expected to make large changes in the Finite Range (FR-DWIA) predictions for $(p, 2p)$ reaction cross sections. Besides this it is hoped that the marked differences one obtains in the behaviour of these $t_L(r)$'s from different realistic N-N interactions through their resulting fits obtained for the $(p, 2p)$ reactions will provide us a tool to select the proper N-N interaction out of the many available in literature.

1. Introduction

For a long time the medium energy $(p, 2p)$ reactions on light medium mass nuclei were considered to be very good tools to obtain the absolute proton spectroscopic factors in those target nuclei. These expectations grew further with the recent advancements made in this field due to the incorporation of the sophisticated relativistic finite range methodology and the development of related computer programs [1, 2, 3, 4, 5, 6]. For performing these advanced calculations one requires not only the scalar and vector components of the optical potentials which in the Dirac phenomenology fit the proton elastic scattering data but also requires the proper p-p t - matrix effective interaction, $t(\vec{r})$. The conventional finite range distorted wave impulse approximation FR-DWIA calculations employ the meson exchange model t-matrix effective interaction by Love and Franey or Horowitz [7, 8, 9] or others in the same category [10, 11] where the direct and exchange contributions to the amplitude are parametrized separately in terms of a number of Yukawa type functions in the first-order Born approximation. Thus far, the use of the Dirac phenomenology in the present relativistic finite range-distorted wave impulse approximation (RFR-DWIA) has not resulted much improvement in agreements with the data in comparison to the non-relativistic FR-DWIA formalisms. The existing studies of the sensitivities of the analyzing power to both the zero range (ZR) and finite range (FR) approximations on ^{40}Ca

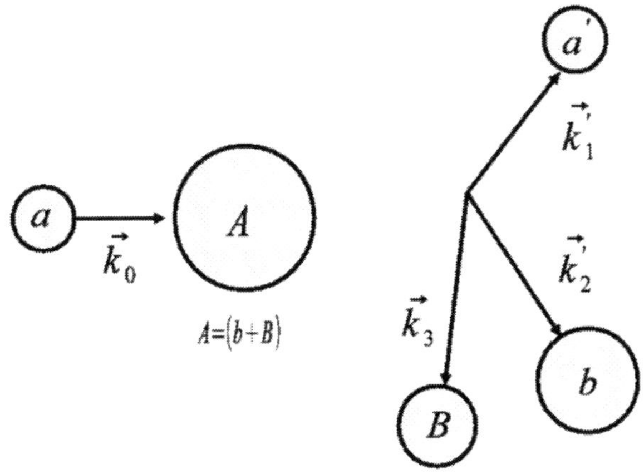

Figure 1. Knockout reaction schematics.

and ^{208}Pb targets have not resulted in any definite conclusion as to which approximation gives a consistently better description of the data [3, 6]. The RFR-DWIA calculations using all the inputs mentioned above sometimes describe the shape of the energy sharing spectra reasonably well but often the analyzing power data fits show significant disagreement. Even the inclusion of the assumed 10% to 20% medium modification of the interaction parameters could not explain the inconsistencies completely indicating thereby that something more basic is missing in the assumed *p*-*p* *t*- matrix effective interaction.

2. Knockout Reactions

The knockout reaction $A(a, a'b)B$ of Fig.1 is akin to a free *a*-*b* scattering [12, 13]. Here the incident proton, a of momentum $\vec{k_o}$ and kinetic energy E_o interacts with the weakly bound proton, b with its momentum, -$\vec{k_3}$ and binding energy, -Q. After scattering, the proton, a' (just another name for a after the scattering) goes out with momentum $\vec{k_1}$ and kinetic energy E_1 and the ejected proton, b now goes out with momentum $\vec{k_2}$ and kinetic energy E_2. The residual nucleus, B behaving as a spectator moves with momentum $\vec{k_3}$ and kinetic energy E_3. Now as the incident energy, E_o is much larger than the binding energy, $-Q$ one can consider it to be negligible in comparison to E_o thus the (a, ab) reaction can be regarded to be close to free $a - b$ scattering ($\vec{k_3} = 0$). The energy momentum conservation equations for the free a - b scattering are:

$$E_o = E'_1 + E'_2 \qquad \text{and} \qquad \vec{k_o} = \vec{k'_1} + \vec{k'_2} \qquad k_0 = k'_1 \cos\theta'_1 + k'_2 \cos\theta'_2.$$

However if the particle b is bound to B by $-Q$ (a few MeV) in the target nucleus, A, and as the momentum k_3 of the recoiling nucleus is small, taking it to be zero one gets:

$$E_o = E_1 + E_2 - Q.$$

As the Q here is negative we have

$$E_1 < E'_1 \qquad \text{and} \qquad E_2 < E'_2 \qquad \text{and} \qquad (k_0 = k_1 \cos\theta_1 + k_2 \cos\theta_2).$$

because, $\quad k_1 = \sqrt{2m_1 E_1}, \qquad k_2 = \sqrt{2m_2 E_2}$

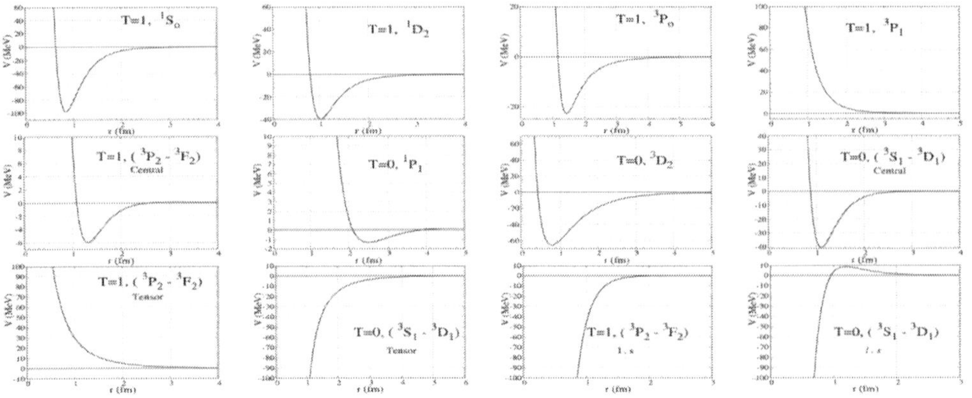

Figure 2. Reid soft core potential, $V^{S,T}(r)$.

Hence k_1' and k_2' are more than k_1 and k_2 respectively. Similarly the angles θ_1 and θ_2 of $\vec{k}_1$ and $\vec{k}_2$ will be slightly less than θ_1' and θ_2' of the free scattering.

Now for the $A(a,a'b)B$ reaction the transition amplitude $T_{fi}^{bL\varphi}$ is written as [12]:

$$T_{fi}^{bL\wedge}\ (\vec{k}_f,\vec{k}_i) = \int \chi_1^{(-)*}(\vec{k}_{aB},\vec{r}_{aB})\chi_2^{(-)*}(\vec{k}_{bB},\vec{R}_{bB})t_{ab}(\vec{r}_{ab})$$
$$\chi_0^{(+)}(\vec{k}_{1-A},\vec{r}_{1-A})\varphi_{b\wedge}(\vec{R}_{bB})d\vec{r}_{aB}d\vec{R}_{bB} \tag{1}$$

The scattering state distorted waves χ_0, χ_1 and χ_2 of this expression are evaluated using the optical potentials for the p_o-A, p_1-B and p_2-B systems with the relative coordinates $\vec{r}_{oA}$, $\vec{r}_{aB}$ and $\vec{R}_{bB}$ respectively [14]. The bound state of b and B is represented by $\varphi_{b\wedge}(\vec{R}_{bB})$ where $\wedge$ represents all the quantum numbers collectively. Finally all these relative coordinates are expressed in terms of $\vec{r}_{ab}(\equiv \vec{r})$ and $\vec{R}_{bB}(\equiv \vec{R})$. While using the zero range approximation the transition matrix element, T_{fi} of this expression is factorized into a half off-shell p-p scattering amplitude and a distorted momentum distribution of the struck proton in the target nucleus. On the other hand when one uses the full finite range $t_{ab}(\vec{r}_{ab})$ the factorization is not possible due to the presence of optical distortions. In the FR-DWIA formalism the chosen relative coordinates $\vec{r}$ and $\vec{R}$ get coupled through the distorted waves $\chi_0^{(+)}(\vec{k}_{oA},\vec{r}_{oA})$ and $\chi_1^{(-)*}(\vec{k}_{aB},\vec{r}_{aB})$ leading to no-factorization of the scattering amplitude of particles a and b [12, 13]. Usually one treats the $t_{ab}(\vec{r}_{ab})$ as a series expansion in terms of Yukawa functions of various range parameters [4] and that too only a function of $|\vec{r_{ab}}|$. The parameters in this expansion are chosen to fit the N-N amplitudes rather than the data directly.

As indicated earlier the quasi-free scattering $(p,\ 2p)$ reaction kinematic is only slightly modified because of the binding of the proton in the target nucleus (-Q, a small value). Therefore the assumption made that one can replace $t_{pp}(\vec{r}_{pp})$ in $(p,\ 2p)$ reaction by free p-p t - matrix effective interaction.

We can obtain the p-p t-matrix effective interaction, $t_{ab}(\vec{r}_{ab})$ through another approach where the N-N interaction potentials obtained directly by fitting N-N scattering data are employed in solving the Schrödinger equation. Our method uses solution of the Schrödinger equation in configuration space and is much different from the conventional method (where this equation is solved in momentum space by matrix inversion only to first order).

To start with we use the Reid soft core realistic N-N potential [15], seen in Fig.2 which is spin and isospin dependent and contains tensor and spin-orbit interaction in triplet states (where $L \neq J$, L and J also $\neq 0$). In such states L is not a good quantum number and $L = J-1$ and $L = J+1$ are present in the same J^π- state. Here one faces the difficulty that there are coupled equations for these triplet states such as ($^3S_1$-$^3D_1$ and $^3P_2$-$^3F_2$ etc.).

In order to resolve this coupling of different L's which lead to coupled second order differential equations a semi-classical treatment has been developed for the N-N scattering problem. This method is similar to the deuteron problem solved earlier with tensor interaction [16]. The coupled wave functions, $\Psi^{(+)}(k,r)$ and $\Psi^{(-)}(k,r)$ are obtained as,

$$\begin{aligned}\Psi^{(+)}(k,r) &= u(k,r)\cos\frac{\theta(k,r)}{2} - w(k,r)\sin\frac{\theta(k,r)}{2}\\ \Psi^{(-)}(k,r) &= u(k,r)\sin\frac{\theta(k,r)}{2} + w(k,r)\cos\frac{\theta(k,r)}{2},\end{aligned} \quad (2)$$

where $\theta(k,r)$ is function of tensor potential, $V_t(r)$, spin -orbit potential-$V_{LS}(r)$ and the difference between the centrifugal repulsions between the two ℓ - states, $u(k,r)$ (for the lower ℓ-value) and $w(k,r)$ (for the larger ℓ-value). As $\theta(k,r) \to 0$ for r beyond the range of $V_t(r)$ these $u(k,r)$ and $w(k,r)$'s are used to get the phase shifts in these J ℓ states.

For the singlet states as well as for the triplet states (with $L \neq J-1$ and $L \neq J+1$ cases) there exist no L-coupling hence the radial Schrödinger equations are easily solvable. These solutions are then matched with the known forms of the wave functions in the external region to obtain the phase shifts.

The phase shifts for the singlet as well as triplet states for the p-p and n-p scattering at various energies have been found to be matching exceedingly well with those obtained by other techniques. We are now assured that the radial distorted wave functions found by this method are correct. We now use these distorted wave functions in the t-matrix equation below for various spin, S and isospin, T values.

3. Evaluation of the t-matrix

The N-N t - matrix effective interaction, $t_{12}(\vec{r}_{01})$ evaluated at the relative energy E, is given by [17, 18]:

$$t^{+}_{12}(S,T,E,\vec{r}) = e^{-ikz}V^{S,T}(\vec{r})\Psi^{+}_{12}(\vec{r}) \equiv \sum_{L=0,1,2,3...} t^{S,T}_{L}(E,r)P_L(\hat{r}) \quad (3)$$

where,

$$\Psi^{+}_{12}(\vec{r}) = \sum_{\ell=0,1,2,3...} i^{\ell}(2\ell+1)\frac{u_\ell(kr)}{kr}e^{i\sigma_\ell}P_\ell(\hat{r}). \quad (4)$$

$$\begin{aligned}t^{S,T}_{L}(E,r) = \frac{(2L+1)}{2}\sum_{\ell,n} V^{S,T}_{\ell}(r)i^{\ell}(2\ell+1)\frac{u_\ell(kr)}{kr}\jmath_n(kr)(-i)^n\\ (2n+1)e^{i\sigma_\ell}\int_{-1}^{+1} P^{*}_{L}(\cos\theta)P_\ell(\cos\theta)P_n(\cos\theta)d(\cos\theta).\end{aligned} \quad (5)$$

where $j_\ell(kr)$'s are the spherical Bessel functions and $P_\ell(x)$ are the Legendre polynomials, σ_ℓ's are the Coulomb phase shifts and $u_\ell(kr)$'s are the radial wave functions in the ℓ channel.

These $t^{ST}_{L}(E,r)$'s form our t -matrix effective interactions. For p-p case the total isospin T = 1, therefore it will be only ($S = 0$, - even) and ($S = 1$, - odd) states present in the realistic interactions $V^{ST}_{\ell}(r)$. Each of these states will provide $t^{ST}_{L}(E,r)$'s which will be complex in general. It is easily seen from Eqs.(3) to (5) that for one ℓ- value of $V^{ST}_{\ell}(r)$, there will be several possible L- values in the effective interaction $t^{ST}_{L}(E,r)$, i.e. L here will vary from 0 to ∞. It is

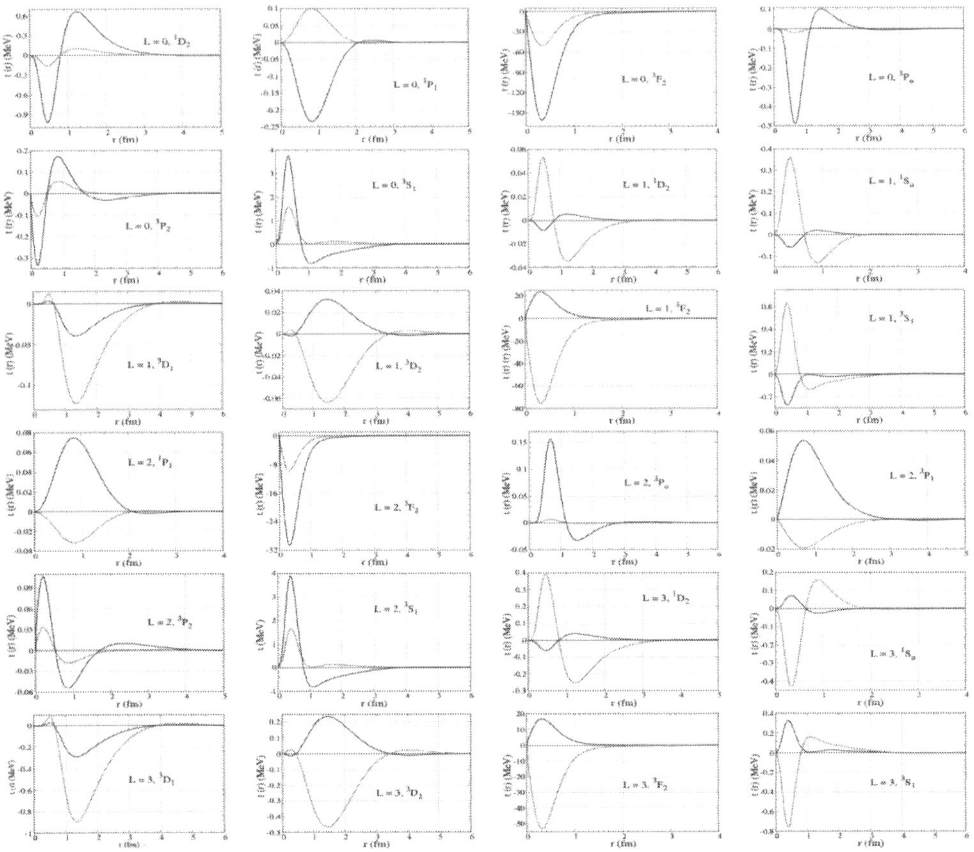

Figure 3. $t_L^{ST}(E,r)$ N-N effective interaction at $E_{inc.} = 200$ MeV for various L values for central, tensor and spin-orbit potentials. Real $t_L^{ST}(E,r)$ (Solid Line), Imaginary $t_L^{ST}(E,r)$ (Dotted Line).

to be emphasized at this stage that even though the interactions $V_\ell^{ST}(r)$'s are real, the t -matrix effective interactions $t_L^{ST}(E,r)$'s could, in general be complex. Moreover for each orbital angular momentum ℓ of the realistic interactions $V_\ell^{ST}(r)$, the L of the effective interaction, $t_L^{ST}(E,r)$ will vary and peak at some other value of L. The observation of different L's for the same ℓ - value indicate that the symmetry character (decided by the ℓ - value) of the state will be modified in the effective interaction, $t_L^{ST}(E,r)$. In fact the derived t- matrix effective interaction will have mixed symmetry which should not be disturbed by any re-symmetrization. This is because all these t_L's are intricately interwoven so as to preserve the symmetry in the overall expansion in terms of the series. It is easy to observe in Eq.(3) that the symmetrization of $t_{12}^+(\vec{r})$ due to symmetry of $\Psi_\ell^{ST}(r)$ is modified due to the multiplication e^{ikz} or $e^{i\vec{k}_o\cdot\vec{r}}$ (which is not symmetrized) in the expression of $t_L^{ST}(E,r)$. One must also realize that as long as one uses the matrix elements of these $t^{ST}(E,\vec{r})$ between the initial plane wave state and the final plane wave state the (effectively taking the Fourier transform of $t^{ST}(E,\vec{r})$) the symmetry is restored. Therefore, after taking the Fourier transform, one can perform any operation what so ever one wants with these matrix element because the symmetry is already restored in this procedure of taking the Fourier transform.

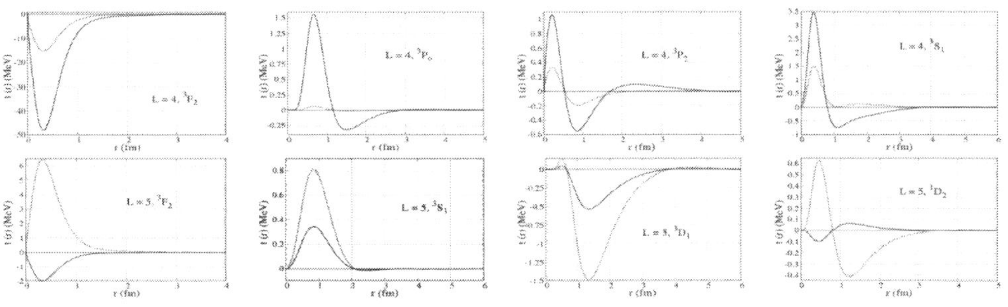

Figure 4. Figure 3. Continued

4. Results and Discussion

The $t_L^{ST}(E,r)$ for 200 MeV incident energy using the Reid soft core N-N potential and the results are shown in Figs.3 and 4. The results are calculated for ${}^{(2S+1)(2T+1)}\ell_J$ states as ${}^{13}S_0$, ${}^{11}P_1$, ${}^{13}D_2$, ${}^{33}P_0$, ${}^{31}S_1$, ${}^{33}P_1$, ${}^{33}P_2$, ${}^{31}D_1$, ${}^{31}D_2$ and ${}^{33}F_2$ for various L - values of $t_L^{ST}(E,r)$. As there are too many L - values contributing to all these states, only certain representative and significant results are presented and discussed here.

It is observed that whatever be the state ${}^{13}S_0$, ${}^{11}P_1$, ${}^{13}D_2$, ${}^{33}P_0$, ${}^{31}S_1$, ${}^{33}P_1$, ${}^{33}P_2$, ${}^{31}D_1$, ${}^{31}D_2$ or ${}^{33}F_2$ and whatever the L - value, the $t_L^{ST}(E,r)$ vanish at $r=0$ which can be understood in terms of vanishing of $\Psi_{12}^{+}(\vec{r})$ at $r=0$. In the case of a hard core, there will be a $\delta(r-r_c)$ at $r=r_c$ as seen in Eq. 53 of [19]. In the case case of a soft repulsive core however, this will result in a $t_L^{ST}(E,r=0)=0$ while there will be a rise in $t_L^{ST}(E,r)$ below $r=r_c$ and then it will fall back to zero as r goes to zero (in the case where there is no mixing due to the tensor interaction), this behaviour is clearly seen in our results in Fig.3. Besides, the real and imaginary components are observed to have large variations in shape and strengths for this energy. Except for the triplet states with $\ell \neq J$ (as well as for L or S or $J \neq 0$) and for the tensor interaction, there is a node at the value of r where the $V_\ell^{ST}(r)$ goes to zero. It is seen that in the case of triplet states with $\ell \neq J$ this node does not appear necessarily. The range of $t_L^{ST}(E,r)$ is not shifted outwards for larger L's as is normally expected from the centrifugal barrier considerations. The general impression one gets is that $t_L^{ST}(E,r)$'s behave much differently in comparison to the corresponding $V_\ell^{ST}(r)$'s. From these one can easily conclude that the N-N t- matrix effective interaction from Love and Franey [7, 8], Satchler and Love [11], Bertsch et al [10] and that of Horowitz [9] which behave like sums of Yukawa functions are not properly behaved as they all peak at $r=0$. We have seen that there are many more multipole (L)-components of the $t_L^{ST}(E,r)$'s instead of just the $L=0$ component. The symmetry character is built-in if large number of $t_L^{ST}(E,r)$'s are incorporated. Moreover the imaginary part of these $t_L^{ST}(E,r)$'s, in the present situation, cannot be interpreted as absorptive potentials because they are present even when the phase shifts are real.

In conclusion one can say that the t- matrix effective interactions have many more multipole components in its structure and their individual components may have symmetry characteristics while the overall symmetry is taken care of in their interwoven behaviour. Because of this as well as because of wide variations in their values and shapes one can not have the L - averaging as it will change their symmetry character as also their rotation characteristics. The tensor interaction makes an order of magnitude large contributions in the $t_L^{ST}(E,r)$'s as compared to the uncoupled multipole components; therefore, this interaction should be treated with caution, without the use of Born approximation. The repulsive core character is also affected by the couplings caused by the tensor interaction, although the t- matrix effective interactions vanishes at $r=0$ the vanishing of $t_L(r)$ where $V_{Central}(r)$ goes to zero may be altered due to the hard core for coupled states.

Journal of Physics: Conference Series **374** (2012) 012009 doi:10.1088/1742-6596/374/1/012009

5. References

[1] M. B. Epstein, *et al* 1980 *Phys. Rev. Lett.* **44** 20
[2] Y. Kudo and M. Miyazaki 1986 *Phys. Rev.* C **34** 1192
[3] E. D. Cooper and O. V. Maxwell 1989 *Nucl. Phys.* A **493** 468, 1990 *Nucl. Phys.* A **513** 584, 1993 *Nucl. Phys.* A **565** 740, 1994 *Nucl. Phys.* A **574** 819, 1996 *Nucl. Phys.* A **603** 441
[4] Y. Ikebata 1995 *Phys. Rev.* C **34** 890
[5] J. Mano and Y. Kudo 1998 *Prog. Theor. Phys.* **100** 91
[6] R. Neveling *et al* 2002 *Phys. Rev.* C **66** 034602
[7] W. G. Love and M. A. Franey 1981 *Phys. Rev.* C **24** 1073
[8] M. A. Franey and W. G. Love 1985 *Phys. Rev.* C **31** 488
[9] C. J. Horowitz 1985 *Phys. Rev.* C **31** 1340
[10] G. Bertsch *et al* 1977 *Nucl. Phys.* A **284** 399
[11] G. R. Satchler and W.G. Love 1979 *Phys. Reports* **55C** 183
[12] Arun K. Jain and Bhushan N. Joshi 2009 *Phys. Rev. Lett.* **103** 132503
[13] B. N. Joshi *et al* 2011 *Phys. Rev. Lett.* **106** 022501
[14] Arun K. Jain and B. N. Joshi 2008 *Phys. Rev.* C **77** 027601
[15] R. V. Reid 1968 *Ann. Phys., NY* **50** 448
[16] S. R. Jain 2004 *J. Phys. G: Nucl. Part. Phys.* **30** 157
[17] D. F. Jackson 1970 *Nuclear Reactions*, (Methuen and Co.) 61
[18] Arun K. Jain and B. N. Joshi 2008 *Prog. Theor. Phys.* **120** 1193
[19] K. A. Brueckner and J. L. Gammel 1958 *Phys. Rev.* **109** 1023

In-medium vector mesons and low mass lepton pairs from heavy ion collisions

Sourav Sarkar and Sabyasachi Ghosh
Theoretical Physics Division, Variable Energy Cyclotron Centre,
1/AF, Bidhannagar, Kolkata 700064, India
E-mail: sourav@vecc.gov.in

Abstract.
The ρ and ω meson self-energy at finite temperature and baryon density have been analysed for an exhaustive set of mesonic and baryonic loops in the real time formulation of thermal field theory. The large enhancement of spectral strength below the nominal ρ mass is seen to cause a substantial enhancement in dilepton pair yield in this mass region. The integrated yield after space-time evolution using relativistic hydrodynamics with quark gluon plasma in the initial state leads to a very good agreement with the experimental data from In-In collisions obtained by the NA60 collaboration.

1. Introduction

Colliding heavy ions at ultra-relativistic energies is the only way to produce and study bulk properties of strongly interacting matter. Systematic efforts, both theoretical and experimental over the last few decades [1, 2] have addressed various facets of the thermodynamics of the underlying theory – QCD. The spectra of hadrons emitted after freeze-out of the fireball produced in heavy ion collisions have provided us with a wealth of information. This includes the recent discovery made by studying the elliptic flow of hadrons that the quark gluon plasma (QGP) produced in Au+Au collisions at RHIC actually behaves as a strongly interacting fluid [3] as opposed to a asymptotically free gas of quarks and gluons. However, by virtue of the fact that electromagnetic probes (real photons and dileptons) are emitted all through the lifetime of the fireball coupled with their low rescattering probability make them penetrating probes capable of mapping the space-time history of the collision [4]. Lepton pairs with both invariant mass and transverse momentum information are in fact preferable to real photons. Large mass pairs produced by Drell-Yan process and from the decays of heavy quarkonia are emitted early whereas pairs with low invariant mass radiated from thermal hadronic matter and Dalitz decays of hadrons are produced late in the collision. The invariant mass spectra of dileptons thus carry time information as displayed explicitly in [5] using the invariant mass dependence of the elliptic flow of lepton pairs.

The rate of production of thermal dileptons is proportional to the two-point correlator of vector currents [6]. In the low invariant mass region which is dominated by lepton pairs produced during the later stages of the collision, dilepton emission takes place due to the decay of vector mesons. Consequently, the spectral properties of vector mesons, the ρ meson in particular has been a subject of intense discussion [7, 4, 2, 8]. We find that it is only for the $\pi-\pi$ loop that one calculates the thermal self-energy loop for the vector mesons directly. In the case of other loops typically involving one heavy and one light particle or both heavy particles one uses in general either the virial formula or the Lindhard function.

The sources modifying the free propagation of a particle find a unified description in terms of contributions from the branch cuts of the self energy function. In addition to the unitary cut present already in vacuum, the thermal amplitude generates a new cut, the so called the Landau cut which provides the effect of collisions with the surrounding particles in the medium. This formalism was applied to obtain the ρ self-energy in hot mesonic [9] and baryonic [10] matter considering an exhaustive set of one-loop diagrams. The framework of real time thermal field theory that we use, enables us to evaluate the imaginary part of the self-energy from the branch cuts for real and positive values of energy and momentum without having to resort to analytic continuation as in the imaginary time approach. To evaluate the baryonic loops we work with the full relativistic baryon propagator in which baryons and anti-baryons manifestly appear on an equal footing. Thus the contributions from all the singularities in the self-energy function including the distant ones coming from the unitary cut of the loops involving heavy baryons are also included. These are not considered in the Lindhard function approach but can contribute appreciably to the real part of the ρ meson self-energy as shown [11] in the case of a $N\Delta$ loop.

The broadening of the vector meson spectral functions leads to an enhancement of lepton pair production in the invariant mass region below the ρ peak. The effect of the evolving matter is handled by relativistic hydrodynamics. The integrated yield is seen to agree very well with the NA60 data [12] from In-In collisions at 17.3 AGeV.

The article is organised as follows. We will begin with a short derivation of the the dilepton emission rate in terms of the current correlation function in section 2. The relation to the spectral function of vector mesons is specified in section 3. This will be followed by a discussion on ρ and ω self-energies in hot and dense matter in section 4. In section 5 a brief account of the space-time evolution and initial conditions will be provided followed by the dilepton invariant mass spectra. We will end with a summary in section 6.

2. Dilepton emission rate and the current correlation function

Let us consider an initial state $|I\rangle$ which goes to a final state $|F\rangle$ producing a lepton pair l^+l^- with momenta p_1 and p_2 respectively. The dilepton multiplicity thermally averaged over initial states is given by [13]

$$N = \sum_I \sum_F |\langle F, l^+l^- | e^{i\int \mathcal{L}_{int} d^4x} | I\rangle|^2 \frac{e^{-\beta E_I}}{Z} \frac{d^3p_1}{(2\pi)^3 2E_1} \frac{d^3p_2}{(2\pi)^3 2E_2} \tag{1}$$

where $Z = Tr[e^{-\beta H}]$ and $\mathcal{L}_{int} = e\bar{\psi}_l(x)\gamma_\mu\psi_l(x)A^\mu(x) + eJ^h_\mu(x)A^\mu(x)$ in which $\psi_l(x)$ is the lepton field operator and $J^h_\mu(x)$ is the electromagnetic current of hadrons. Following [6, 13, 4] this expression can be put in the form

$$\frac{dN}{d^4x d^4q} = \frac{e^4 L(q^2)}{3(2\pi)^5 q^4} e^{-\beta q_0} W^>_{\mu\nu}(q)(q^\mu q^\nu - q^2 g^{\mu\nu}) \tag{2}$$

where, $W^>_{\mu\nu} = \int d^4x\, e^{iq\cdot x}\langle J^h_\mu(x) J^h_\nu(0)\rangle_\beta$ is the Fourier transform of the thermal expectation value of the two-point correlator of the hadronic currents and $L(q^2) = (1 + \frac{2m_l^2}{q^2})\sqrt{1 - \frac{4m_l^2}{q^2}}$. We now define $W^<_{\mu\nu}$ by interchanging the order of the currents getting $W^<_{\mu\nu} = e^{-\beta q_0} W^>_{\mu\nu}$. Using these to define the commutator $W_{\mu\nu} = W^>_{\mu\nu} - W^<_{\mu\nu} = \int d^4x\, e^{iq\cdot x}\langle [J^h_\mu(x), J^h_\nu(0)]\rangle_\beta$ we finally have

$$\frac{dN}{d^4x d^4q} = -\frac{\alpha^2}{6\pi^3} \frac{g^{\mu\nu}}{q^2} L(q^2) f_{BE}(q_0) W_{\mu\nu}(q_0, \vec{q}) \tag{3}$$

for conserved hadronic currents. This expression appears in different forms in the literature. Replacing $f_{BE}(q_0)W_{\mu\nu}$ by (i) $W^M_{\mu\nu} = \int d^4x\, e^{iq\cdot x}\langle J^h_\mu(x) J^h_\nu(0)\rangle_\beta$ yields the expression in [6], by (ii) $2\text{Im}W^R_{\mu\nu} f_{BE}(q_0)$ where $W^R_{\mu\nu} = i\int d^4x\, e^{iq\cdot x}\theta(t-t')\langle [J^h_\mu(x), J^h_\nu(0)]\rangle_\beta$ yields the rate in [14] by (iii) $2\text{Im}W^T_{\mu\nu}/(1+e^{\beta q_0})$ where $W^T_{\mu\nu} = i\int d^4x\, e^{iq\cdot x}\langle T J^h_\mu(x) J^h_\nu(0)\rangle_\beta$ gives the rate in [15].

The rate given by eq. (3) is to leading order in electromagnetic interactions but exact to all orders in the strong coupling encoded in the current correlator $W_{\mu\nu}$. The q^2 in the denominator indicates the exchange of a single virtual photon and the Bose distribution implies the thermal weight of the source.

3. Current correlator and the spectral function of vector mesons

Thermal field theory is the appropriate framework to carry out perturbative calculations in the medium. In the real time version of this formalism two point functions assume a 2×2 matrix structure on account of the shape of the contour in the complex time plane [16, 17]. It is convenient to begin with the quantity,

$$T^{ab}_{\mu\nu} = i\int d^4x e^{iq\cdot x}\langle T_c J^h_\mu(x) J^h_\nu(0)\rangle_\beta \tag{4}$$

where T_c denotes ordering along the contour and a, b are the thermal indices which take values 1 and 2. This quantity can be diagonalised by means of a matrix U so that

$$T^{ab}_{\mu\nu} = U\begin{pmatrix} \overline{T}_{\mu\nu} & 0 \\ 0 & -\overline{T}^*_{\mu\nu}\end{pmatrix}U\ ; \quad U = \begin{pmatrix}\sqrt{1+n} & \sqrt{n} \\ \sqrt{n} & \sqrt{1+n}\end{pmatrix}, \quad n = \frac{1}{e^{\beta|q_0|}-1} \tag{5}$$

where $\overline{T}_{\mu\nu}$ is an analytic function. Equating both sides it follows that this function is obtainable from any one (thermal) component of $T^{ab}_{\mu\nu}$. It is related e.g. to the 11-component as,

$$\text{Re}\overline{T}_{\mu\nu}(q_0,\vec{q}) = \text{Re}T^{11}_{\mu\nu}(q_0,\vec{q})\ ; \quad \text{Im}\overline{T}_{\mu\nu}(q_0,\vec{q}) = \coth(\frac{\beta|q_0|}{2})\text{Im}T^{11}_{\mu\nu}(q_0,\vec{q})\ . \tag{6}$$

Furthermore, $\overline{T}_{\mu\nu}$ has the spectral representation [18, 16]

$$\overline{T}_{\mu\nu}(q_0,\vec{q}) = \int\frac{dq_0'}{2\pi}\frac{W_{\mu\nu}(q_0',\vec{q})}{q_0'-q_0-i\eta\epsilon(q_0)} \tag{7}$$

which immediately leads to

$$W_{\mu\nu}(q_0,\vec{q}) = 2\epsilon(q_0)\text{Im}\overline{T}_{\mu\nu}(q_0,\vec{q})\ . \tag{8}$$

In the QGP where quarks and gluons are the relevant degrees of freedom, the time ordered correlation function $T^{11}_{\mu\nu}$ can be directly evaluated by writing the hadron current in terms of quarks of flavour f i.e. $J^h_\mu = \sum_f e_f\overline{\psi}_f\gamma_\mu\psi_f$. To leading order we obtain using relations (6) and (8),

$$g^{\mu\nu}W_{\mu\nu} = -\frac{3q^2}{2\pi}\sum_f e_f^2(1-\frac{4m_q^2}{q^2})\ . \tag{9}$$

The rate in this case corresponds to dilepton production due to process $q\bar{q} \to \gamma^* \to l^+l^-$. To obtain the rate of dilepton production from hadronic interactions it is convenient to break up the quark current J^h_μ into parts with definite isospin

$$\begin{aligned} J^h_\mu &= \frac{1}{2}(\bar{u}\gamma_\mu u - \bar{d}\gamma_\mu d) + \frac{1}{6}(\bar{u}\gamma_\mu u + \bar{d}\gamma_\mu d) + \cdots \\ &= J^V_\mu + J^S_\mu + \cdots \\ &= J^\rho_\mu + J^\omega_\mu/3 + \cdots \end{aligned} \tag{10}$$

where V and S denote iso-vector and iso-scalar currents and the dots denote currents comprising of quarks with strangeness and heavier flavours. These currents couple to individual hadrons as well as multiparticle states with the same quantum numbers and are usually labelled by the lightest meson in the corresponding channel [19]. We thus identify the isovector and isoscalar currents with the ρ and ω mesons respectively. Using eq. (10) in eq. (4) and neglecting possible mixing between the isospin states, we write

$$T^{ab}_{\mu\nu} = T^{(\rho)ab}_{\mu\nu} + T^{(\omega)ab}_{\mu\nu}/9 + \cdots \tag{11}$$

where

$$T^{(\rho)ab}_{\mu\nu} = i\int d^4x e^{iq\cdot x}\langle T_c J^\rho_\mu(x) J^\rho_\nu(0)\rangle_\beta \tag{12}$$

and similarly for the scalar current. The current commutator in the isospin basis follows as

$$W_{\mu\nu} = W^\rho_{\mu\nu} + W^\omega_{\mu\nu}/9 + ... \tag{13}$$

The correlator of vector-isovector currents $W^\rho_{\mu\nu}$ have in fact been measured [20] in vacuum along with the axial-vector correlator by studying τ decays into even and odd number of pions. The former is found to be dominated at lower energies by the prominent peak of the ρ meson followed by a continuum at high energies. The axial correlator, on the other hand, is characterised by the broad hump of the a_1. The distinctly different shape in the two spectral densities is an experimental signature of the fact that chiral symmetry of QCD is dynamically broken by the ground state [21]. It is expected that this symmetry may be restored at high temperature and/or density and will be signalled by a complete overlap of the vector and axial-vector correlators [22].

In the medium, both the pole and the continuum structure of the correlation function gets modified [4, 23]. We will first evaluate the modification of the pole part due to the self-energy of vector mesons in the following. Using Vector Meson Dominance the isovector and scalar currents are written in terms of dynamical field operators for the mesons allowing us to express the correlation function in terms of the exact(full) propagators of the vector mesons in the medium. Writing $J^\rho_\mu(x) = F_\rho m_\rho \rho_\mu(x)$ and $J^\omega_\mu(x) = 3F_\omega m_\omega \omega_\mu(x)$ in eq. (13) and using eq. (8) the current commutator becomes

$$W_{\mu\nu} = 2\epsilon(q_0)F_\rho^2 m_\rho^2 \mathrm{Im}\overline{D}^\rho_{\mu\nu} + 2\epsilon(q_0)F_\omega^2 m_\omega^2 \mathrm{Im}\overline{D}^\omega_{\mu\nu} + \cdots \tag{14}$$

where $\overline{D}_{\mu\nu}$ is the diagonal element of the thermal propagator matrix which is a two point function of the fields of vector mesons and is diagonalisable as in (5). The exact propagator is obtained in terms of the in-medium self-energies using the Dyson equation. Following [24] this is given by

$$\overline{D}_{\mu\nu}(q) = -\frac{P_{\mu\nu}}{q^2 - m_\rho^2 - \overline{\Pi}_t(q)} - \frac{Q_{\mu\nu}/q^2}{q^2 - m_\rho^2 - q^2\overline{\Pi}_l(q)} - \frac{q_\mu q_\nu}{q^2 m_\rho^2} \tag{15}$$

where $P_{\mu\nu}$ and $Q_{\mu\nu}$ are the transverse and longitudinal projections. The imaginary part is then put in eqs. (14) and then in eq. (3) to arrive at the dilepton emission rate [24]

$$\frac{dN}{d^4q d^4x} = \frac{\alpha^2}{\pi^3 q^2} L(q^2) f_{BE}(q_0)\left[F_\rho^2 m_\rho^2 A_\rho(q_0,\vec{q}) + F_\omega^2 m_\omega^2 A_\omega(q_0,\vec{q}) + \cdots\right] \tag{16}$$

where e.g. $A_\rho(= -g^{\mu\nu}\mathrm{Im}\overline{D}^\rho_{\mu\nu}/3)$ is given by

$$A_\rho = -\frac{1}{3}\left[\frac{2\sum \mathrm{Im}\Pi_t^R}{(q^2 - m_\rho^2 - \sum \mathrm{Re}\Pi_t^R)^2 + (\sum \mathrm{Im}\Pi_t^R)^2} + \frac{q^2\sum \mathrm{Im}\Pi_l^R}{(q^2 - m_\rho^2 - q^2\sum \mathrm{Re}\Pi_l^R)^2 + q^4(\sum \mathrm{Im}\Pi_l^R)^2}\right] \tag{17}$$

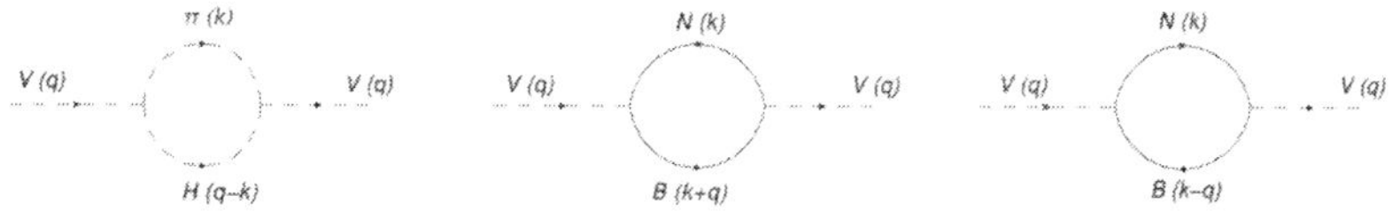

Figure 1. One-loop Feynman diagrams for ρ or ω self-energy involving mesons (first figure) and baryons (second and third figures). V stands for the ρ or ω in the external line. In the internal lines, h stands for π, ω, a_1 and h_1 mesons. For the baryonic loops, N and B indicate respectively nucleon and baryonic internal lines.

the sum running over all mesonic and baryonic loops. Thus, the dilepton emission rate in the present scenario actually boils down to the evaluation of the self energy graphs (shown in Fig. 1). The self-energy is also a 2×2 matrix and is diagonalisable by the matrix U^{-1}. The real and imaginary parts of the self energy function can then be obtained from the 11-component as [17, 10]

$$\begin{aligned} \mathrm{Re}\Pi^R_{\mu\nu}(q_0,\vec{q}) &= \mathrm{Re}\overline{\Pi}_{\mu\nu}(q_0,\vec{q}) = \mathrm{Re}\Pi^{11}_{\mu\nu}(q_0,\vec{q}) \\ \mathrm{Im}\Pi^R_{\mu\nu}(q_0,\vec{q}) &= \epsilon(q_0)\mathrm{Im}\overline{\Pi}_{\mu\nu}(q_0,\vec{q}) = \tanh(\beta q_0/2)\mathrm{Im}\Pi^{11}_{\mu\nu}(q_0,\vec{q}) \end{aligned} \tag{18}$$

where Π^R denotes the retarded self-energy.

As indicated earlier, coupling of the hadronic current to multiparticle states gives rise to a continuum structure in the current correlation function $W^{\mu\nu}$. Following Shuryak [19] we take a parametrised form for this contribution and augment the dilepton emission rate with

$$\frac{dN}{d^4q d^4x} = \frac{\alpha^2}{\pi^3} L(q^2) f_{BE}(q_0) \sum_{V=\rho,\omega} A_V^{\mathrm{cont}}. \tag{19}$$

where

$$A_\rho^{\mathrm{cont}} = \frac{1}{8\pi}\left(1+\frac{\alpha_s}{\pi}\right)\frac{1}{1+\exp(\omega_0 - q_0)/\delta} \tag{20}$$

with $\omega_0 = 1.3, 1.1$ GeV for ρ, ω and $\delta = 0.2$ for both ρ and ω. The continuum contribution for the ω contains an additional factor of $\frac{1}{9}$.

4. Spectral function and self-energy of vector mesons

Obtaining the in-medium spectral functions for the ρ and ω mesons essentially involves the evaluation of the self-energies. For the ρ meson these have been evaluated recently for mesonic [9] and baryonic [10] loops. In the following we summarise these results for the ρ followed by those for the ω meson.

4.1. ρ meson

The 11-component of the self-energy of vector mesons for loops containing mesons or baryons can be generically expressed as

$$\Pi^{11}_{\mu\nu}(q) = i\int \frac{d^4k}{(2\pi)^4} L_{\mu\nu}(k,q)\Delta^{11}(k,m_k)\Delta^{11}(p,m_p) . \tag{21}$$

where the term $L_{\mu\nu}(k,q)$ contains factors from the numerators of the two propagators in the loop as well as from the two vertices and

$$\Delta^{11}(k,m_k) = \frac{-1}{k^2 - m_k^2 + i\epsilon} + a2\pi i\delta(k^2 - m_k^2)N_1 . \tag{22}$$

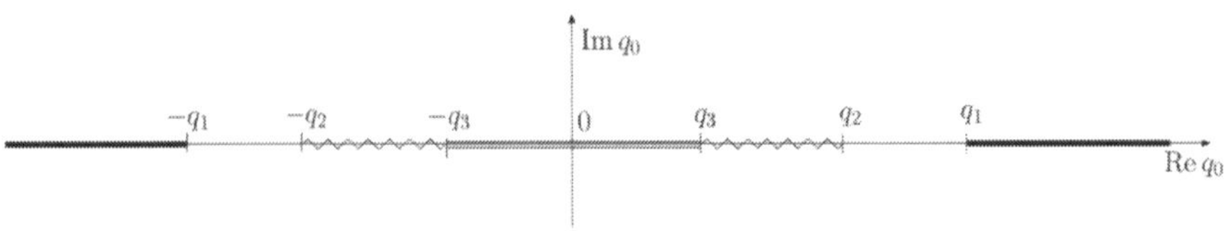

Figure 2. Branch cuts of self-energy function in q_0 plane for fixed $\vec{q}$ given by πh loop. The quantities $q_{1,2,3}$ denote the end points of cuts discussed in the text : $q_1 = \sqrt{(m_h + m_\pi)^2 + |\vec{q}|^2}$, $q_2 = \sqrt{(m_h - m_\pi)^2 + |\vec{q}|^2}$ and $q_3 = |\vec{q}|$.

For bosons, $a = 1$ and $N_1 = n$ with $n(\omega_k) = \frac{1}{e^{\beta\omega_k}-1}$ whereas for fermions, $a = -1$ and $N_1 = n_+\theta(k_0) + n_-\theta(-k_0)$ with $n_\pm(\omega_k) = \frac{1}{e^{\beta(\omega_k \mp \mu)}+1}$. The loop momentum p in (21) for the various cases are as shown in Fig. (1).

We begin with mesonic loops. The ρ self-energy for four possible π-h loops, where $h = \pi, \omega, h_1, a_1$ have been evaluated. The imaginary part of the retarded self-energy (18) is given by,

$$\begin{aligned}\mathrm{Im}\Pi_R^{\mu\nu}(q_0,\vec{q}) &= -\pi \int \frac{d^3\vec{k}}{(2\pi)^3 4\omega_\pi\omega_h} \times \\ &[L_1^{\mu\nu}\{(1+n_\pi+n_h)\delta(q_0-\omega_\pi-\omega_h) - (n_\pi - n_h)\delta(q_0-\omega_\pi+\omega_h)\} \\ &+L_2^{\mu\nu}\{(n_\pi - n_h)\delta(q_0+\omega_\pi-\omega_h) - (1+n_\pi+n_h)\delta(q_0+\omega_\pi+\omega_h)\}] \,. \end{aligned} \tag{23}$$

where the Bose distribution functions $n_\pi \equiv n(\omega_\pi)$ with $\omega_\pi = \sqrt{\vec{k}^2 + m_\pi^2}$ and $n_h \equiv n(\omega_h)$ with $\omega_h = \sqrt{(\vec{q}-\vec{k})^2 + m_h^2}$. $L_i^{\mu\nu}(i = 1, 2)$ are the values of $L^{\mu\nu}(k_0)$ for $k_0 = \omega_\pi, -\omega_\pi$ respectively and the vertices used in $L^{\mu\nu}(k_0)$ have been obtained from the chiral Lagrangians which are specified in [9].

The regions, in which the four terms of eq. (23) are non-vanishing, give rise to cuts in the self-energy function (Fig. 2). These regions are controlled by the respective δ-functions. Thus, the first and the fourth terms are non-vanishing for $q^2 \geq (m_h + m_\pi)^2$, giving the unitary cut, while the second and the third are non-vanishing for $q^2 \leq (m_h - m_\pi)^2$, giving the so-called Landau cut. The unitary cut arises from the states, which can communicate with the ρ. These states are, of course, the same as in vacuum, but, as we see above, the probabilities of their occurrence in the medium are modified by the distribution functions. On the other hand, the Landau cut appears only in medium and arises from scattering of ρ with particles present there. We note that this contribution appears as the first term in the virial expansion of the self-energy function.

Integrating over the angle and restricting to the kinematic region $q_0, q^2 > 0$ the contribution to the imaginary part coming from the unitary cut is given by [9]

$$\mathrm{Im}\Pi_R^{\mu\nu} = -\frac{1}{16\pi|\vec{q}|}\int_{\omega_\pi^-}^{\omega_\pi^+} d\omega_\pi L_1^{\mu\nu}\{1 + n(\omega_\pi) + n(q_0 - \omega_\pi)\} \tag{24}$$

and that from the Landau cut is given by

$$\mathrm{Im}\Pi_R^{\mu\nu} = -\frac{1}{16\pi|\vec{q}|}\int_{\omega_\pi^{-\prime}}^{\omega_\pi^{+\prime}} d\omega_\pi L_2^{\mu\nu}\{n(\omega_\pi) - n(q_0 + \omega_\pi)\} \tag{25}$$

where $\omega_{\pi\pm} = \frac{S_\pi^2}{2q^2}(q_0 \pm |\vec{q}|W_\pi)$, $\omega_{\pi\pm}{}' = \frac{S_\pi^2}{2q^2}(-q_0 \mp |\vec{q}|W_\pi)$ with $W_\pi = \sqrt{1 - \frac{4q^2m_\pi^2}{S_\pi^4}}$ and $S_\pi^2 = q^2 - m_h^2 + m_\pi^2$.

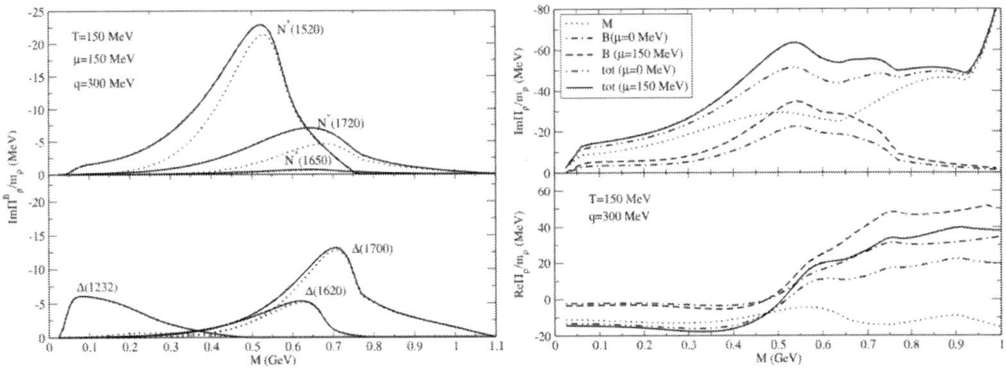

Figure 3. Left panel shows the imaginary part from baryonic loops and the right panel shows the total contribution from meson and baryon loops.

The real part can be obtained from the imaginary part by a dispersion relation or can be evaluated directly from the graphs.

Let us now turn to the baryonic loops in the ρ meson self-energy. Using $p = k - q$ for the third diagram of Fig. 1 in eq. (21) the retarded self-energy is evaluated for NB loops including all spin one-half and three-half 4−star resonances listed by the Particle Data Group so that B stands for the $N^*(1520)$, $N^*(1650)$, $N^*(1700)$, $N^*(1720)$ $\Delta(1230)$, $\Delta^*(1620)$, as well as the $N(940)$ itself.

As before, the imaginary part can be evaluated from the discontinuities of the self-energy. However, the threshold for the unitary cut for the baryon loops being far away from the ρ pole we only consider the Landau part. The expression of the self-energy corresponding to the second and third diagrams of Fig. 1 can be obtained from one another by inverting the sign of q. Adding the contributions coming from the two diagrams, the imaginary part of the self energy is given by

$$\mathrm{Im}\Pi^{\mu\nu}_R = -\frac{1}{16\pi|\vec{q}|}\int_{\omega_N^+}^{\omega_N^-} d\omega_N[L_1^{\mu\nu}(-q)\{n_+(q_0+\omega_N)-n_+(\omega_N)\}+L_2^{\mu\nu}(q)\{n_-(q_0+\omega_N)-n_-(\omega_N)\}] \tag{26}$$

where $\omega_N^{\pm} = \frac{S_N^2}{2q^2}(-q_0 \pm |\vec{q}|W_N)$ with $W_N = \sqrt{1-\frac{4q^2m_N^2}{S_N^4}}$, $S_N^2 = q^2 - m_B^2 + m_N^2$. The factors $L_i^{\mu\nu}(i = 1, 2)$ in eq. (26) are the values of $L^{\mu\nu}(k_0)$ for $k_0 = \omega_N, -\omega_N$ respectively and is obtained using gauge invariant interactions details of which are provided in [10].

The calculations described above treats the heavy mesons and baryon resonances in the narrow width approximation. We have included their widths by folding with their vacuum spectral functions as done in [25].

We now present the results of numerical evaluation. We plot in the left panel of Fig. 3 the imaginary part of the ρ self-energy for baryon loops as a function of the invariant mass $\sqrt{q^2} \equiv M$ for $\vec{q} = 300$. The transverse (solid line) and longitudinal (dashed line) components $\mathrm{Im}\Pi_t$ and $q^2\mathrm{Im}\Pi_l$ have been shown separately. The $NN^*(1520)$ loop makes the most significant contribution followed by the $N^*(1720)$ and $\Delta(1700)$. On the right panel is plotted the spin-averaged ρ self-energy defined by $\Pi = \frac{1}{3}(2\Pi_t + q^2\Pi_l)$ showing contributions from the baryon and meson loops for two values of the baryonic chemical potential. The small positive contribution from the baryon loops to the real part is partly compensated by the negative contributions from the meson loops. The substantial baryon contribution at vanishing baryonic chemical potential reflects the importance of anti-baryons.

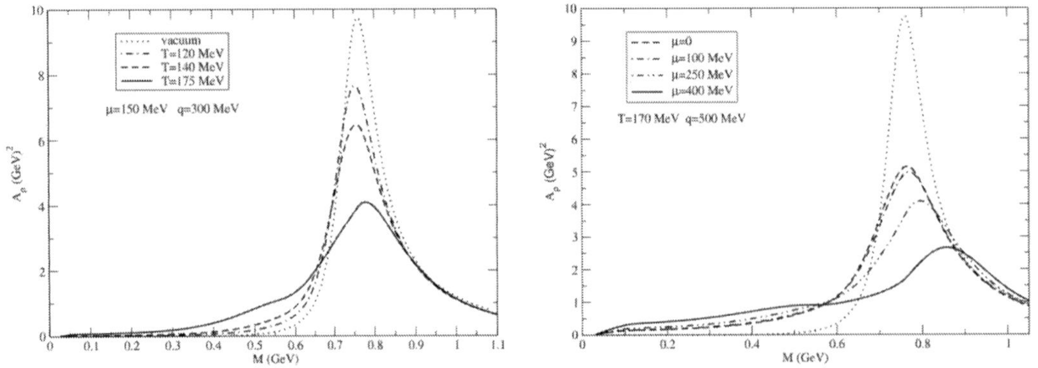

Figure 4. The spectral function of the ρ meson for (left) different values of the temperature T and (right) different values of the baryonic chemical potential (μ).

We now turn to the spin averaged spectral function defined in eq. (17). First, in the left panel of Fig.4 we plot the spectral function at fixed values of the baryonic chemical potential and three-momentum for various representative values of the temperature. We observe an increase of spectral strength at lower invariant masses resulting in broadening of the spectral function with increase in temperature. This is purely a Landau cut contribution from the baryonic loop arising from the scattering of the ρ from baryons in the medium. We then plot in the right panel of Fig. 4, the spectral function for various values of the baryonic chemical potential for a fixed temperature. For high values of μ we observe an almost flattened spectral density of the ρ.

4.2. ω meson

The ω self-energy is evaluated along similar lines [26]. The ω meson decays mostly into three pions. Assuming this decay to proceed via an intermediate ρ meson i.e. $\omega \to \rho\pi \to 3\pi$, the dominant contribution to the ω self-energy in meson matter can be expressed as [27]

$$\Pi^{\mu\nu}_{R(3\pi)}(q) = \frac{1}{N_\rho}\int_{4m_\pi^2}^{(q-m_\pi)^2} dM^2[\Pi^{\mu\nu}_{R(\pi\rho)}(q,M)]A_\rho(M) \tag{27}$$

where $N_\rho = \int_{4m_\pi^2}^{(q-m_\pi)^2} dM^2 A_\rho(M^2)$, A_ρ being the vacuum spectral function of the ρ. Here $\Pi^{\mu\nu}_{R(\pi\rho)}(q,M)$ can be obtained by evaluating the first diagram of Fig. (1) with $h=\rho$.

For ω self-energy due to baryons we have evaluated NB loops where $B = N^*(1440), N^*(1520), N^*(1535), N^*(1650), N^*(1720), N(940)$. The calculation proceeds similarly as in case of the ρ.

Left panel of Fig. (5) shows the transverse and longitudinal components of the imaginary part coming from the baryonic loops and the right panel shows the individual mesonic and baryonic loop contributions to the ω self-energy at two different chemical potentials. We see that the $N^*(1535)$ plays the dominating role mainly due to the strong coupling compared to the other baryonic resonances [27]. The spin averaged spectral function at different temperatures and chemical potentials are plotted in the left and right panels of Fig. (6) respectively. We observe a slight positive shift in the peak position.

Having obtained the in-medium spectral functions of the two most important low-lying vector mesons, we are now in a position to evaluate the static rate of dilepton production using eq. (16). Integrating over the transverse momentum q_T and rapidity y of the electron pairs we plot dR/dM^2 vs M in Fig. 7 for T=175 MeV. Because of the kinematical factors multiplying the ρ

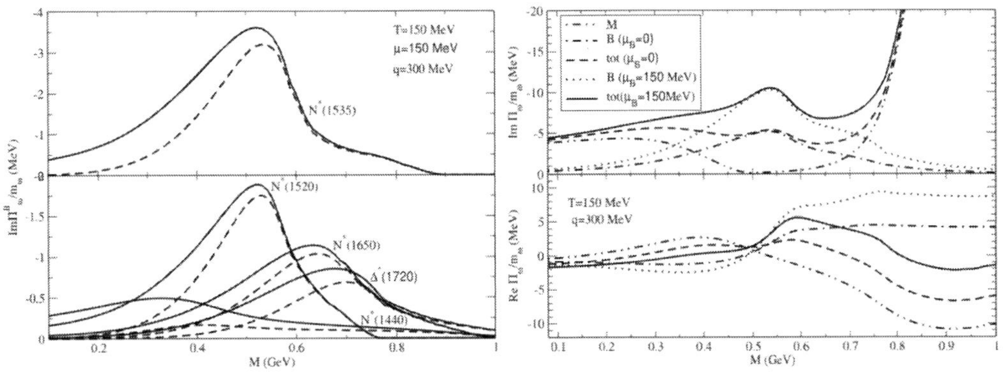

Figure 5. Same as Fig. (3) for the ω.

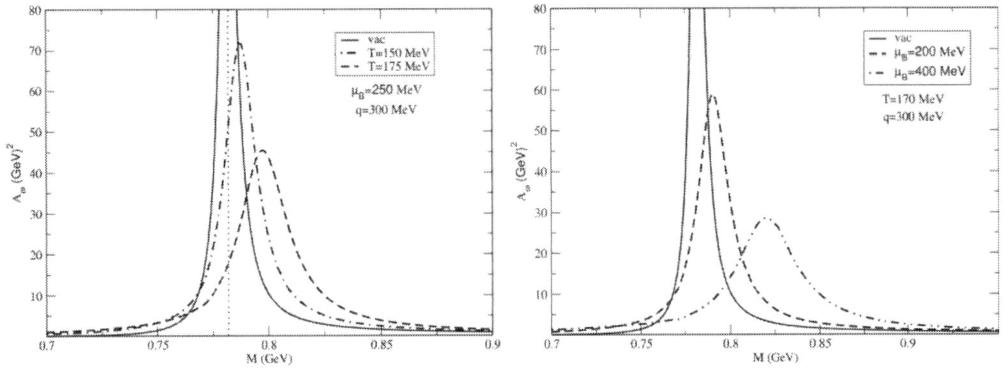

Figure 6. The spectral function of the ω for different values of T (left panel) and μ (right panel).

spectral function the broadening appears magnified in the dilepton emission rate. A significant enhancement is seen in the low mass lepton production rate due to baryonic loops over and above the mesonic ones shown by the dot-dashed line. The substantial contribution from baryonic loops even for vanishing chemical potential points to the important role played by antibaryons in thermal equilibrium in systems created at RHIC and LHC energies.

5. Space time evolution and dilepton spectra in In-In collisions

Since dileptons are produced at all stages of the collision it is necessary to integrate the emission rates over the space-time volume from creation to freeze-out. We assume that quark gluon plasma having a temperature T_i is produced at an initial time τ_i. Hydrodynamic expansion and cooling follows up to a temperature T_c where QGP undergoes a transition to hadronic matter. Subsequent cooling leads to freeze-out of the fluid element into observable hadrons. In the present work the fireball is taken to undergo an azimuthally symmetric transverse expansion along with a boost invariant longitudinal expansion [28]. The local temperature of the fluid element and the associated flow velocity as a function of the radial coordinate and proper time is obtained by solving the the energy momentum conservation equation $\partial_\mu T^{\mu\nu} = 0$ where $T^{\mu\nu} = (\epsilon + P)u^\mu u^\nu + g^{\mu\nu} P$ is the energy momentum tensor for ideal fluid. This set of equations are closed with the Equation of State (EoS); typically a functional relation between the pressure P and the energy density ϵ. It is a crucial input which essentially controls the profile of expansion

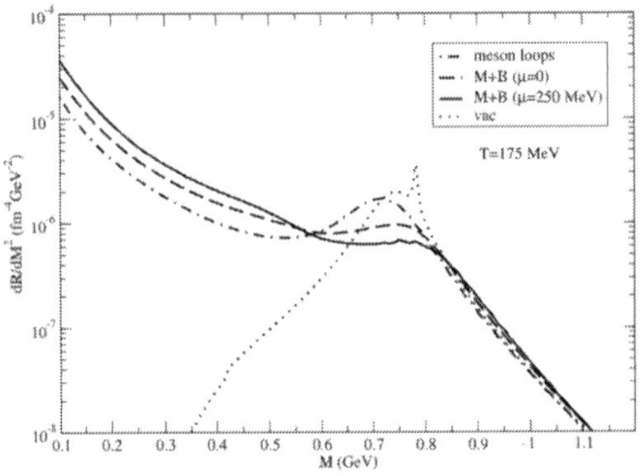

Figure 7. The lepton pair emission rate at $T = 175$ MeV with and without baryon (B) loops in addition to the meson (M) loops.

of the fireball.

The initial temperature is constrained by the experimentally measured hadron multiplicity through entropy conservation [29],

$$T_i^3(b_m)\tau_i = \frac{2\pi^4}{45\zeta(3)\pi R_\perp^2 4a_k}\langle\frac{dN}{dy}(b_m)\rangle \tag{28}$$

where $\langle dN/dy(b_m)\rangle$ is the hadron (predominantly pions) multiplicity for a given centrality class with maximum impact parameter b_m, $R_\perp$ is the transverse dimension of the system and a_k is the degeneracy of the system created. The initial radial velocity, $v_r(\tau_i, r)$ and energy density, $\epsilon(\tau_i, r)$ profiles are taken as [30], $v_r(\tau_i, r) = 0$ and $\epsilon(\tau_i, r) = \epsilon_0/(e^{\frac{r-R_A}{\delta}} + 1)$ where the surface thickness, $\delta = 0.5$ fm. In the present work we assume $T_c = 175$ MeV [31]. In a quark gluon plasma to hadronic matter transition scenario, we use the bag model EoS for the QGP phase and all resonances with mass ≤ 2.5 GeV for the hadronic gas. The transition region is parametrized as [32]

$$s = f(T)s_q + (1 - f(T))s_h \quad \text{with} \quad f(T) = \frac{1}{2}(1 + \tanh(\frac{T - T_c}{\Gamma})) \ . \tag{29}$$

where s_q (s_h) is the entropy density of the quark (hadronic) phase at T_c. The value of the parameter Γ can be varied to make the transition strong first order or continuous. We take $\Gamma = 20$ MeV in this work.

The ratios of various hadrons measured experimentally at different $\sqrt{s_{\rm NN}}$ indicate that the system formed in heavy ion collisions chemically decouple at a temperature ($T_{\rm ch}$) which is higher than the temperature for kinetic freeze-out (T_f)determined by the transverse spectra of hadrons [33]. Therefore, the system remains out of chemical equilibrium from $T_{\rm ch}$ to T_f. The chemical non-equilibration affects the dilepton yields through (a) the emission rate through the phase space factor and (b) the space-time evolution of the matter via the equation of state. The value of the chemical potential and its inclusion in the EoS has been taken into account following [34].

Finally, we have obtained the dimuon yield (dN/dM) in In-In collisions at SPS at a center of mass energy of 17.3 AGeV. The initial energy density is taken as 4.5 GeV/fm^3 corresponding to a thermalisation time $\tau_i = 0.7$ fm. We take the QGP to hadronic matter transition temperature $T_c = 175$ MeV and the freeze-out temperature T_f =120 MeV which can reproduce the slope of the hadronic spectra measured by the NA60 Collaboration. In Fig. (8) we have shown the

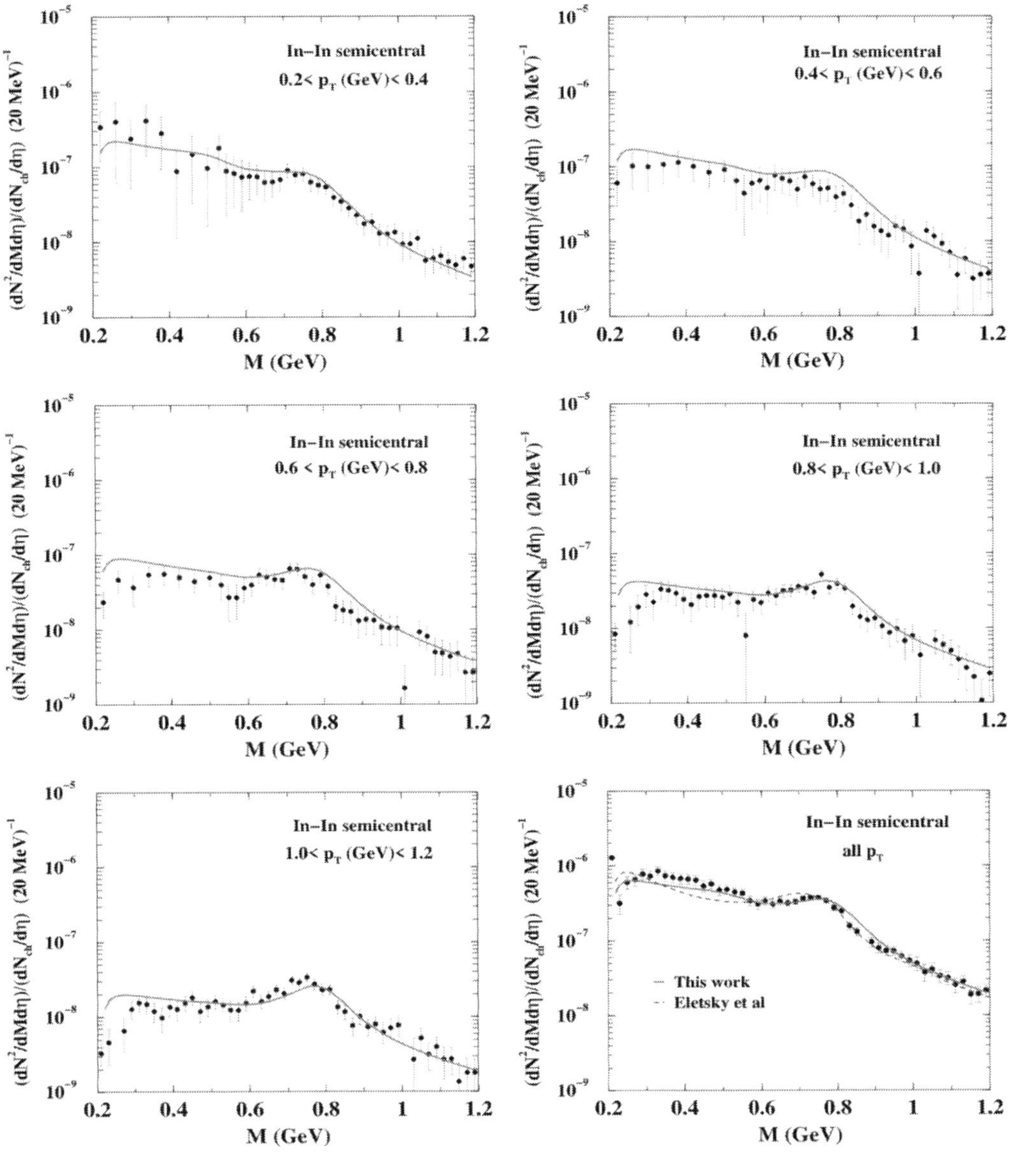

Figure 8. Dilepton invariant mass spectra for different p_T-bins compared with the NA60 data.

invariant mass spectra for different transverse momentum (p_T) windows. The theoretical curves agree quite well with the experimental data [12] for all the p_T ranges. The strong enhancement in the low M domain is clearly due to the large broadening of the ρ (and ω) in the thermal medium which comes entirely from the Landau cut in the self-energy diagrams. In the last panel we also plot for comparison the spectra calculated in [35] where the self-energy due to baryons has been evaluated following the approach of [36]. It is seen that this approach depicted by the dashed curve does not produce the required enhancement to explain the data in the range $0.35 \leq M \leq 0.65$ GeV.

6. Summary

The self-energy of ρ and ω mesons have been computed in nuclear matter at finite temperature and baryon density. Loop graphs involving mesons, nucleons and 4-star N^* and Δ resonances up to spin 3/2 were calculated using gauge invariant interactions in the framework of real time thermal field theory to obtain the correct relativistic expressions for the self-energy. The singularities in the complex energy plane were analysed and the imaginary part obtained from the Landau cut contribution. Results for the real and imaginary parts at non-zero three-momenta for various values of temperature and baryonic chemical potential were shown for the individual loop graphs. The spectral function of the ρ was observed to undergo a significant modification at and below the nominal rho mass which was seen to bring about a large enhancement of lepton pair yield in this region. After a space-time evolution using relativistic hydrodynamics, the invariant mass spectra for various p_T windows was found to be in very good agreement with the experimental data obtained in In-In collisions at 17.3 AGeV.

References

[1] "The Physics Of The Quark-Gluon Plasma," eds. S. Sarkar, H. Satz and B. Sinha, Lect. Notes Phys. **785**, 1 (2010).
[2] "The CBM Physics Book," eds. B. Friman et al, Lect. Notes Phys. **814**, 1 (2011).
[3] L. P. Csernai, J. I. Kapusta and L. D. McLerran, Phys. Rev. Lett. **97** (2006) 152303.
[4] J. Alam, S. Sarkar, P. Roy, T. Hatsuda, B. Sinha, Ann. Phys. 286, 159 (2000).
[5] P. Mohanty, V. Roy, S. Ghosh, S. K. Das, B. Mohanty, S. Sarkar, J. Alam and A. Chaudhuri, arXiv:1111.2159 [nucl-th].
[6] L. D. McLerran and T. Toimela, Phys. Rev. D **31**, 545 (1985).
[7] R. Rapp and J. Wambach, Adv. Nucl. Phys. **25**, 1 (2000)
[8] S. Leupold, V. Metag and U. Mosel, Int. J. Mod. Phys. E **19**, 147 (2010)
[9] S. Ghosh, S. Sarkar and S. Mallik, Eur. Phys. J. C **70**, 251 (2010)
[10] S. Ghosh and S. Sarkar, Nucl. Phys. **A 870-871**, 94 (2011)
[11] S. Ghosh, S. Sarkar and S. Mallik, Phys. Rev. C **83**, 018201 (2011).
[12] R. Arnaldi et al for the NA60 collaboration, Phys. Rev. Lett. **100** 022302 (2008); Eur. Phys. J. C **61**, 711 (2009); S. Damjanovic et al for the NA60 Collaboration, J. Phys. G: Nucl. Part. Phys. **35**, 104036 (2008).
[13] H. A. Weldon, Phys. Rev. D **42**, 2384 (1990).
[14] R. Rapp, J. Wambach and H. van Hees, arXiv:0901.3289 [hep-ph].
[15] J. V. Steele, H. Yamagishi and I. Zahed, Phys. Lett. B **384** (1996) 255.
[16] S. Mallik and S. Sarkar, Eur. Phys. J. C **61**, 489 (2009)
[17] M. Le Bellac, *Thermal Field Theory* (Cambridge University Press, Cambridge, 1996).
[18] S. Sarkar, B. K. Patra, V. J. Menon and S. Mallik, Indian J. Phys. **76A**, 385 (2002).
[19] E.V. Shuryak, Rev. Mod. Phys. 65, 1 (1993)
[20] R. Barate *et al.* [ALEPH Collaboration], Eur. Phys. J. C **4**, 409 (1998), K. Ackerstaff *et al.* [OPAL Collaboration], Eur. Phys. J. C **7**, 571 (1999)
[21] S. Sarkar, Nucl. Phys. A **862-863** (2011) 13.
[22] J. I. Kapusta and E. V. Shuryak, Phys. Rev. D **49**, 4694 (1994).
[23] S. Mallik and S. Sarkar, Eur. Phys. J. C **25**, 445 (2002).
[24] S. Ghosh, S. Sarkar, J. Alam Eur. Phys. J. C **71**, 9 (2011).
[25] S. Sarkar, E. Oset and M. J. Vicente Vacas, Nucl. Phys. A **750**, 294 (2005)
[26] S. Ghosh and S. Sarkar, (work in progress)
[27] P. Muehlich, V. Shklyar, S. Leupold, U. Mosel, M. Post, Nucl. Phys. **A 780**, 187 (2006)
[28] J. D. Bjorken, Phys. Rev. D **27**, 140 (1983).
[29] R. C. Hwa and K. Kajantie, Phys. Rev. D **32**,1109 (1985).
[30] H. von Gersdorff, M. Kataja, L. D. McLerran and P. V. Ruskanen, Phys. Rev. D **34** 794(1986).
[31] S. Borsanyi *et al.*, Jour. High Eener. Phys. **1011**, 077 (2010).
[32] M. Asakawa and T. Hatsuda Phys. Rev. D **55** 4488 (1997).
[33] P. Braun-Munzinger, J. Stachel, J. P. Wessels and N. Xu, Phys. Lett. B **344** 43 (1995); **365** 1 (1996).
[34] T. Hirano and K. Tsuda, Phys. Rev. C **66** 054905 (2002).
[35] J. K. Nayak, J. e. Alam, T. Hirano, S. Sarkar and B. Sinha, arXiv:0902.0446 [nucl-th].
[36] V. L. Eletsky, M. Belkacem, P. J. Ellis and J. I. Kapusta, Phys. Rev. C **64** (2001) 035202

5th DAE-BRNS Workshop on Hadron Physics (Hadron 2011) IOP Publishing
Journal of Physics: Conference Series **374** (2012) 012011 doi:10.1088/1742-6596/374/1/012011

Constraining Form Factors with the Method of Unitarity Bounds

B. Ananthanarayan[a,*] **Irinel Caprini**[b,†]

[a]Centre for High Energy Physics, Indian Institute of Science, Bangalore 560 012, India
[b] Horia Hulubei National Institute for Physics and Nuclear Engineering, P.O.B. MG-6, 077125, Magurele, Romania

E-mail: *anant@cts.iisc.ernet.in, †caprini@theory.nipne.ro

Abstract. The availability of a reliable bound on an integral involving the square of the modulus of a form factor on the unitarity cut allows one to constrain the form factor at points inside the analyticity domain and its shape parameters, and also to isolate domains on the real axis and in the complex energy plane where zeros are excluded. In this lecture note, we review the mathematical techniques of this formalism in its standard form, known as the method of unitarity bounds, and recent developments which allow us to include information on the phase and modulus along a part of the unitarity cut. We also provide a brief summary of some results that we have obtained in the recent past, which demonstrate the usefulness of the method for precision predictions on the form factors.

1. Introduction

Form factors are basic observables in strong interaction dynamics. They provide information on the nature of the strong force and confinement. The pion electromagnetic form factor is one such observable. The weak form factors are of crucial importance for the determination of standard model parameters such as the elements of the Cabibbo-Kobayashi-Maskawa (CKM) matrix.

Before the advent of the modern theory of strong interactions, mathematical methods were advanced to obtain bounds on the form factors when a suitable integral of the modulus squared along the unitarity cut (t_+, ∞) was available [1, 2] (for a topical review of the results at that time, see [3]). Using the methods of complex analysis, this condition leads to constraints on the values at points inside the analyticity domain or on the derivatives at $t = 0$, such as the slope and curvature. Also of interest was whether or not the form factors can vanish on the real energy or in the complex energy plane (in the context of the pion electromagnetic form factor, see, e.g. [4]). Mathematically, these problems belong to a class referred to as the Meiman problem [5].[1]

At present, effective theories of strong interactions, like Chiral Perturbation Theory (ChPT) or Heavy Quark Effective Theories (HQET), as well as lattice calculations or various types of QCD sum rules, allow us to make predictions, sometimes very precise, on the form factors at particular kinematical regions. However, a full calculation of the hadronic form factors from first principles is not yet possible. On the other hand, experimental information on the form factors has improved considerably in recent years. In these circumstances, analytic techniques as those

[1] Stated differently, the problem belongs to the standard analytic interpolation theory for functions in the Hardy class H^2.

mentioned above prove to be very useful, providing rigorous correlations and consistency checks of various approaches and in improving the phenomenological analyses of the form factors. In a recent review, ref. [6] we have presented a complete treament of the mathematical background required for studying these problems and a comprehensive bibliography on the subject.

The integral condition was provided either from an observable (like muon's $g-2$ in the case of the electromagnetic form factor of the pion) or, in the modern approach first applied in [7], from the dispersion relation satisfied by a suitable correlator calculated by perturbative QCD in the spacelike region, and whose positive spectral function has, by unitarity, a lower bound involving the modulus squared of the relevant form factor. Therefore, the constraints derived in this framework are often referred to as "unitarity bounds". Information about the phase of the form factor, available by Watson's theorem from the associated elastic scattering, can be used to improve the results in a systematic fashion [8, 9]. In some cases the modulus is also measured independently along a part of the unitarity cut.

Recent applications in the modern approach concerned mainly the form factors relevant for the $B \to D^{(*)}$ semileptonic decays, or the so-called Isgur-Wise function, where heavy quark symmetry provided strong additional constraints at interior points [10, 11, 12, 13, 14, 15]. More recent applications revisited the electromagnetic form factor of the pion [16, 17, 18, 19, 20], the strangeness changing $K\pi$ form factors [21, 22, 23, 24, 25], the $B\pi$ vector form factor [26, 27, 28], and also the $D\pi$ form factors [29]. The results confirm that the approach represents a useful tool in the study of the form factors, complementary and free of additional assumptions inherent in standard dispersion relations. The techniques presented here are the framework of including inputs coming from disparate sources and effectively testing their consistency. In this lecture note we will describe the mathematical machinery that has been developed to address these issues. We will also review some recent results obtained by us using these methods.

In Sec. 2 we present our notation and describe the general Meiman interpolation problem, for an arbitrary number of derivatives at the origin and an arbitrary number of values at points inside the analyticity domain. The solution is obtained by Lagrange multipliers, and is written as a positivity condition of a determinant written down in terms of the input values. Alternatively, using techniques of analytic interpolation theory, the result can be written as a compact convex quadratic form, given in ref. [6]. We present the complete treatment of the inclusion of the phase on a part of the unitarity cut, $(t_+, t_{\rm in})$, along with an arbitrary number of constraints of the Meiman type in Sec. 3. The general treatment was only recently provided in entirety despite the lengthy literature on the problem. Two equivalent sets of integral equations have been found and here we provide the result obtained from using the method of Lagrange multipliers. The result obtained from analytic interpolation theory can be found in [6]. In Sec. 4 we treat a modified analytic optimization problem, where the input on the cut consists from the phase below a certain threshold $t_{\rm in}$ and a weighted integral over the modulus squared along $t > t_{\rm in}$. This is a mathematically more complicated problem solved for the first time in [16]. The method uses the fact that the knowledge of the phase allows one to describe exactly the elastic cut of the form factor by means of the Omnès function. The problem is thus reduced to a standard Meiman problem on a larger analyticity domain. In practice, the integral along $t > t_{\rm in}$ can be obtained by subtracting from the integral along the whole cut the low-energy contribution, $t < t_{\rm in}$, which can be estimated if some information on the modulus on that part of the cut is available from an independent source. The integral can be evaluated also if precise data on the modulus are available up to high energies, as is the case with the pion electromagnetic form factor [30, 31]. In Sec. 5 we present for illustration most recent results of these methods applied to the pion electromagnetic form factor [20]. We conclude with an afterword, Sec. 6.

Journal of Physics: Conference Series **374** (2012) 012011 doi:10.1088/1742-6596/374/1/012011

2. Meiman Problem

In order to set up the notation, let us begin by letting $F(t)$ denote a form factor, which is real analytic (*i.e.* $F(t^*) = F^*(t)$) in the complex t-plane with a cut along the positive real axis from the lowest unitarity branch point t_+ to ∞. The essential condition considered in the present context is an inequality:

$$\int_{t_+}^{\infty} dt\ \rho(t)|F(t)|^2 \leq I, \tag{1}$$

where $\rho(t) \geq 0$ is a positive semi-definite weight function and I is a known quantity. As mentioned in the Introduction, such inequalities can be obtained starting from a dispersion relation satisfied by a suitable correlator, evaluated in the deep Euclidean region by perturbative QCD, and whose spectral function is bounded from below by a term involving the modulus squared of the relevant form factor.

Of interest in the analysis of form factors are the shape parameters that appear in the expansion around $t = 0$. For instance, in the case of the pion electromagnetic form factor the expansion is customarily written as

$$F(t) = 1 + \frac{1}{6}\langle r_\pi^2\rangle t + ct^2 + dt^3 + \cdots, \tag{2}$$

whereas in the analysis of the semileptonic decays, one is interested in the shape parameters appearing in

$$F(t) = F(0)\left[1 + \lambda' \frac{t}{M^2} + \lambda'' \frac{t^2}{2M^4} + \cdots\right], \tag{3}$$

where M is a suitable mass and λ' and λ'' denote the dimensionless slope and curvature, respectively.

Low energy theorems, reflecting the chiral symmetry of the strong interaction, and lattice calculations can provide information on $F(t)$ at several special points inside the analyticity domain. The standard unitarity bounds exploit analyticity of the form factor and the inequality (1) in order to correlate in an optimal way these values and the expansion parameters in (3).

In order to set up the stage, the problem is brought to a canonical form by making the conformal transformation

$$\tilde{z}(t) = \frac{\sqrt{t_+} - \sqrt{t_+ - t}}{\sqrt{t_+} + \sqrt{t_+ - t}}, \tag{4}$$

that maps the cut t-plane onto the unit disc $|z| < 1$ in the $z \equiv \tilde{z}(t)$ plane, such that t_+ is mapped onto $z = 1$, the point at infinity to $z = -1$ and the origin to $z = 0$. After this mapping, the inequality (1) is written as

$$\frac{1}{2\pi}\int_0^{2\pi} \mathrm{d}\theta |g(\mathrm{e}^{i\theta})|^2 \leq I, \tag{5}$$

where the analytic function $g(z)$ is defined as

$$g(z) = F(\tilde{t}(z))w(z). \tag{6}$$

Here $\tilde{t}(z)$ is the inverse of (4) and $w(z)$ is an *outer function*, *i.e.* a function analytic and without zeros in $|z| < 1$, such that its modulus on the boundary is related to $\rho(\tilde{t}(\mathrm{e}^{i\theta}))$ and the Jacobian of the transformation (4). In particular cases of physical interest, the outer functions $w(z)$ have a simple analytic form. In general, an outer function is obtained from its modulus on the boundary by the integral

$$w(z) = \exp\left[\frac{1}{2\pi}\int_0^{2\pi} \mathrm{d}\theta \frac{e^{i\theta} + z}{e^{i\theta} - z} \ln|w(e^{i\theta})|\right]. \tag{7}$$

The function $g(z)$ is analytic within the unit disc and can be expanded as:

$$g(z) = g_0 + g_1 z + g_2 z^2 + \cdots, \tag{8}$$

and (5) implies

$$\sum_{k=0}^{\infty} g_k^2 \leq I. \tag{9}$$

Using (6), the real numbers g_k are expressed in a straightforward way in terms of the coefficients of the Taylor expansion (3). The inequality (9), with the sum in the left side truncated at some finite order, represents the simplest "unitarity bound" for the shape parameters defined in (3). In what follows we shall improve it by including additional information on the form factor.

We consider the general case when the first K derivatives of $g(z)$ at $z = 0$ and the values at N interior points are assumed to be known:

$$\begin{aligned}\left[\frac{1}{k!}\frac{d^k g(z)}{dz^k}\right]_{z=0} &= g_k, \quad 0 \leq k \leq K-1; \\ g(z_n) &= \xi_n, \quad 1 \leq n \leq N,\end{aligned} \tag{10}$$

where g_k and ξ_n are given numbers. They are related, by (6), to the derivatives $F^{(j)}(0)$, $j \leq k$ of $F(t)$ at $t = 0$, and the values $F(\tilde{t}(z_n))$, respectively. For simplicity and in view of phenomenological inputs that we will use, we assume the points z_n to be real, so ξ_n are also real. The Meiman problem requires to find the optimal constraints satisfied by the numbers defined in (10) if (5) holds. One can prove that the most general constraint satisfied by the input values appearing in (10) is given by the inequality:

$$\mu_0^2 \leq I, \tag{11}$$

where μ_0^2 is the solution of the minimization problem:

$$\mu_0^2 = \min_{g \in \mathcal{G}} ||g||_{L^2}^2. \tag{12}$$

Here $||g||_{L^2}^2$ denotes the L^2 norm, *i.e.* the quantity appearing in the l.h.s. of (5) or (9), and the minimum is taken over the class $\mathcal{G}$ of analytic functions which satisfy the conditions (10).

The minimization problem (12) can be solved by using the Lagrange multiplier method. The Lagrangian may be written as

$$\mathcal{L} = \frac{1}{2}\sum_{k=0}^{\infty} g_k^2 + \sum_{n=1}^{N} \alpha_n (\xi_n - \sum_{k=0}^{\infty} g_k z^k), \tag{13}$$

where α_n are real Lagrange multipliers. Solving the Lagrange equations obtained by varying with respect to g_k for all $k \geq K$, and eliminating the Lagrange multipliers yields the solution of the minimization problem (12). For purposes of illustration, when $N = 2$, the Lagrange equations yield

$$g_k = \alpha_1 z_1^k + \alpha_2 z_2^k, \qquad k \geq K, \tag{14}$$

and the inequality (11) can be expressed in terms of the two Lagrange multipliers as:

$$\alpha_1 \bar{\xi}_1 + \alpha_2 \bar{\xi}_2 \leq \bar{I}, \tag{15}$$

where $\bar{\xi}_n$ are known numbers defined as

$$\bar{\xi}_n = \xi_n - \sum_{k=0}^{K-1} g_k z_n^k, \tag{16}$$

and

$$\bar{I} = I - \sum_{k=0}^{K-1} g_k^2. \tag{17}$$

The constraint conditions themselves are

$$\begin{aligned} \alpha_1 \frac{z_1^{2K}}{1-z_1^2} + \alpha_2 \frac{(z_1 z_2)^K}{1-z_1 z_2} &= \bar{\xi}_1, \\ \alpha_1 \frac{(z_1 z_2)^K}{1-z_1 z_2} + \alpha_2 \frac{z_2^{2K}}{1-z_2^2} &= \bar{\xi}_2. \end{aligned} \tag{18}$$

The consistency of eqs. (15) and (18) as a system of equations for α_i can be written as:

$$\begin{vmatrix} \bar{I} & \bar{\xi}_1 & \bar{\xi}_2 \\ \bar{\xi}_1 & \frac{z_1^{2K}}{1-z_1^2} & \frac{(z_1 z_2)^K}{1-z_1 z_2} \\ \bar{\xi}_2 & \frac{(z_1 z_2)^K}{1-z_1 z_2} & \frac{z_2^{2K}}{1-z_2^2} \end{vmatrix} \geq 0. \tag{19}$$

This can be readily extended to the case of N constraints:

$$\begin{vmatrix} \bar{I} & \bar{\xi}_1 & \bar{\xi}_2 & \cdots & \bar{\xi}_N \\ \bar{\xi}_1 & \frac{z_1^{2K}}{1-z_1^2} & \frac{(z_1 z_2)^K}{1-z_1 z_2} & \cdots & \frac{(z_1 z_N)^K}{1-z_1 z_N} \\ \bar{\xi}_2 & \frac{(z_1 z_2)^K}{1-z_1 z_2} & \frac{(z_2)^{2K}}{1-z_2^2} & \cdots & \frac{(z_2 z_N)^K}{1-z_2 z_N} \\ \vdots & \vdots & \vdots & \vdots & \vdots \\ \bar{\xi}_N & \frac{(z_1 z_N)^K}{1-z_1 z_N} & \frac{(z_2 z_N)^K}{1-z_2 z_N} & \cdots & \frac{z_N^{2K}}{1-z_N^2} \end{vmatrix} \geq 0. \tag{20}$$

Alternatively, the solution can be obtained by introducing Lagrange multipliers also for the given coefficients g_k, $k = 0, ..., K-1$ in (13). This leads to an inequality, equivalent to (20), written as [32]:

$$\begin{vmatrix} I & g_0 & g_1 & \cdots & g_{K-1} & \xi_1 & \xi_2 & \cdots & \xi_N \\ g_0 & 1 & 0 & 0 & 0 & 1 & 1 & \cdots & 1 \\ g_1 & 0 & 1 & 0 & 0 & z_1 & z_2 & \cdots & z_N \\ \vdots & \vdots & \vdots & \vdots & \vdots & \vdots & \vdots & \vdots \\ g_{K-1} & 0 & 0 & \cdots & 1 & z_1^{K-1} & z_2^{K-1} & \cdots & z_N^{K-1} \\ \xi_1 & 1 & z_1 & \cdots & z_1^{K-1} & \frac{1}{1-z_1^2} & \frac{1}{1-z_1 z_2} & \cdots & \frac{1}{1-z_1 z_N} \\ \xi_2 & 1 & z_2 & \cdots & z_2^{K-1} & \frac{1}{1-z_1 z_2} & \frac{1}{1-z_2^2} & \cdots & \frac{1}{1-z_2 z_N} \\ \vdots & \vdots & \vdots & \vdots & \vdots & \vdots & \vdots & \vdots \\ \xi_N & 1 & z_N & \cdots & z_N^{K-1} & \frac{1}{1-z_1 z_N} & \frac{1}{1-z_2 z_N} & \cdots & \frac{1}{1-z_N^2} \end{vmatrix} \geq 0. \tag{21}$$

The conditions (20) or (21) can be expressed in a straightforward way in terms of the values of the form factor $F(t)$ at $t_i = \tilde{t}(z_i)$ and the derivatives at $t = 0$, using eqs. (4) and (6). It can be shown that these inequalities define convex domains in the space of the input parameters.

3. Inclusion of Phase Information

Additional information on the unitarity cut can be included in the formalism. According to the Fermi-Watson theorem, below the inelastic threshold $t_{\rm in}$ the phase of $F(t)$ is equal (modulo π) to the phase $\delta(t)$ of the associated elastic scattering process. Thus,

$$F(t+i\epsilon) = |F(t)|e^{i\delta(t)}, \qquad t_+ < t < t_{\rm in}, \tag{22}$$

where $\delta(t)$ is known.

In order to impose the constraint (22), we define first the Omnès function

$$\mathcal{O}(t) = \exp\left(\frac{t}{\pi}\int_{t_+}^{\infty} dt' \frac{\delta(t')}{t'(t'-t)}\right), \tag{23}$$

where $\delta(t)$ is known for $t \leq t_{\rm in}$, and is an arbitrary function, sufficiently smooth (*i.e.* Lipschitz continuous) for $t > t_{\rm in}$. From (22) and (23) it follows that

$$\mathrm{Im}\left[\frac{F(t+i\epsilon)}{\mathcal{O}(t+i\epsilon)}\right] = 0, \qquad t_+ \leq t \leq t_{\rm in}. \tag{24}$$

Expressed in terms of the function $g(z)$ this condition becomes

$$\mathrm{Im}\left[\frac{g(e^{i\theta})}{W(\theta)}\right] = 0, \qquad \theta \in (-\theta_{\rm in}, \theta_{\rm in}). \tag{25}$$

Here $\theta_{\rm in}$ is defined by $\tilde{z}(t_{\rm in}) = \exp(i\theta_{\rm in})$ and the function $W(\theta)$ is defined as:

$$W(\theta) = w(e^{i\theta})O(e^{i\theta}), \tag{26}$$

where $w(z)$ is the outer function and

$$O(z) = \mathcal{O}(\tilde{t}(z)). \tag{27}$$

The constraint (25) can be imposed by means of a generalized Lagrange multiplier, while constraints at interior points can be treated with standard real-valued Lagrange multipliers. The Lagrangian of the minimization problem (12) with the constraints (10) and (25) reads

$$\begin{aligned}\mathcal{L} &= \frac{1}{2}\sum_{k=0}^{\infty} g_k^2 + \sum_{n=1}^{N}\alpha_n(\xi_n - \sum_{k=0}^{\infty} g_k z^k) \\ &+ \frac{1}{\pi}\sum_{k=0}^{\infty} g_k \lim_{r\to 1}\int_{-\theta_{\rm in}}^{\theta_{\rm in}} \lambda(\theta')|W(\theta')|\mathrm{Im}[[W(\theta')]^{-1} r^k e^{ik\theta'}]d\theta'.\end{aligned} \tag{28}$$

The Lagrange multiplier $\lambda(\theta)$ is an odd function, $\lambda(-\theta) = -\lambda(\theta)$ and the factor $|W(\theta')|$ was introduced in the integrand for convenience.

We minimize $\mathcal{L}$ by brute force method with respect to the free parameters g_k with $k \geq K$. The Lagrange multipliers $\lambda(\theta)$ and α_n are found in the standard way by imposing the constraints (10) and (25).

The calculations are straightforward. In order to write the equations in a simple form, it is convenient to define the phase $\Phi(\theta)$ of the function $W(\theta)$ by

$$W(\theta) = |W(\theta)|e^{i\Phi(\theta)}. \tag{29}$$

From (26) we have

$$\Phi(\theta) = \phi(\theta) + \delta(\tilde{t}(e^{i\theta})), \tag{30}$$

where $\phi(\theta)$ is the phase of the outer function $w(e^{i\theta})$ and $\delta(t)$ is the elastic scattering phase shift. We introduce also the functions β_n for $n = 1, ...N$, by

$$\beta_n(\theta) = z_n^K \frac{\sin[K\theta - \Phi(\theta)] - z_n \sin[(K-1)\theta - \Phi(\theta)]}{1 + z_n^2 - 2z_n \cos\theta}. \tag{31}$$

Then the equations for the Lagrange multipliers $\lambda(\theta)$ and α_n take the form:

$$\sum_{k=0}^{K-1} g_k \sin[k\theta - \Phi(\theta)] = \lambda(\theta) - \sum_{n=1}^{N} \alpha_n \beta_n(\theta) \tag{32}$$

$$-\frac{1}{2\pi} \int_{-\theta_{\rm in}}^{\theta_{\rm in}} {\rm d}\theta' \lambda(\theta')) \mathcal{K}_\Phi(\theta, \theta'), \qquad \theta \in (-\theta_{\rm in}, \theta_{\rm in}),$$

$$-\frac{1}{\pi} \int_{-\theta_{\rm in}}^{\theta_{\rm in}} \lambda(\theta)\beta_n(\theta){\rm d}\theta + \sum_{n'=1}^{N} \alpha_{n'} \frac{(z_n z_{n'})^K}{1 - z_n z_{n'}} = \xi_n, \tag{33}$$

where $n = 1, \ldots N$.

The integral kernel in (32), defined as

$$\mathcal{K}_\Phi(\theta, \theta') \equiv \frac{\sin[(K - 1/2)(\theta - \theta') - \Phi(\theta) + \Phi(\theta')]}{\sin[(\theta - \theta')/2]}, \tag{34}$$

is of Fredholm type if the phase $\Phi(\theta)$ is Lipschitz continuous Then the above system can be solved numerically in a straightforward manner. Finally, the inequality (11) takes the form:

$$\frac{1}{\pi} \sum_{k=0}^{K-1} g_k \int_{-\theta_{\rm in}}^{\theta_{\rm in}} {\rm d}\theta \lambda(\theta) \sin[k\theta - \Phi(\theta)] + \sum_{n=1}^{N} \alpha_n \xi_n \leq \bar{I}, \tag{35}$$

with $\bar{I}$ defined in (17). Using the relation (6), the above inequality defines an allowed domain for the values of the form factor and its derivatives at the origin. It must be emphasized that the theory for arbitrary number of constraints was presented for the first time in ref. [6]. It is easy to see that, if $t_{\rm in}$ is increased, the allowed domain defined by the inequality (35) becomes smaller. The reason is that by increasing $t_{\rm in}$ the class of functions entering the minimization (12) becomes gradually smaller, leading to a larger value for minimum μ_0^2 entering the definition (11) of the allowed domain.

4. A Modified Optimization Problem

In this section we shall present the solution of a modified interpolation problem, where the input consists from the phase condition (22) and the inequality

$$\int_{t_{\rm in}}^{\infty} {\rm d}t \rho(t) |F(t)|^2 \leq I', \tag{36}$$

where I' is known. Unlike the previous condition (1), which involved the whole unitarity cut, we assume now that an integral of the modulus squared along $t > t_{\rm in}$ (or a reliable upper bound on it) is known.

In some cases, the quantity I' can be obtained from the inequality (1) and the modulus of the form factor along the elastic part of the unitarity cut, if the latter is available from an independent source, for instance from experiment. Then I' is given by

$$I' = I - \int_{t_+}^{t_{\rm in}} {\rm d}t \rho(t)|F(t)|^2 < I. \tag{37}$$

In other cases, like for the pion electromagnetic form factor, I' can be estimated directly from experimental data on the modulus at high energies, taking into account also the asymptotic decrease $|F(t)| \sim 1/t$ predicted by perturbative QCD.

Once the conditions (22) and (36) are adopted, one can find the optimal domain allowed for the values and the derivatives of the form factor defined in (10). Below we present the solution of the problem as first described in [16]. We start with the remark that the knowledge of the phase was implemented in the previous section by the relation (24), which says that the function $f(t)$ defined through

$$F(t) = f(t)\mathcal{O}(t), \tag{38}$$

is real in the elastic region, $t_+ \leq t \leq t_{\rm in}$. In fact, since the Omnès function $\mathcal{O}(t)$ fully accounts for the elastic cut of the form factor, the function $f(t)$ has a larger analyticity domain, namely the complex t-plane cut only for $t > t_{\rm in}$. Moreover, (37) implies that f satisfies the condition

$$\int_{t_{\rm in}}^{\infty} {\rm d}t \rho(t)|\mathcal{O}(t)|^2 |f(t)|^2 \leq I'. \tag{39}$$

This inequality leads, through the techniques presented in section 2, to constraints on the values of f inside the analyticity domain. It is easy to see that the problem differs from the standard one described there only by the appearance of the additional factor $|\mathcal{O}(t)|^2$ in the integral, and the fact that the cut starts now at $t_{\rm in} > t_+$.

The problem is brought into a canonical form by the new transformation

$$\tilde{z}(t) = \frac{\sqrt{t_{\rm in}} - \sqrt{t_{\rm in} - t}}{\sqrt{t_{\rm in}} + \sqrt{t_{\rm in} - t}}, \tag{40}$$

which maps the complex t-plane cut for $t > t_{\rm in}$ on the unit disc in the z-plane defined by $z = \tilde{z}(t)$. Then (39) can be written as

$$\frac{1}{2\pi}\int_0^{2\pi} {\rm d}\theta |g(\exp(i\theta))|^2 \leq I', \tag{41}$$

where the function g is now

$$g(z) = w(z)\,\omega(z)\,F(\tilde{t}(z))\,[O(z)]^{-1}. \tag{42}$$

Here $w(z)$ is the outer function related to the weight $\rho(t)$ and the Jacobian of the new mapping (40) and $O(z)$ is defined as

$$O(z) = \mathcal{O}(\tilde{t}(z)), \tag{43}$$

where $\tilde{t}(z)$ is the inverse of $z = \tilde{z}(t)$ with $\tilde{z}(t)$ defined in (40), and

$$\omega(z) = \exp\left(\frac{\sqrt{t_{\rm in} - \tilde{t}(z)}}{\pi}\int_{t_{\rm in}}^{\infty} {\rm d}t' \frac{\ln|\mathcal{O}(t')|}{\sqrt{t' - t_{\rm in}}(t' - \tilde{t}(z))}\right). \tag{44}$$

The inequality (41) has exactly the same form as (5), and leads to the constraints (20) or (21) for the values and derivatives of $g(z)$ at interior points. Using (42), these constraints are expressed in terms of the physically interesting values of the form factor $F(t)$.

A remark on the uniqueness is of interest: we recall that the Omnès function $\mathcal{O}(t)$ defined in (23) is not unique, as it involves the arbitrary function $\delta(t)$ for $t > t_{\rm in}$. In section 3 we have seen that the results are not affected by this arbitrariness, as the integral equations involved only the known phase below $t_{\rm in}$. This is true also for the results here: the reason is that a change of the function $\delta(t)$ for $t > t_{\rm in}$ is equivalent with a multiplication of $g(z)$ by a function analytic and without zeros in $|z| < 1$ (*i.e.* an outer function). According to the general theory of analytic functions of Hardy class, the multiplication by an outer function does not change the class of functions used in minimization problems. In our case, the arbitrary function $\delta(t)$ for $t > t_{\rm in}$ enters in both the functions $O(z)$ and $\omega(z)$ appearing in (42), and their ambiguities compensate each other exactly. The independence of the results on the choice of the phase for $t > t_{\rm in}$ is confirmed numerically, for functions $\delta(t)$ that are Lipschitz continuous. The constraints provided by the technique of this section are expected to be quite strong since they result from a minimization on a restricted class of analytic functions, where the second Riemann sheet of the form factor is accounted for explicitly by the Omnès function.

On the other hand, it is easy to see that the fulfillment of the condition (39) does not automatically imply that the original condition (1) is satisfied, and both conditions should be applied in order to reduce the allowed range of the parameters of interest.

The mathematical techniques presented can be adapted in a straightforward way to the problem of zeros. Let us assume that the form factor $F(t)$ has a simple zero on the real axis, $F(t_0) = 0$. We shall use this information in the determinant condition: if the zero is compatible with the remaining information, the inequality can be satisfied. If, on the contrary, the inequality is violated, the zero is excluded. It follows that we can obtain a rigorous condition for the domain of points z_0 (or t_0) where the zeros are excluded.

For illustration, assume first that we use as input only the value of the form factor at $t = 0$. Then from the argument given above, it follows that the domain of real points t_0 where the form factor cannot have zeros is described by the inequality [20]

$$\begin{vmatrix} I' - g_0^2 & -g_0 \\ -g_0 & \frac{\tilde{z}(t_0)^2}{1-\tilde{z}(t_0)^2} \end{vmatrix} \leq 0. \tag{45}$$

Here, I' is the known bound appearing in (36), $g_0 = g(0)$ is calculated in terms of the input phase from (42), and $\tilde{z}(t)$ is defined in (40). If we include in addition the value $F(t_1)$ of the form factor at some point t_1, the condition reads [20]:

$$\begin{vmatrix} I' - g_0^2 & g(\tilde{z}(t_1)) - g_0 & -g_0 \\ g(\tilde{z}(t_1)) - g_0 & \frac{\tilde{z}(t_1)^2}{1-\tilde{z}(t_1)^2} & \frac{\tilde{z}(t_1)\tilde{z}(t_0)}{1-\tilde{z}(t_1)\tilde{z}(t_0)} \\ -g_0 & \frac{\tilde{z}(t_1)\tilde{z}(t_0)}{1-\tilde{z}(t_1)z(t_0)} & \frac{\tilde{z}(t_0)^2}{1-\tilde{z}(t_0)^2} \end{vmatrix} \leq 0. \tag{46}$$

5. Recent Applications to the Pion Electromagnetic Form Factor

We will now describe some applications of the methods discussed above. Whilst there has been a long series of applications of these methods to the pion electromagnetic form factor, ref. [20] captures the most stringent results. This is based on the fact that the phase of the form factor is determined up to the $\pi\omega$ threshold ($\sqrt{t_{\rm in}}$=917 MeV) in terms of the $I = 1$, $l = 1$ partial wave of $\pi\pi$ elastic scattering. We have used the recent parametrization given in ref. [33], which agrees well with the solutions of the Roy equations [34] between threshold and 800 MeV. The modulus $|F(t)|$ has been measured also recently with improved precision by BaBar [30] and KLOE [31] collaborations.

Journal of Physics: Conference Series **374** (2012) 012011 doi:10.1088/1742-6596/374/1/012011

Table 1. Spacelike data from [37, 38].

t	Value[GeV^2]	$F(t)$
t_1	-1.60	$0.243 \pm 0.012^{+0.019}_{-0.008}$
t_2	-2.45	$0.167 \pm 0.010^{+0.013}_{-0.007}$

We consider the two-pion contribution to the muon $g-2$, when the weight $\rho(t)$ has the form

$$\rho(t) = \frac{\alpha^2 M_\mu^2}{12\pi} \frac{(t-t_+)^{3/2}}{t^{7/2}} K(t),$$
$$K(t) = \int_0^1 du \, \frac{(1-u)u^2}{1-u+M_\mu^2 u^2/t}. \quad (47)$$

The two-pion contribution to muon anomaly was evaluated recently with great precision [35] from the accurate BaBar data on the modulus [30]. In ref. [20] the value of the integral I' defined in (36) for the weight (47) and the above choice of $\sqrt{t_{\rm in}}$ was evaluated to be 22.17×10^{-10}.

A concrete application of these methods is one of constraining the shape parameters appearing in (2) using as input the conditions (22) and (36) and the technique described in section 4. We use as input also the precise estimate of the coefficient of the linear term $\langle r_\pi^2 \rangle = (0.435 \pm 0.005)\,{\rm fm}^2$ given in [36], and additional spacelike data coming from [37, 38], which are given in Table 1, where the first error is statistical and the second is systematic.

An illustrative example of the results on the shape parameters c and d is given in Fig. 1. Varying $F(t_1)$ given in Table 1 inside the error bars, we obtain the allowed domain of the c and

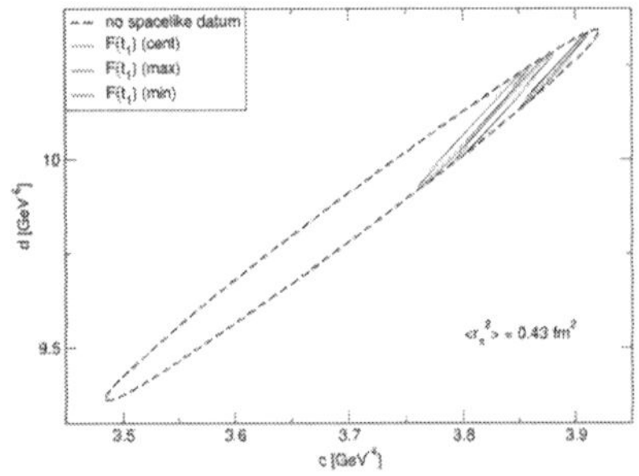

Figure 1. Allowed domain in the $c-d$ plane calculated with $t_{\rm in} = (0.917~{\rm GeV})^2$ and $\langle r_\pi^2 \rangle = 0.43\,{\rm fm}^2$, for three values of $F(t_1)$ at the spacelike point $t_1 = -1.6~{\rm GeV}^2$ (central value in Table 1 and the extreme values obtained from the error intervals). Also shown is the large ellipse when no spacelike datum is included.

d parameters at the present level of knowledge as the union of the three small ellipses in Fig. 1. By varying also $\langle r_\pi^2 \rangle$ in the range given above the allowed ranges read

$$3.75~{\rm GeV}^{-4} \lesssim c \lesssim 3.98~{\rm GeV}^{-4},$$
$$9.91~{\rm GeV}^{-6} \lesssim d \lesssim 10.45~{\rm GeV}^{-6}, \quad (48)$$

with a strong correlation between the two coefficients. Similar results are obtained when the second datum in Table 1 is used in place of the first.

Turning now to the issue of zeros, and using the machinery as described in the preceding section, we find that, for with $\langle r_\pi^2 \rangle = 0.43\,\text{fm}^2$, simple zeros are excluded from the interval $-1.93\ \text{GeV}^2 \leq t \leq 0.83\ \text{GeV}^2$ of the real axis. If we impose the additional constraint at a spacelike point $t_1 = -1.6\ \text{GeV}^2$, the interval for the excluded zeros is much bigger. The left end of the range is sensitive to the input value $F_1 = F(t_1)$. Using the central value $F_1 = 0.243$ given in Table 1, we find that the form factor cannot have simple zeros in the range $-5.56\ \text{GeV}^2 \leq t \leq 0.84\ \text{GeV}^2$. By varying F_1 inside the error interval given in Table 1 (with errors added in quadrature), we find that zeros are excluded from the range $-4.46\ \text{GeV}^2 \leq t \leq 0.84\ \text{GeV}^2$.

The machinery may now be used to find regions in the complex plane where no zeros are allowed. Complex zeros occur simultaneously at complex conjugate points due to the reality condition. In Fig. 2 we present the exclusion region obtained with central values of the radius and the first datum in Table1. As in the case of the simple zeros on the real axis, there is significant sensitivity to the experimental uncertainty at that point, the allowed region being obtained as the intersection of the regions corresponding to the extreme values. The net effect is a somewhat reduced region compared to that given in Fig. 2. A detailed discussion on the

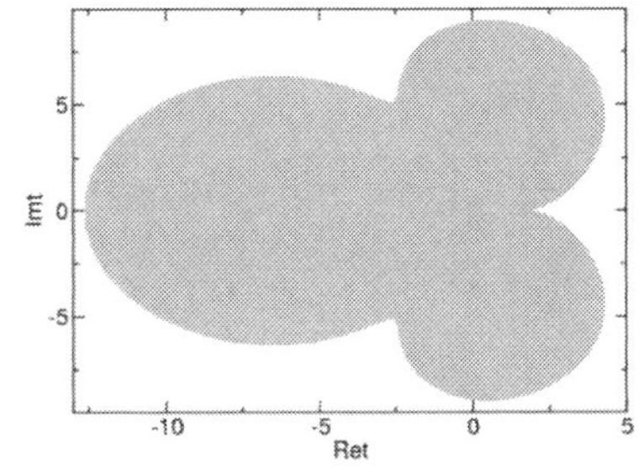

Figure 2. Domain without zeros obtained with $t_{\text{in}} = (0.917\ \text{GeV})^2$ and $\langle r_\pi^2 \rangle = 0.43\,\text{fm}^2$, using in addition the central experimental value $F(t_1) = 0.243$ at the spacelike point $t_1 = -1.6\ \text{GeV}^2$.

sensitivity is given in ref. [20].

6. Afterword

In this lecture note we have provided a comprehensive introduction to the modern theory of unitarity bounds for hadronic form factors. As an illustration we have shown how these methods can be exploited to improve the knowledge on the electromagnetic form factor of the pion at low energies. In [23, 24, 25] the techniques were applied to the pion-kaon form factors, which represent an important input for the extraction of the CKM matrix element V_{us}, while in [29] the techniques were applied the $D\pi$ form factors, of interest for the extraction of the CKM matrix element V_{cd}. The results of these investigations are presented in the contributions [39] and [40] to these Proceedings, respectively. An important conclusion of these analyses is that there is a remarkable coherence between various theoretical approaches to the strong interactions, including perturbative QCD, chiral symmetries, lattice evaluations, and also experimental information which is now available at high precision.

Acknowledgments

We wish to thank the organizers of the workshop for inviting us to present our work. We thank Gauhar Abbas, I. Sentitemsu Imsong and Sunethra Ramanan for discussions.

References

[1] S. Okubo, Phys. Rev. D **3**, 2807 (1971).
[2] S. Okubo, Phys. Rev. D **4**, 725 (1971).
[3] V. Singh and A. K. Raina, Fortsch. Phys. **27**, 561 (1979).
[4] I. Raszillier, W. Schmidt and I. S. Stefanescu, J. Math. Phys. **17**, 1957 (1976).
[5] N. N. Meiman, JETP **17**, 830 (1963).
[6] G. Abbas, B. Ananthanarayan, I. Caprini, I. Sentitemsu Imsong and S. Ramanan, Eur. Phys. J. A **45**, 389 (2010) [arXiv:1004.4257 [hep-ph]].
[7] C. Bourrely, B. Machet and E. de Rafael, Nucl. Phys. B **189**, 157 (1981).
[8] M. Micu, Phys. Rev. D **7**, 2136 (1973).
[9] G. Auberson, G. Mahoux and F. R. A. Simao, Nucl. Phys. B **98**, 204 (1975).
[10] E. de Rafael and J. Taron, Phys. Lett. B **282**, 215 (1992).
[11] E. de Rafael and J. Taron, Phys. Rev. D **50**, 373 (1994) [arXiv:hep-ph/9306214].
[12] C. G. Boyd, B. Grinstein and R. F. Lebed, Phys. Lett. B **353**, 306 (1995) [arXiv:hep-ph/9504235].
[13] C. G. Boyd, B. Grinstein and R. F. Lebed, Nucl. Phys. B **461**, 493 (1996) [arXiv:hep-ph/9508211].
[14] I. Caprini and M. Neubert, Phys. Lett. B **380**, 376 (1996) [arXiv:hep-ph/9603414].
[15] I. Caprini, L. Lellouch and M. Neubert, Nucl. Phys. B **530**, 153 (1998) [arXiv:hep-ph/9712417].
[16] I. Caprini, Eur. Phys. J. C **13**, 471 (2000) [arXiv:hep-ph/9907227].
[17] B. Ananthanarayan and S. Ramanan, Eur. Phys. J. C **54**, 461 (2008) [arXiv:0801.2023 [hep-ph]].
[18] B. Ananthanarayan and S. Ramanan, Eur. Phys. J. C **60**, 73 (2009) [arXiv:0811.0482 [hep-ph]].
[19] G. Abbas, B. Ananthanarayan and S. Ramanan, Eur. Phys. J. A **41**, 93 (2009) [arXiv:0903.4297 [hep-ph]].
[20] B. Ananthanarayan, I. Caprini and I. S. Imsong, Phys. Rev. D **83**, 096002 (2011) [arXiv:1102.3299 [hep-ph]].
[21] C. Bourrely and I. Caprini, Nucl. Phys. B **722**, 149 (2005) [arXiv:hep-ph/0504016].
[22] R. J. Hill, Phys. Rev. D **74**, 096006 (2006) [arXiv:hep-ph/0607108].
[23] G. Abbas and B. Ananthanarayan, Eur. Phys. J. A **41**, 7 (2009) [arXiv:0905.0951 [hep-ph]].
[24] G. Abbas, B. Ananthanarayan, I. Caprini, I. Sentitemsu Imsong and S. Ramanan, Eur. Phys. J. A **44**, 175 (2010) [arXiv:0912.2831 [hep-ph]].
[25] G. Abbas, B. Ananthanarayan, I. Caprini and I. Sentitemsu Imsong, Phys. Rev. D **82**, 094018 (2010) [arXiv:1008.0925 [hep-ph]].
[26] L. Lellouch, Nucl. Phys. B **479**, 353 (1996) [arXiv:hep-ph/9509358].
[27] T. Becher and R. J. Hill, Phys. Lett. B **633**, 61 (2006) [arXiv:hep-ph/0509090].
[28] C. Bourrely, I. Caprini and L. Lellouch, Phys. Rev. D **79**, 013008 (2009) [Erratum-ibid. D **82**, 099902 (2010)] [arXiv:0807.2722 [hep-ph]].
[29] B. Ananthanarayan, I. Caprini and I. S. Imsong, Eur. Phys. J. A **47**, 147 (2011) [arXiv:1108.0284 [hep-ph]].
[30] B. Aubert *et al.* [BABAR Collaboration], Phys. Rev. Lett. **103**, 231801 (2009) [arXiv:0908.3589 [hep-ex]].
[31] F. Ambrosino *et al.* [KLOE Collaboration], Phys. Lett. B **670**, 285 (2009) [arXiv:0809.3950 [hep-ex]].
[32] A. K. Raina and V. Singh, J. Phys. G **3**, 315 (1977).
[33] R. Kaminski, J. R. Pelaez and F. J. Yndurain, Phys. Rev. D **77**, 054015 (2008) [arXiv:0710.1150 [hep-ph]].
[34] B. Ananthanarayan, G. Colangelo, J. Gasser and H. Leutwyler, Phys. Rept. **353**, 207 (2001) [arXiv:hep-ph/0005297].
[35] M. Davier, A. Hoecker, B. Malaescu, C. Z. Yuan and Z. Zhang, Eur. Phys. J. C **66**, 1 (2010) [arXiv:0908.4300 [hep-ph]].
[36] G. Colangelo, Nucl.Phys.Proc.Suppl. 131 (2004) 185-191 [arXiv:hep-ph/0312017]
[37] T. Horn *et al.* [Jefferson Lab F(pi)-2 Collaboration], Phys. Rev. Lett. **97**, 192001 (2006) [arXiv:nucl-ex/0607005].
[38] G.M. Huber *et al.* [Jefferson Lab Collaboration], Phys. Rev. C **78**, 045203 (2008) [arXiv:0809.3052 [nucl-ex]].
[39] Gauhar Abbas, B.Ananthanarayan, I.Caprini and I.Sentitemsu Imsong, *The $K\pi$ form factors from analyticity and unitarity*, these Proceedings
[40] B.Ananthanarayan, I.Caprini and I.Sentitemsu Imsong, *The $D\pi$ form factors from analyticity and unitarity*, these Proceedings

The $K\pi$ form factors from Analyticity and Unitarity

Gauhar Abbas[a,*], **B.Ananthanarayan**[a,*], **I.Caprini**[b,†], **I.Sentitemsu Imsong**[a,‡]

[a] Centre for High Energy Physics, Indian Institute of Science, Bangalore 560 012, India
[b] Horia Hulubei National Institute for Physics and Nuclear Engineering, P.O.B. MG-6, 077125 Magurele, Romania

E-mail: *gabbas@cts.iisc.ernet.in
*anant@cts.iisc.ernet.in
†caprini@theory.nipne.ro
‡senti@cts.iisc.ernet.in

Abstract. Analyticity and unitarity techniques are employed to obtain bounds on the shape parameters of the scalar and vector form factors of semileptonic K_{l3} decays. For this purpose we use vector and scalar correlators evaluated in pQCD, a low energy theorem for scalar form factor, lattice results for the ratio of kaon and pion decay constants, chiral perturbation theory calculations for the scalar form factor at the Callan-Treiman point and experimental information on the phase and modulus of $K\pi$ form factors up to an energy $t_{\rm in} = 1{\rm GeV}^2$. We further derive regions on the real axis and in the complex-energy plane where the form factors cannot have zeros.

1. Introduction

K_{l3} decays are important for determining the matrix element V_{us}, for recent reviews, see [1, 2] of the Cabibbo-Kobayashi-Maskawa (CKM) which in turn is crucial for testing the unitarity of the CKM matrix. Information on the experimental sector is rich, in particular for the $K_{\ell 3}$ decay rates, which were measured by a number of experiments BNL-E865 [3], KLOE [4], ISTRA+ [5], KTEV [6] and more recently a new analysis by NA48 [7]. Recent lattice studies have also been carried out, see refs.[8, 9, 10, 11].

The decay of a kaon to a pion, a charged lepton and a neutrino is described by the matrix element

$$\langle \pi^0(p')|\bar{s}\gamma_\mu u|K^+(p)\rangle = \frac{1}{\sqrt{2}}[(p'+p)_\mu f_+(t) + (p-p')_\mu f_-(t)], \tag{1}$$

where $f_+(t)$ is the vector form factor and the combination

$$f_0(t) = f_+(t) + \frac{t}{M_K^2 - M_\pi^2} f_-(t) \tag{2}$$

is known as the scalar form factor. The expansion at $t = 0$

$$f_k(t) = f_+(0)\left(1 + \lambda'_k \frac{t}{M_\pi^2} + \frac{1}{2}\lambda''_k \frac{t^2}{M_\pi^4} + \cdots\right), \tag{3}$$

Published under licence by IOP Publishing Ltd

Journal of Physics: Conference Series **374** (2012) 012012 doi:10.1088/1742-6596/374/1/012012

defines the slope λ'_k and the curvature λ''_k parameters where $k = 0$ denotes the scalar and $k = +$ denotes the vector form factor. The precise determination of the element $|V_{us}|$ depends on how accurate the parametrization of the form factor is. To improve the precision and to provide bounds on the shape parameters of the form factors, we use inputs coming from certain low energy theorems, perturbative QCD, lattice computations and chiral perturbation theory. Our techniques allow us to incorporate the phase and modulus information of the form factors. We also apply the technique to find regions on the real axis and in the complex t-plane where zeros are excluded. The knowledge of zeros is of interest, for instance, for the dispersive methods (Omnès-type representations) and for testing specific models of the form factors. For more details, see refs. [12, 13, 14, 15].

2. Formalism

The formalism described in [12, 13, 14] and in the contribution to these Proceedings [16] , exploits an integral of the type

$$\int_{t_+}^{\infty} dt\ \rho_{+,0}(t)|f_{+,0}(t)|^2 \leq I_{+,0}, \tag{4}$$

along the unitarity cut, whose upper bound is known from a dispersion relation, satisfied by a certain QCD correlator. For the scalar form factor this reads

$$\chi_0(Q^2) \equiv \frac{\partial}{\partial q^2}\left[q^2\Pi_0\right] = \frac{1}{\pi}\int_{t_+}^{\infty} dt\, \frac{t\mathrm{Im}\Pi_0(t)}{(t+Q^2)^2}, \tag{5}$$

$$\mathrm{Im}\Pi_0(t) \geq \frac{3}{2}\frac{t_+t_-}{16\pi}\frac{[(t-t_+)(t-t_-)]^{1/2}}{t^3}|f_0(t)|^2, \tag{6}$$

with $t_\pm = (M_K \pm M_\pi)^2$. Analogues expression, involving a suitable correlator denoted by $\chi_1(Q^2)$, can be written down for the vector form factor. We can now use the conformal map $t \to z(t)$

$$z(t) = \frac{\sqrt{t_+} - \sqrt{t_+ - t}}{\sqrt{t_+} + \sqrt{t_+ - t}}, \tag{7}$$

that maps the cut t-plane onto the unit disc $|z| < 1$ in the $z \equiv z(t)$ plane, with t_+ mapped onto $z = 1$, the point at infinity to $z = -1$ and the origin to $z = 0$. Using this map, we cast eqn.(4) into a canonical form, incorporate phase and modulus information as well as the Callan-Treiman relations and finally employ a determinant for obtaining bounds on the shape parameters and for finding regions of excluded zeros in the complex t-plane.

3. Inputs

The essential inputs of our formalism, the vector $\chi_1(Q^2)$ and the scalar $\chi_0(Q^2)$ correlators, can be calculated in perturbative QCD up to order α_s^4 for $Q^2 >> \Lambda_{QCD}^2$ [17, 18]. We get $\chi_1(2\,\mathrm{GeV}) = (343.8 \pm 51.6) \times 10^{-5}\,\mathrm{GeV}^{-2}$ and $\chi_0(2\,\mathrm{GeV}) = (253 \pm 68) \times 10^{-6}$, see also [12]. An improvement can be achieved when we implement theoretical and experimental information into the formalism of unitarity bounds. The first improvement comes when we use the the value of vector form factor at zero momentum transfer. Recent determinations from the lattice give $f_+(0) = 0.964(5)$ [19]. We can also use two low energy theorems, namely soft pion and soft kaon theorems, for the improvement of the bounds on the slope and curvature parameters in the scalar case. The soft pion theorem relates the value of scalar form factor at first Callan-Treiman piont $\Delta_{K\pi} \equiv M_K^2 - M_\pi^2$ to the ratio F_K/F_π of the decay constants [20, 21]:

$$f_0(\Delta_{K\pi}) = F_K/F_\pi + \Delta_{CT}. \tag{8}$$

Recent lattice evaluations with $N_f = 2 + 1$ flavors of sea quarks give $F_K/F_\pi = 1.193 \pm 0.006$ [8, 9]. In the isospin limit, $\Delta_{CT} = -3.1 \times 10^{-3}$ to one loop [22] and $\Delta_{CT} \simeq 0$ to two-loops in chiral perturbation theory [23, 24, 25].

At $\bar{\Delta}_{K\pi}(= -\Delta_{K\pi})$, a soft-kaon result [26] relates the value of the scalar form factor to F_π/F_K

$$f_0(-\Delta_{K\pi}) = F_\pi/F_K + \bar{\Delta}_{CT}. \tag{9}$$

A calculation in ChPT to one-loop in the isospin limit [22] gives $\bar{\Delta}_{CT} = 0.03$, but the higher order ChPT corrections are expected to be larger in this case. As discussed in [14], due to the poor knowledge of $\bar{\Delta}_{CT}$, the low-energy theorem eqn.(9) is not useful for further constraining the shape of the $K_{\ell 3}$ form factors at low energies. On the other hand, we obtain from the same machinery, the stringent bound on the quantity $\bar{\Delta}_{CT}$ which is $-0.046 \leq \bar{\Delta}_{CT} \leq 0.014$.

Further improvement of the bounds can be achieved if the phase of the form factor along the elastic part of the unitarity cut is known from an independent source. In our calculations we use below $t_{\rm in}$ the phases from [27, 28] for the scalar form factor, and from [29, 30] for the vector form factor. Above $t_{\rm in}$ we take $\delta(t)$ Lipschitz continuous, i.e., a smooth function approaching π at high energies. The results are independent of the choice of the phase for $t > t_{\rm in}$. We can further improve the bounds if the modulus of the form factor is known along the unitarity cut, $t \leq t_{\rm in}$: we can shift the branch point from $t_\pm$ to t_{in} by subtracting the low energy integral from the integral Eq. (4). In order to estimate the low-energy integral, which is the value of the integral contribution from t_+ to $t_{\rm in}$, see for expressions ref.[13], we use the Breit-Wigner parameterizations of $|f_+(t)|$ and $|f_0(t)|$ in terms of the resonances given by the Belle Collaboration [31] for fitting the rate of $\tau \to K\pi\nu$ decay. The above leads to the value $31.4 \times 10^{-5}\,{\rm GeV}^{-2}$ for the vector form factor and 60.9×10^{-6} for the scalar form factor. By combining with the values $I_{+,0}$, we obtain the new upper bound on the integral Eq. (4) from $t_{\rm in}$ to ∞, $I'_+ = (312 \pm 69) \times 10^{-5}{\rm GeV}^{-2}$ and $I'_0 = (192 \pm 90) \times 10^{-6}$.

4. Results

In Figs. 1 and 2, the constraints for the scalar form factor are represented together with experimental information from various experiments. As shown in Fig. 1, the slope λ'_0 of the scalar form factor, predicted by NA48 (2007) is not consistent with our predictions (yellow band) which are obtained by taking into account the phase, modulus as well as the CT constraint. Nevertheless our predicted range for the slope is well-respected by the recent 2011 analysis by NA48 [7] .

The value of λ'_0 for this new determination by the NA48 reads $\lambda'_0 = (15.6 \pm 1.2 \pm 0.9) \times 10^{-3}$. On the theoretical side, the prediction of ChPT to two loops gives $\lambda'_0 = (13.9^{-0.4}_{+1.3} \pm 0.4) \times 10^{-3}$ and $\lambda''_0 = (8.0^{-1.7}_{+0.3}) \times 10^{-4}$ which are consistent with our results within errors as shown in Fig. 2. For the central value of the slope λ'_0 given above, the range of λ''_0 is $(8.24 \times 10^{-4}, 8.42 \times 10^{-4})$. The corresponding theoretical predictions are $\lambda'_0 = (16.00 \pm 1.00) \times 10^{-3}$, $\lambda''_0 = (6.34 \pm 0.38) \times 10^{-4}$ obtained from dispersion relations.

Comparison of the experimental results with our constraints for the vector form factor is shown in Fig. 3. We find that except for the results from NA48 (2007) and KLOE, which have curvatures slightly larger than the allowed values, the experimental data satisfy our constraints. The new results from NA48 [7] provide a curvature which overlaps with our constraints while the slope lies completely within our domain as seen in the figure. We also note that the theoretical

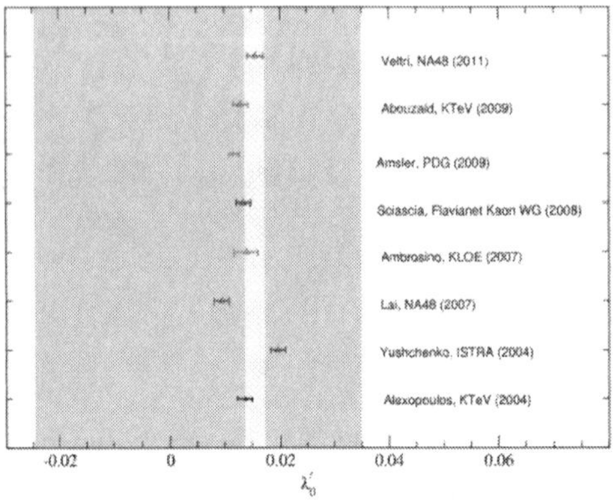

Figure 1. The allowed range for the slope of the scalar form factor, when we include phase, modulus and the CT constraint (yellow band). The grey band shows the range without the CT constraint.

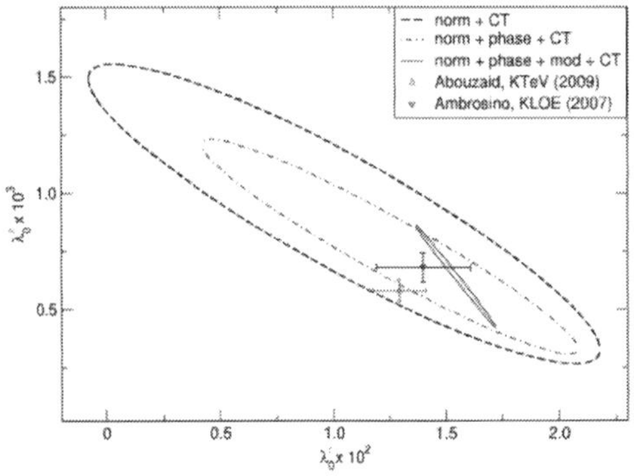

Figure 2. Allowed domain for the slope and curvature of the scalar form factor, using the normalization $f_+(0) = 0.962$, the value $f_0(\Delta_{K\pi}) = 1.193$, and phase and modulus information up to $t_{\rm in} = (1\,{\rm GeV})^2$.

predictions $\lambda'_+ = (24.9 \pm 1.3) \times 10^{-3}$, $\lambda''_+ = (1.6 \pm 0.5) \times 10^{-3}$ obtained from ChPT to two loops, and $\lambda'_+ = (26.05^{+0.21}_{-0.51}) \times 10^{-3}$, $\lambda''_+ = (1.29^{+0.01}_{-0.04}) \times 10^{-3}$, and $\lambda'_+ = (25.49 \pm 0.31) \times 10^{-3}$, $\lambda''_+ = (1.22 \pm 0.14) \times 10^{-3}$ obtained from dispersion relations are consistent with the constraint. For more results, see [12].

We can also extend our technique to derive regions in the complex plane where the form factors can not have zeros. For the $K\pi$ form factors, the influence of possible zeros in the context of Omnès dispersive representations has been analyzed in [30]. The absence of zeros is assumed in the recent analysis of KTeV data reported in [6]. In Fig. 4 we show the region where zeros of the scalar form factors are excluded. The formalism rules out zeros in the physical region of the kaon semileptonic decay. In the case of complex zeros, we have obtained a rather large region where they cannot be present. In the case of the vector form factor, the analysis of [30] using data from τ decay concludes that complex zeros cannot be excluded, due to the lack

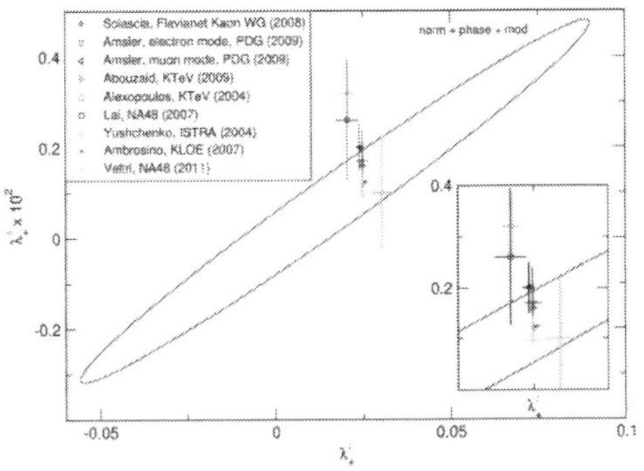

Figure 3. The best constraints for the slope and curvature of the vector form factor in the slope-curvature plane, where the allowed region is the interior of the ellipse.

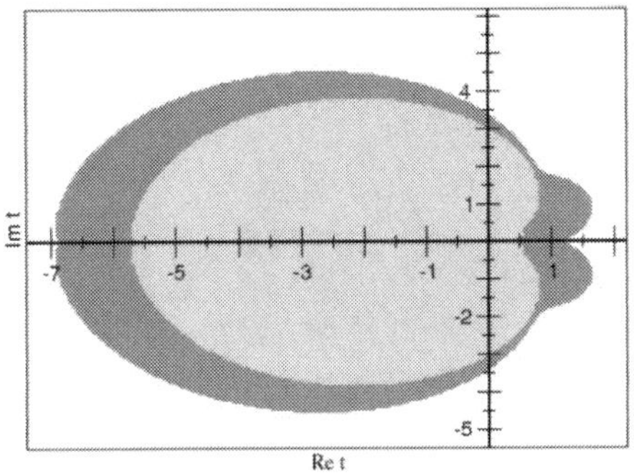

Figure 4. Domain without zeros for the scalar form factor: the small domain is obtained without including phase and modulus in the elastic region, the bigger one using phase, modulus and CT constraint.

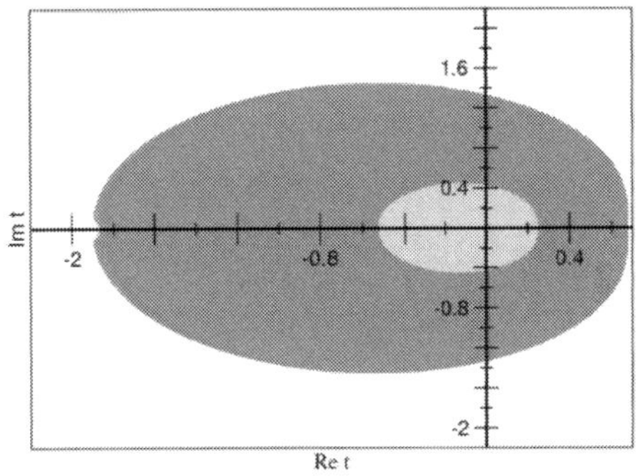

Figure 5. Domain without zeros for the vector form factor: the small domain is obtained without including phase and modulus in the elastic region, the bigger one using phase and modulus.

of information on the phase of the form factor in the inelastic region. However our formalism is independent of phase information in the inelastic region and leads without any assumptions to a rather large domain where complex zeros are excluded. For the vector form factor, fig. 5 shows the region where complex zeros are excluded. For more results, see [12].

5. Conclusion

We have derived stringent constraints on the shape parameters of the form factors of K_{l3} decay which is the best source for the extraction of CKM matrix element V_{us}. The results are promising and stringent especially in the case of the scalar form factor. The most recent results from NA48 [7] is consistent with our prediction for the slope of scalar form factor and restricts the range of the slope to $\sim 0.01 - 0.02$. We have also excluded zeros in a rather large domain at low energies both for the scalar and vector form factor. The Callan-Treiman input provides an additional constraint in the case of the scalar form factor and as a result excludes a larger domain of the energy plane where zeros can exist. Thus, this work represents a powerful application of the theory of unitarity bounds, which relies not so much on experimental information, but on theoretical inputs from perturbative QCD, low energy theorems and lattice calculations. It provides a powerful consistency check on determinations of shape parameters from phenomenology and experimental analyses.

References

[1] M. Antonelli, V. Cirigliano, G. Isidori, F. Mescia, M. Moulson, H. Neufeld, E. Passemar and M. Palutan *et al.*, Eur. Phys. J. C **69**, 399 (2010). [arXiv:1005.2323 [hep-ph]].
[2] V. Cirigliano, G. Ecker, H. Neufeld, A. Pich and J. Portoles, arXiv:1107.6001 [hep-ph].
[3] A. Sher *et al.*, Phys. Rev. Lett. **91**, 261802 (2003) [arXiv:hep-ex/0305042].
[4] F. Ambrosino *et al.* [KLOE Collaboration], Phys. Lett. B **636**, 166 (2006). [arXiv:hep-ex/0601038].
[5] V. I. Romanovsky *et al.*, arXiv:0704.2052 [hep-ex].
[6] E. Abouzaid *et al.* [KTeV collaboration], Phys. Rev. D **81**, 052001 (2010). [arXiv:0912.1291 [hep-ex]].
[7] M. Veltri, arXiv:1101.5031 [hep-ex].
[8] L. Lellouch, PoS LATTICE **2008**, 015 (2009). [arXiv:0902.4545 [hep-lat]].
[9] S. Durr *et al.*, Phys. Rev. D **81**, 054507 (2010). [arXiv:1001.4692 [hep-lat]].
[10] G. Colangelo *et al.*, Eur. Phys. J. C **71**, 1695 (2011). [arXiv:1011.4408 [hep-lat]].
[11] A. Ramos *et al.*, arXiv:1101.3968 [hep-lat].
[12] G. Abbas, B. Ananthanarayan, I. Caprini and I. Sentitemsu Imsong, Phys. Rev. D **82**, 094018 (2010). [arXiv:1008.0925 [hep-ph]].
[13] G. Abbas, B. Ananthanarayan, I. Caprini, I. Sentitemsu Imsong and S. Ramanan, Eur. Phys. J. A **45**, 389 (2010). [arXiv:1004.4257 [hep-ph]].
[14] G. Abbas, B. Ananthanarayan, I. Caprini, I. Sentitemsu Imsong and S. Ramanan, Eur. Phys. J. A **44**, 175 (2010). [arXiv:0912.2831 [hep-ph]].
[15] G. Abbas, B. Ananthanarayan, I. Caprini and I. S. Imsong, arXiv:1112.4270 [hep-ph].
[16] B. Ananthanarayan, I. Caprini, *Constraining Form Factors with the Method of Unitarity Bounds,* these Proceedings.
[17] P. A. Baikov, K. G. Chetyrkin and J. H. Kuhn, Phys. Rev. Lett. **96**, 012003 (2006). [hep-ph/0511063].
[18] P. A. Baikov, K. G. Chetyrkin and J. H. Kuhn, Phys. Rev. Lett. **101**, 012002 (2008). [arXiv:0801.1821 [hep-ph]].
[19] P. A. Boyle, A. Juttner, R. D. Kenway, C. T. Sachrajda, S. Sasaki, A. Soni, R. J. Tweedie and J. M. Zanotti, Phys. Rev. Lett. **100**, 141601 (2008). [arXiv:0710.5136 [hep-lat]].
[20] C. G. Callan and S. B. Treiman, Phys. Rev. Lett. **16**, 153 (1966).
[21] R. F. Dashen and M. Weinstein, Phys. Rev. Lett. **22**, 1337 (1969).
[22] J. Gasser and H. Leutwyler, Nucl. Phys. B **250**, 517 (1985).
[23] A. Kastner and H. Neufeld, Eur. Phys. J. C **57**, 541 (2008). [arXiv:0805.2222 [hep-ph]].
[24] J. Bijnens and P. Talavera, Nucl. Phys. B **669**, 341 (2003). [hep-ph/0303103].
[25] J. Bijnens and K. Ghorbani, arXiv:0711.0148 [hep-ph].
[26] R. Oehme, Phys. Rev. Lett. **16**, 215 (1966).
[27] P. Buettiker, S. Descotes-Genon and B. Moussallam, Eur. Phys. J. C **33**, 409 (2004). [hep-ph/0310283].
[28] B. El-Bennich, A. Furman, R. Kaminski, L. Lesniak, B. Loiseau and B. Moussallam, Phys. Rev. D **79**, 094005 (2009). [Erratum-ibid. D **83**, 039903 (2011)] [arXiv:0902.3645 [hep-ph]].
[29] B. Moussallam, Eur. Phys. J. C **53**, 401 (2008). [arXiv:0710.0548 [hep-ph]].
[30] V. Bernard, M. Oertel, E. Passemar and J. Stern, Phys. Rev. D **80**, 034034 (2009). [arXiv:0903.1654 [hep-ph]].
[31] D. Epifanov *et al.* [Belle Collaboration], Phys. Lett. B **654**, 65 (2007). [arXiv:0706.2231 [hep-ex]].

Journal of Physics: Conference Series **374** (2012) 012013 doi:10.1088/1742-6596/374/1/012013

The $D\pi$ form factors from analyticity and unitarity

B.Ananthanarayan[a,*], **I.Caprini**[b,†], **I.Sentitemsu Imsong**[a,‡]

[a] Centre for High Energy Physics, Indian Institute of Science, Bangalore 560 012, India
[b] Horia Hulubei National Institute for Physics and Nuclear Engineering, P.O.B. MG-6, 077125 Magurele, Romania

E-mail: *anant@cts.iisc.ernet.in
†caprini@theory.nipne.ro
‡senti@cts.iisc.ernet.in

Abstract. We study the shape parameters of the $D\pi$ scalar and vector form factors using as input dispersion relations and unitarity for the moments of suitable heavy-light correlators evaluated with Operator Product Expansions, including $O(\alpha_s^2)$ terms in perturbative QCD. For the scalar form factor, a low energy theorem and phase information on the unitarity cut are implemented to further constrain the shape parameters. We finally determine points on the real axis and isolate regions in the complex energy plane where zeros of the form factors are excluded.

1. Introduction

The knowledge of shape parameters of the $D\pi$ form factors is of interest for the determination of the element $|V_{cd}|$ of the Cabibbo-Kobayashi-Maskawa (CKM) matrix entering precision tests of the Standard Model. The branching fractions of the semileptonic decays $D \to \pi l\nu$ and $D \to Kl\nu$ have recently been analysed by the CLEO collaboration [1, 2], which renewed the interest in the theoretical study of these processes.

The knowledge on the form factors can be further improved using analyticity and unitarity techniques, in particular, the method of unitarity bounds [3, 4]. Using this framework, one can obtain bounds on the form factors using as input an integral of the modulus squared of the form factor along the unitarity cut, known from unitarity and a dispersion relation for a suitable correlator of the same current, which can be evaluated by Operator Product Expansion (OPE) in the spacelike region. Using standard mathematical techniques, one can then correlate the values of the form factor and its derivatives at different points inside the analyticity domain. A review of the method was presented recently in [5]. In this work, we apply the above technique to the $D\pi$ form factors. The work is motivated in part by the progress in perturbative QCD calculations, which now yield the heavy-light correlators of interest to order α_s^2 [6] for a massless light quark and partly form the new experimental analyses of the semileptonic D decays.

We investigate the shape parameters entering the Taylor expansion around $t = 0$,

$$f_k(t) = f_k(0)\left(1 + \lambda'_k \frac{t}{M_\pi^2} + \frac{1}{2}\lambda''_k \frac{t^2}{M_\pi^4} + \cdots\right), \quad k = +, 0, \tag{1}$$

and derive allowed ranges for the slopes λ'_k and the curvatures λ''_k. Here, $f_+(t)$ and $f_0(t)$ are respectively the vector and scalar form factors and are related to each other by the relation,

$$f_0(t) = f_+(t) + \frac{t}{M_D^2 - M_\pi^2} f_-(t), \quad t = q^2 = (p - p')^2. \tag{2}$$

We further find points on the real t-axis and isolate regions in the complex t-plane where zeros of the form factors are excluded. The knowledge of the zeros are of interest, for instance, for the dispersive methods based on phase (Omnès-type representations) and for testing specific models of the form factors.

2. Review of the method

To start with, we have the heavy-light invariant amplitudes $\Pi_+(q^2)$ and $\Pi_0(q^2)$ defined by the vector-vector correlation function,

$$-(q^2 g^{\mu\nu} - q^\mu q^\nu)\Pi_+(q^2) + q^\mu q^\nu \Pi_0(q^2) = i \int d^4x e^{iqx} \langle 0|TV^\mu(x)V^{\nu\dagger}(0)|0\rangle, \tag{3}$$

where $V_\mu = \bar{d}\gamma_\mu c$.

We then consider the moments of the invariant amplitudes at $q^2 = 0$,

$$\chi_k^{(n)} \equiv \frac{1}{n!} \frac{d^n}{(dq^2)^n} \left[\Pi_k(q^2)\right]_{q^2=0}, \quad k = +, 0, \tag{4}$$

which satisfy dispersion relations of the form,

$$\chi_k^{(n)} = \frac{1}{\pi} \int_{t_+}^{\infty} dt \frac{\mathrm{Im}\Pi_k(t + i\epsilon)}{t^{n+1}}, \quad k = +, 0, \tag{5}$$

where $t_\pm = (M_D \pm M_\pi)^2$.

It is well-known from QCD that the amplitude $\Pi_+(q^2)$ satisfies a once subtracted dispersion relation, while for $\Pi_0(q^2)$ an unsubtracted relation converges. The quantities $\chi_+^{(n)}$ and $\chi_0^{(n)}$ are hence defined for $n \geq 1$ and $n \geq 0$, respectively. These relations are connected to the form factors $f_+(t)$ and $f_0(t)$ by means of unitarity,

$$\mathrm{Im}\Pi_+(t + i\epsilon) \geq \frac{3}{2} \frac{1}{48\pi} \frac{[(t - t_+)(t - t_-)]^{3/2}}{t^3} |f_+(t)|^2, \tag{6}$$

$$\mathrm{Im}\Pi_0(t + i\epsilon) \geq \frac{3}{2} \frac{t_+ t_-}{16\pi} \frac{[(t - t_+)(t - t_-)]^{1/2}}{t^3} |f_0(t)|^2, \tag{7}$$

respectively for the vector and scalar form factor, which hold for $t > t_+$. The above expressions are the unitarity sums for the spectral functions, restricted to the contribution of the $D\pi$ states in the isospin limit.

OPE on the other hand allows for the calculation of $\chi_k^{(n)}$ as the sum of perturbative and nonperturbative contributions. While the perturbative parts of the moments of heavy-light correlators for $n \leq 7$ were calculated up to order α_s^2 in [6], the leading non-perturbative contribution of the quark and gluon condensates were obtained from [7], the corresponding expressions of which are presented in [8]. For the vector and scalar form factors, we specifically use the dispersion relations for the moments $\chi_+^{(1)}, \chi_+^{(2)}, \chi_+^{(3)}$ and $\chi_0^{(0)}, \chi_0^{(1)}, \chi_0^{(2)}$ respectively, as a result of which we obtain, for each form factor, a family of three different constraints. The final allowed domain for the parameters of interest will be the intersection of the three individual domains.

From the dispersion relations (5) and the unitarity conditions given in eqns.(6) and (7), it follows that each form factor $f_k(t)$ satisfies a set of integral inequalities written as,

$$\frac{1}{\pi}\int_{t_+}^{\infty} dt\ \rho_k^{(n)}(t)|f_k(t)|^2 \le \chi_k^{(n)}, \qquad k = +, 0, \tag{8}$$

where the weights $\rho_k^{(n)}(t)$ are the product of $1/t^{n+1}$ with the phase space factors entering the unitarity relations. Using the fact that the form factors $f_+(t)$ and $f_0(t)$ are analytic functions in the complex t-plane cut along the real axis from t_+ to ∞, we apply standard techniques to derive from eqn.(8) constraints on their values, in particular on the shape parameters and on the regions in the real and complex energy plane where zeros are excluded. Specifically, we cast the problem into a canonical form via a conformal map, and derive a determinant which is central to our investigations for obtaining bounds on the shape-parameters [5, 8, 9] and for finding the exclusion regions of the zeros. The quantity $f_k(0)$ defined in eqn.(1) goes as an important input to our work. Besides, additional information for the scalar form factor provided by a low-energy soft-pion theorem namely the Callan-Treiman relation [10], is exploited to obtain more stringent bounds on the shape parameters of the scalar form factor For the $D\pi$ case, the corresponding expression is given below,

$$f_0(\Delta_{D\pi}) = f_D/f_\pi, \tag{9}$$

where $\Delta_{D\pi} = M_D^2 - M_\pi^2$ is the relevant Callan-Treiman point and f_D and f_π are the meson decay constants.

Even more stringent constraints on the shape parameters can be obtained if we also have some information on $f_k(t)$ on the unitarity cut, in particular, if the phase $\delta_k(t)$ defined as,

$$f_k(t + i\epsilon) = |f_k(t)|e^{i\delta_k(t)}, \quad k = +, 0, \tag{10}$$

is known along a low-energy interval, $t_+ \le t \le t_{\text{in}}$. This information on the phase can be implemented using the technique of generalized Lagrange multipliers and involves the solution of an integral equation. A review of the method and more references are given in [5].

3. Inputs and Results

The perturbative and nonperturbative contributions to the moments χ_k^n have been tabulated in [8]. The value of the inputs used for $f_k(0)$ and $f_0(\Delta_{D\pi})$ which reads $0.67^{+0.10}_{-0.07}$ and 1.58 ± 0.07 respectively are obtained from a recent work based on analyticity [11]. We roughly estimate the phase of the $D\pi$ form factors from the masses and widths of the resonances dominant at low energies, namely, from the relativistic Breit-Wigner parametrization given by,

$$\delta(t) = \arctan\left(\frac{M_R\Gamma(t)}{M_R^2 - t}\right), \tag{11}$$

where M_R is the mass and $\Gamma(t)$ the energy-dependent width -

$$\Gamma(t) = \left(\frac{q(t)}{q(M_R^2)}\right)^{2J+1}\frac{M_R}{\sqrt{t}}\Gamma_R, \tag{12}$$

written in terms of the angular momentum J, the width Γ_R and the centre of mass momentum $q(t) = \sqrt{(t - t_-)(t - t_+)/4t}$.

The central values of the masses and widths of the lowest charged $D\pi$ vector and scalar resonances, namely the D^* and D_0^* respectively are listed in [12]. We note that the vector resonance D^* is very close to the threshold so that a reasonable expression of the phase cannot be obtained from a Breit-Wigner parametrization, but in the scalar case we can assume that the phase $\delta_0(t)$ is reliably described by the expression (11) with $J = 0$. For our analysis, we consider the phase up to the point $t_{\rm in} = (2.6\,{\rm GeV})^2$, close to the first inelastic $D\eta$ channel opening at 2.42 GeV.

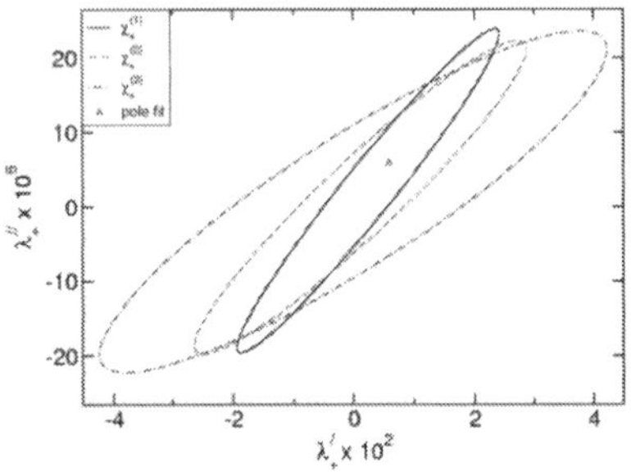

Figure 1. Constraints on the slope and curvature of the vector form factor obtained using as input different moments of the correlator. The point indicates the slope and curvature of the pole ansatz given in [13].

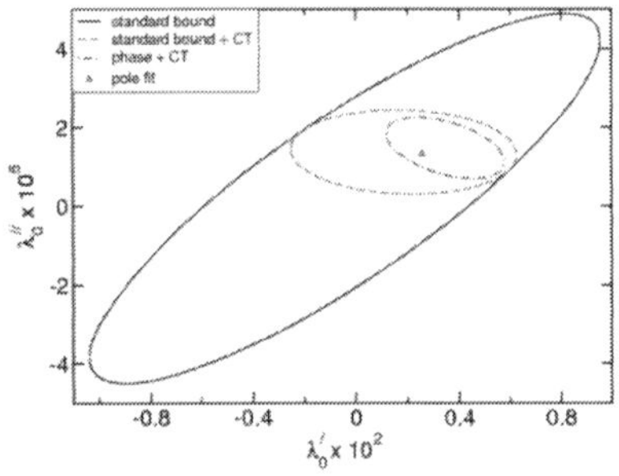

Figure 2. Constraints on the slope and curvature of the scalar form factor obtained with the moment $\chi_0^{(0)}$, from the standard bound and by including the phase and the low-energy theorem. The point indicates the slope and curvature of the pole ansatz given in [13].

The results of our analysis are presented in Fig.1, where the the interior of the ellipses represent the allowed domains in the slope-curvature plane for the vector form factor, obtained with three moments $\chi_+^{(n)}$ of the vector correlator with just $f_+(0)$ and no other input. While the smallest or in other words the best domain is obtained from the lowest moment, we find that the higher moments tend to slightly reduce the domain, since one must take the intersection of all the domains in order to fulfill simultaneously the constraints. We also indicate the slope and curvature of the simple pole ansatz [13], which is seen to satisfy the unitarity constraints coming from our analysis. In Fig.2, we show the bounds derived from the lowest moment $\chi_0^{(0)}$ of the scalar form factor where we also use information on the phase and the soft pion theorem. The effect of these additional constraints can clearly be seen in the figure. We also show the slope and curvature from the pole ansatz [13] which satisfies our constraints. It may be noted that

Journal of Physics: Conference Series **374** (2012) 012013 doi:10.1088/1742-6596/374/1/012013

the above bounds were obtained with the central values of the input parameters. By varying simultaneouly all the input values we can obtain more conservative regions, which are slightly larger than the domains shown in the figures.

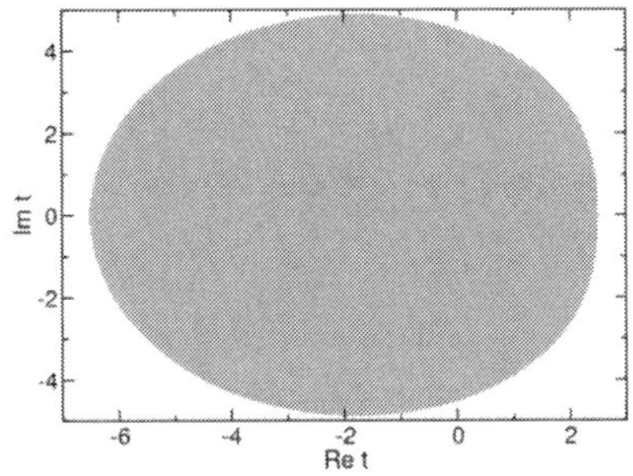

Figure 3. Domain without zeros for the vector form factor, obtained from the lowest moment $\chi_{+}^{(1)}$ and the input $f_{+}(0)$.

We now present results on the zeros of the scalar and vector $D\pi$ form factors. We have considered in the following only the constraints imposed by the lowest moments, corresponding to $n = 1$ in the vector case and $n = 0$ in the scalar one, using $f_{+}(0)$ as input quantity. For the vector form factor,real zeros are excluded in the range $(-1.0, 0.80)\,\mathrm{GeV}^2$. while for the scalar case it is $(-2.51, 1.55)\,\mathrm{GeV}^2$ for the standard bounds, With the inclusion of the low-energy constraint in the case of the scalar form factor, the range is increased to $(-3.55, 3.89)\,\mathrm{GeV}^2$. We present in Figs. 3 and 4 the regions of excluded zeros in the complex t-plane for the vector and scalar form factors respectively. In the scalar case the zeros are excluded in a larger domain if the low-energy constraint is also imposed.

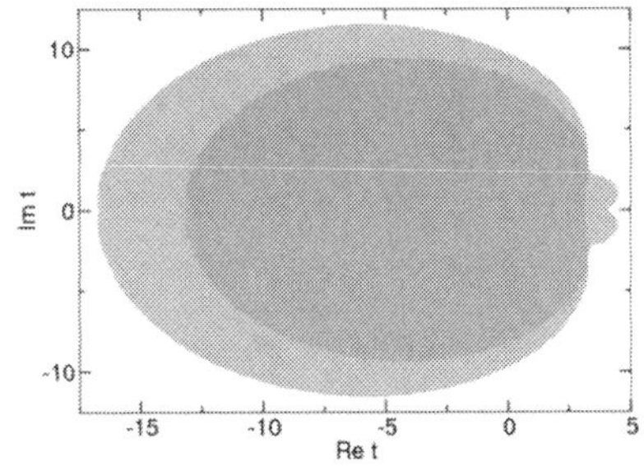

Figure 4. Domain without zeros for the scalar form factor, obtained from the lowest moment $\chi_{0}^{(0)}$ and the input $f_0(0)$ (smaller region) and using in addition the CT constraint (larger region).

4. Conclusions

In this work, we have used analyticity and unitarity techniques to derive information on the $D\pi$ form factors, which might be of interest for the determination of the element $|V_{cd}|$ of the CKM matrix. We have applied the formalism of unitarity bounds and have derived a family of constraints to be satisfied by the shape parameters of both the scalar and vector form factors at

$t = 0$ and explored regions on the real axis as well as in the complex plane where the form factors cannot have zeros. The exclusion regions for the zeros that we have isolated basically cover a significant part of the entire low energy region. They are of practical interest, for instance, for the dispersive methods based on phase (Omnès-type representations) and for testing specific models of the form factors. The technique can be extended also for deriving new parameterizations for the $D\pi$ form factors in the semileptonic region, similar to those proposed in [14] for the semileptonic $B \to \pi l \nu$ decay. The new parametrizations properly implement the singularities related to the lowest charmed resonances and are useful for the description of the semileptonic data (see Ref.[8] for more details).

References

[1] J.Y. Ge *et al.* [CLEO Collaboration], Phys. Rev. D **79**, 052010 (2009) [arXiv:0810.3878 [hep-ex]].
[2] D. Besson *et al.* [CLEO Collaboration], Phys. Rev. D **80**, 032005 (2009) [arXiv:0906.2983 [hep-ex]].
[3] S. Okubo, Phys. Rev. D **3** (1971) 2807.
[4] V. Singh and A.K. Raina, Fortsch. Phys. **27** (1979) 561.
[5] G. Abbas, B. Ananthanarayan, I. Caprini, I. Sentitemsu Imsong and S. Ramanan, Eur. Phys. J. A **45**, 389 (2010) [arXiv:1004.4257 [hep-ph]].
[6] K.G. Chetyrkin and M. Steinhauser, Eur. Phys. J. C **21**, 319 (2001) [arXiv:hep-ph/0108017].
[7] L. Lellouch, Nucl. Phys. B **479**, 353 (1996). [arXiv:hep-ph/9509358].
[8] B. Ananthanarayan, I. Caprini and I. S. Imsong, Eur. Phys. J. A **47**, 147 (2011) [arXiv:1108.0284 [hep-ph]].
[9] B. Ananthanarayan, I. Caprini, I.Sentitemsu Imsong, Phys. Rev. D**83** 096002 (2011) [arXiv:1102.3299 [hep-ph]]
[10] C.G. Callan and S.B. Treiman, Phys. Rev. Lett. **16**, 153 (1966).
[11] A. Khodjamirian, C. Klein, T. Mannel and N. Offen, Phys. Rev. D **80**, 114005 (2009) [arXiv:0907.2842 [hep-ph]].
[12] K. Nakamura *et al.* [Particle Data Group], J. Phys. G **37**, 075021 (2010).
[13] D. Becirevic and A.B. Kaidalov, Phys. Lett. B **478**, 417 (2000) [arXiv:hep-ph/9904490].
[14] C. Bourrely, I. Caprini and L. Lellouch, Phys. Rev. D **79**, 013008 (2009), (E) D**82** 099902 (2010) [arXiv:0807.2722 [hep-ph]].

Journal of Physics: Conference Series **374** (2012) 012014 doi:10.1088/1742-6596/374/1/012014

Development of detector and physics analysis tools for PANDA experiments at FAIR

B. J.Roy, V.Jha, H.Kumawat, A.Chatterjee, S.Kailas
Nuclear Physics Division, B.A.R.C., Mumbai- 400 085, India

Manoj Jadhav, Raghava Varma
Department of Physics, Indian Institute of Technology Bombay, Mumbai, India

C. Schwarz, H.Orth, L. Schmitt, K.Peters
HAD-1, GSI, Darmstadt, Germany

Abstract.
The PANDA collaboration at FAIR facility will address a rich variety of physics programs for studies of strong interaction and perturbative QCD. In India, initiatives have been taken up for the design and construction of some components of the PANDA detector. A silicon photo multiplier based scintillation detector is being planned to be built in India in collaboration with GSI, Darmstadt. Simulation studies and R&D work are being carried out in designing this detector which can provide a fast trigger signal for the PANDA experiments. The other component of the PANDA detector which the Indian group is interested to develop is luminosity monitor detector(LMD). Currently two possible solutions are being considered for LMD: (i)a double sided silicon micro strip detector and (ii) radiation hard Gallium Nitride(GaN)/Diamond sensors with a pitch of 50 μm. Simulations are being carried out. The analysis tools of vertex and kinematic fitting have been developed within the software framework of Pandaroot. The vertex fitting for the primary and secondary vertex based on kinematic constraints have been implemented for the full decay tree reconstruction. Simulation studies have been performed using a few benchmark channels to evaluate the performance of these methods.

1. Introduction

The PANDA experiment at FAIR is one of the key projects at FAIR(Facility for Anti proton and Ion Research), Darmstadt, Germany. The High Energy Storage Ring - HESR- at FAIR will provide intense anti proton beams of momentum in the range of 1 - 15 GeV/c. One of the main topics of study at PANDA[1] is charmonium spectroscopy and search for glueballs and hybrids in the charmonium mass region. Glueballs are excited state of pure gluons whereas hybrids are resonances consisting of a quark, an antiquark and excited gluons. These glueballs and hybrids, in contrast to normal mesons and other fermion - anti fermion systems, can have spin-exotic quantum numbers as the gluons carry additional degrees of freedom. Lattice QCD calculations predict a whole spectrum of bound gluon states and hybrids, whereas experimental information is scarce. Bound gluon systems offer a unique way to study one of the long standing problems in hadron physics- the origin of mass of strongly interacting particles. As we know, the Higgs mechanism might be responsible for creation of mass of elementary particles. But for proton, the Higgs mechanism accounts for only a few percent of its mass while the rest of it, is believed to be, created by the strong interaction. The gluons which are massless carry color charge i.e., charges of strong interaction and thereby interact strongly between themselves. This mutual gluon attraction may allow formation of meson-like bound states of gluon even if no quarks are present. Thus glueballs acquire mass which arises solely from the strong interaction. At PANDA it should be possible to search for all glueballs and hybrids upto mass 5.5 GeV and make a high precision study by detecting both hadronic and electromagnetic decay modes revealing their true nature.

For the charmonium spectroscopy (both hidden and open charm), the goal at PANDA is to make a comprehensive measurement of spectroscopy of charmonium systems and hence to provide a detailed experimental information of the QCD confining forces in the charm region to complement theoretical investigation. It should be highlighted that most of the charmonium states have very narrow width and low production cross section, only in experiments like PANDA with its phase space cooled anti proton beam and high luminosity it is possible to obtain their mass, width and a systematic information of decay modes.

Investigation of medium modification of hadron mass in nuclear matter is another interesting topic that has been planned at PANDA. These studies, so far, have focused on the light quark sector eg., pion and kaon mesons due to the limitation in available energy. The in-medium pion mass is observed to shift as compared to its vacuum value and pion-nucleus potential has been deduced from experimental studies of deeply bound pionic atoms at GSI[2]. For kaons, repulsive mass shifts for K+ and attractive mass shifts for negative kaons are observed experimentally[3]. At PANDA, the medium modification of hadron mass studies can be extended to the charm sector. For example, experiments are being planned to study sub-threshold production of D-Dbar mesons in anti proton - nucleus collision. If the D meson mass gets reduced in the nuclear environment, the in-medium DDbar threshold would be lowered resulting in an enhancement of production cross section at sub-threshold energies.

The PANDA collaboration has an extensive program on single and double hyper nuclear physics. Such studies will provide us information on hyperon-nucleus and hyperon-hyperon interaction, the present knowledge of which is very limited. So far only few double hyper nuclear events have been observed. Hypernuclei are nuclear systems in which one or more nucleons are replaced by hyperons. A hyperon in a nucleus is not constrained by the Pauli exclusion principle, unlike a neutron or a proton. As a result it can populate all possible nuclear states which are not accessible otherwise. Certainly studies of hyper nuclei will provide a sensitive probe to the nuclear structure.

The PANDA detector is a complex detector designed to achieve 4π acceptance, high resolution for tracking, particle identification, calorimetry, and high rate capabilities (about 2 x 10^7 annihilations / s). The details of the PANDA detector are described elsewhere[1]. The Indian group in the PANDA collaboration is involved [4] in detector construction, detector and physics simulation and development of analysis tools for PANDA experiments. In collaboration with the German group at GSI, Darmstadt and at FZ-Juelich, R&D work has been initiated. The present article reports on the initial simulation studies and R&D work towards the design and construction of i)silicon photomultiplier(SiPM) based fast scintillation detector known as SciTil detector and ii) luminosity monitor detector(LMD). The development of algorithm for vertex reconstruction that will be used in physics analysis has also been described.

2. The SciTil hodoscope detector

The detector, SciTil, will serve for precision time measurements for triggering and determination of time-of-flight. The detector, in the present design, consists of about 5700 scintillator tiles readout by Silicon Photomultiplier (SiPM) and will be mounted in the space between the electromagnetic calorimeter (EMC) and DIRC Cherenkov detector (DIRC stands for Detection of Internally Reflected Cherenkov light). The concept of SciTil provides the use of minimum material (so as not to deteriorate the performance of other surrounding detectors) and a good spatial resolution due to its granularity. The timing detector concept is based on 2.85 x 2.85 x 0.5 cm^3 scintillator tiles matching the front face of the calorimeter crystals. The scintillation photons produced in the scintillator tiles due to the passage of charged particles will be collected and readout by SiPMs. The number and size of the SiPMs and position of SiPM that will be coupled to the tile will be optimized based on a detailed simulation and R&D studies. In addition to timing and position information, the hodoscope will allow clean detection of gamma-conversions in front of the EMC in particular within the region of DIRC. In addition, it is best suited for discrimination between charged and neutral particles. The conceptual design[5] of the SciTil is shown in the Figs.1 and 2.

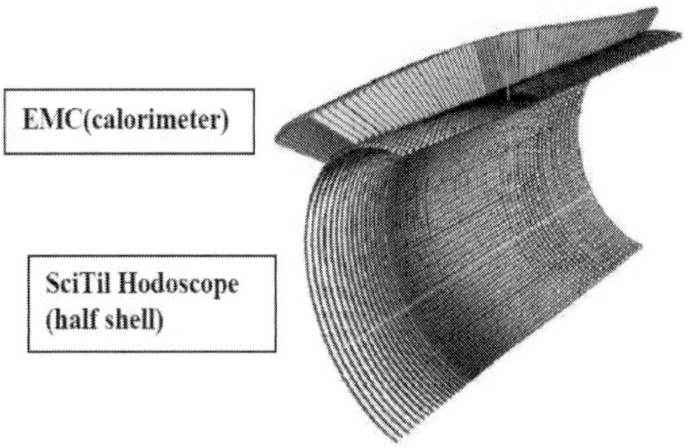

Figure 1. A conceptual design of half shell of the SciTil is shown along with the EMC crystal.

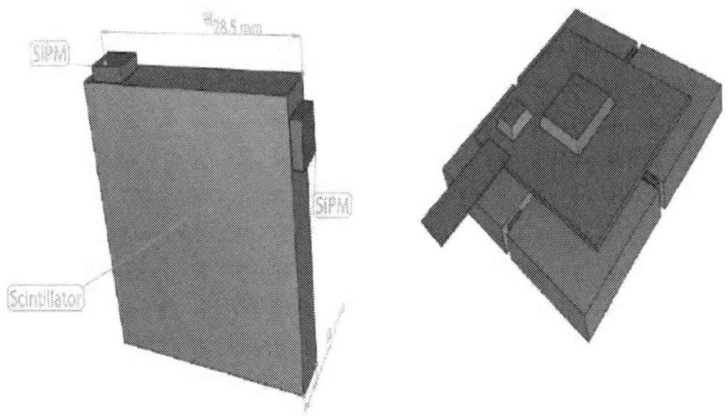

Figure 2. (left) Design view of a single scintillator tile with SiPM attached. (Right) a quad SciTil module with planned PCB readout.

2.1. R&D studies with Silicon Photomultiplier

Silicon photomultiplier (SiPMs) are very new type of photon counting devices that show great promise to be used as detection device in combination with scintillators/Cherenkov radiators. SiPM is essentially an avalanche photo-diode operated in limited Geiger mode. The SiPM module is a photon counting device capable of low light level detection. It is essentially an opto-semiconductor device with excellent photon counting capability and possesses great advantages over the conventional PMTs because of low voltage operation and insensitivity to magnetic fields. In many of the high energy physics experiments, the photon sensors are required to operate in high magnetic fields precluding the use of conventional PMTs. This problem can be overcome with the use of SiPMs. SiPM operating in Geiger mode, a very large gain ($\sim 10^6$), magnitude of which is determined by the internal diode capacitance and applied over-bias voltage, comparable to that of PMTs can be achieved. A SiPM consists of matrix of micro cells (known as pixels), typically between 100 and 10000 per mm^2. Each micro cell acts as digital device where the output signal is independent of the number of photons absorbed. When all the cells are connected in parallel, the SiPM becomes an analog device thereby allowing the number of incident photons to be counted. Detailed R&D studies with SiPM are needed for the use of this device in PANDA experiments. SiPM test facility has been developed both at GSI, Darmstat and at the Nuclear Physics Division, BARC. Mumbai. Several commercially available SiPMs (Hamamatsu make MPPC with active area 1 x 1 mm^2 and different pixel density, Zecotek make MAPD3N with dimension 3 x 3 mm^2 and pixel density as large as 15000) have been tested for their performance[6, 7, 8, 9] and comparison with other photon counting devices. Studies are also being done with an array (2 x 2) of silicon photomultiplier, Hamamatsu make, having a total active area 6 x 6 mm^2.

During the measurement, all devices were mounted in a light tight box and were illuminated by a pico-second pulsed diode laser (make Pico-Quant) of 660 nm wavelength as well as LED with $\lambda = 460$ nm. In order to be able to distinguish between single and multi-photon peaks, the laser intensity was controlled. The voltage and current on the SiPM were measured by high precision multimeter and I-V characteristics of the photo-diode was studied. The preamplifier used was "Photonique SA" make of two varieties: one with high gain(20x.....60x) but relatively slow rise time(5ns) and the other one with lower gain(10x.....20x) but having faster rise time ($\sim$ 700ps). We have also measured the particle detection efficiency (PDE) of SiPM as a function of wavelength of the incident photons. For this, a monochromator that spans wavelength from 200 to 800 nm was used. Different intensity filters were used for light intensity attenuation. The photo-sensitivity of different SiPMs was normalized with a PIN diode which itself was calibrated by the producer. The dark count of the MPPCs (Multi Pixel Photon Counter) were measured at 0.5thr and 1.5thr and found to be in agreement with the specifications provided by the supplier.

In fig.3, we plot the photon detection efficiency (PDE) distribution as a function of the wavelength λ for Zecotek make MAPD3N and Hamamatsu make MPPC. For MAPD3N, the distribution has been normalized with PDE=24.5% at $\lambda = 450$ nm[10] and for MPPC, PDE=32.4% at $\lambda = 450$ nm[11]. It is to be noted that the present data for MPPC shows much broader distribution extended over larger wavelength range as compared to the report of Hamamatsu.

For the SciTil hodoscope, in order to optimize the geometry of SiPM and light collection efficiency, simulations are being performed with a Monte Carlo simulation program SLitrani. SLitrani stands for light transmission in anisotropic media, with 'S' having the meaning of super. It is a general purpose Monte Carlo program that simulates light propagation in isotropic media (it can also be used for anisotropic media) and is built upon ROOT. The main emphasis of this hodoscope detector is to provide very fast timing ($\sim$ 100 - 200 ps). A detailed simulation studies

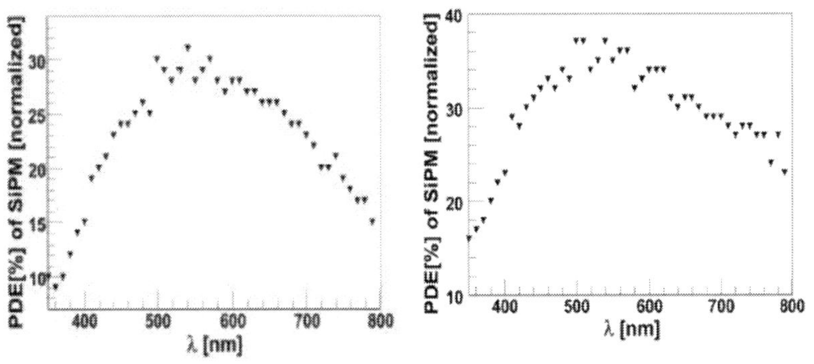

Figure 3. left part) Photon detection efficiency distribution of MAPD3N as a function of wave length normalized with data from ref.[10] at $\lambda = 450$ nm. (Right part) Photon detection efficiency distribution of MPPC normalized with data from ref.[11]at $\lambda = 450$ nm.

and prototype development work are in progress[12] to achieve this sub-nanosecond timing.

3. Luminosity Monitor

The basic concept of the luminosity monitor is to reconstruct the angle (and thus the momentum transfer 't') of the scattered anti protons in the polar angle range of 3-8 mrad with respect to the beam axis. Due to the large transverse dimensions of the interaction region when using the pellet target, there is only a weak correlation of the position of the anti proton at e.g. z = +10.0 m to the recoil angle. Therefore, it is necessary to reconstruct the angle of the anti proton at the luminosity monitor.

As a result the luminosity monitor will consist of a sequence of four planes of double sided silicon strip detectors located as far downstream and as close to the beam axis as possible. The planes are separated by 10 cm along the beam direction. Each plane consists of 4 sensors arranged radially to the beam axis. Four planes are required for sufficient suppression of redundancy and background. The use of 4 sensors (up, down, right, left) in each plane allows systematic errors to be strongly suppressed.

The Sensors are Trapezoidal in shape with dimensions 3.35 cm × 7.8 cm × 7 cm. The thickness of each sensor is 150 μm and pitch of 50 μm. Strips are oriented parallel to side walls of trapezoids and the overlapping of front side strips with strips on back side form a diamond shape structure. Figure 4 and 5 show conceptual design of luminosity sensor and graphical 3D-view, respectively, of the LMD detector.

The silicon wafers will be placed inside the vacuum chamber to minimize scattering of the anti protons. The acceptance for the anti proton beam in the HESR is about 3 mrad, corresponding to the 89 mm inner diameter of the beam pipe at the quadrupole position which is located at about 15 meter downstream of the interaction point. The luminosity monitor can be located between z = +10 m and z = +12 m downstream of the target. At this distance from the target point, the luminosity monitor needs to measure particles at a radial distance of between 3 and 8 cm from the beam axis. As simulations show, at a beam momentum of 6.2 GeV/c the proposed detector measures anti protons elastically scattered in the range 0.0006 GeV^2 < -t < 0.0035 GeV^2, which spans the Coulomb-nuclear interference region. Based upon the granularity of the readout the resolution of t could reach $\sigma_t \approx 0.0001\ GeV^2$. In reality this value is expected to degrade to $\sigma_t \approx 0.0005\ GeV^2$ when taking small-angle scattering into account. At the nominal

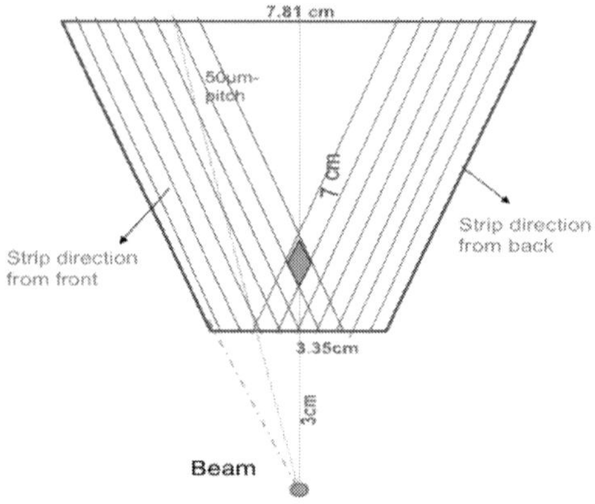

Figure 4. A schematic diagram of silicon sensor for LMD detector showing strips on both sides.

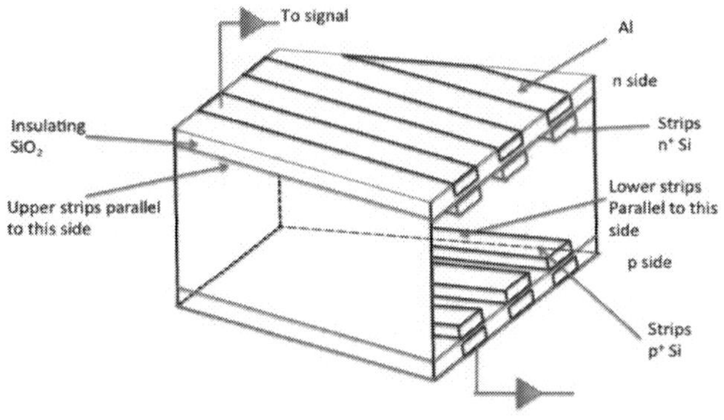

Figure 5. Same as Fig.4 but in 3D view.

PANDA interaction rate of $2 \times 10^7/s$ there will be an average of 10 kHz$/cm^2$ in the sensors. In comparison with other experiments an absolute precision of about 3% is considered feasible for this detector concept at PANDA, which will be verified.

3.1. Radiation Induced Defects

Damage caused by radiation can be divided into two groups, surface damage and bulk damage. Due to the interest of electronic industry, the surface damage is better understood and can be controlled to certain extent by proper design and manufacturing process. Surface damage mainly manifests as charge accumulation in the oxide and subsequent breakdown. Charge accumulation at the silicon-oxide interface significantly decreases the inter-strip resistance of micro strip detectors. Therefore the signal has to be collected very quickly to minimize the loss due to leakage to neighboring strips.

Charge trapping at defect centers and Enhancement in generation or recombination currents causes reduction in charge collection efficiency (CCE). Electrostatic potential within the device changes due to modification in space-charge profile. Also the depletion bias voltage is a strong function of radiation fluences.

To overcome all these defect the simulation of GaN and Diamond sensors is done to build the radiation hard detector. Gallium Nitride (GaN) is Binary III/V direct band gap semiconductor detector with energy gap of 3.4 eV (average e-h creation energy 8-10 eV) and energy gap of a Diamond is 5.5 with avrage e-h creation energy 13 eV. It can be used at high temperature making it promising material for application in detection of ionizing radiation. In table 1, the properties of some wide band gap materials like GaN and Diamond are listed.

Property	Diamond	Si	GaN
E_g (eV)	5.5	1.12	3.39
μ_e ($cm^2 V s^{-1}$)	1800	1500	1000
μ_h ($cm^2 V s^{-1}$)	1200	450	30
e-h energy (eV)	13	3.6	~8-10
Density ($g\ cm^{-3}$)	3.52	2.33	6.15
Radiation length, χ_0 (cm)	12.2	9.4	2.7
e-h pairs/χ_0 ($106\ cm^{-1}$)	4.4	10.1	~2-3

Table 1. Properties of wide band gap materials and Silicon

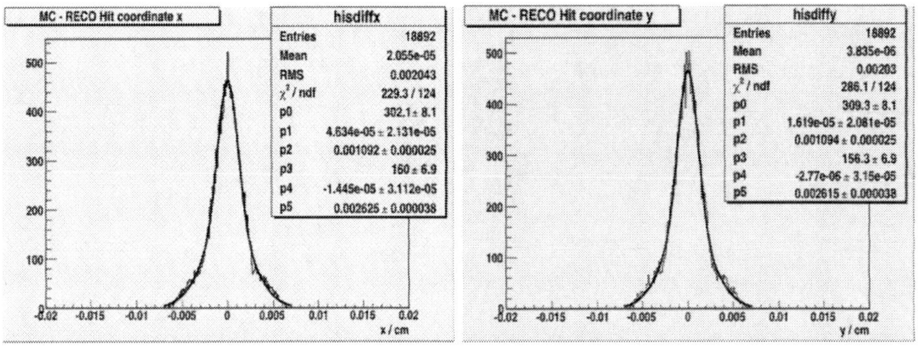

Figure 6. The resolution of the reconstructed hit position after clustering with respect to the simulated value along x and y direction.

3.2. Technology

To form Double Sided Strips from GaN or Diamond material the growth technology is needed. GaN layer deposition on both sides polished sapphire substrate needs Atomic Layer Deposition (ALD) growth setup. Lithography technique can be implemented to form the strips of GaN on Sapphire.

3.3. Simulation and Analysis

The simulation is done using Pandaroot environment and particle tracking through the complete PANDA detector is done by using the GEANT4 transport code. The Monte Carlo simulation is done for forward going anti proton with energy 8.9 GeV and single particle event generator is used to create events. Digitization which models the signals of the individual detectors and their processing in the front-end-electronics (APV25) provides detector response of interaction with incident particle.

The LMD provides very precise space point measurements as a basis for the track and vertex reconstruction. The hit resolution of individual LMD measurements is shown in Fig.6. along x and y direction. The distribution in Fig. 6 shows the difference between the reconstructed position on the sensor and the generated Monte Carlo value. The contribution from hits where two strips are fired (i.e. multiplicity = 2) are more preferable as it gives precise measurement of hit position. Dividing total energy deposited by energy required to create one e-h pair gives total charge create by hit. Fig 7 shows effect of solenoid field on particle propagation, for

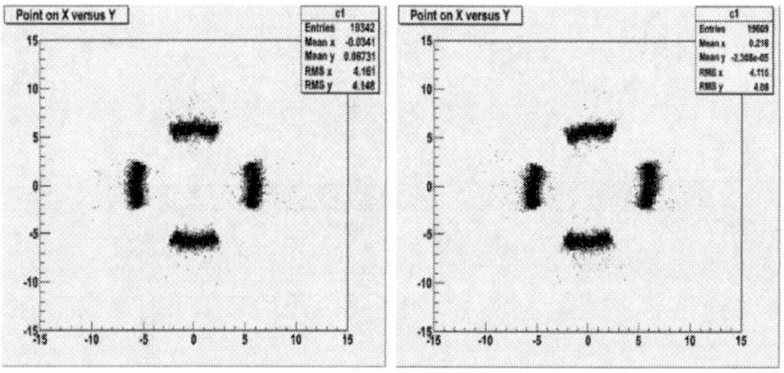

Figure 7. Simulation results showing the effect of solenoid magnet field. Left panel shows the result when the magnetic field is off and the right panel with field on

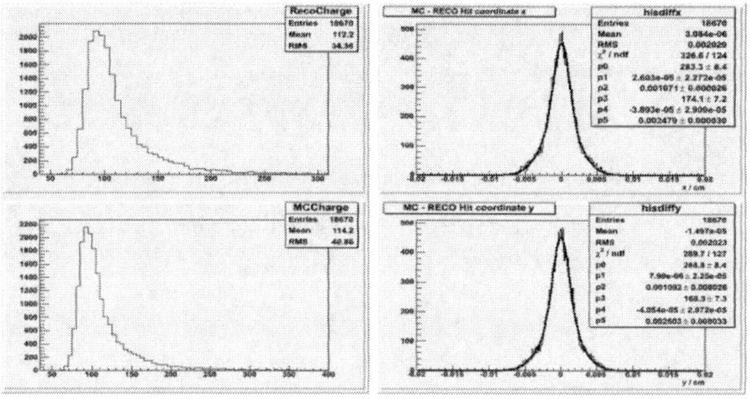

Figure 8. Simulation result for GaN detector, (left panel) Charge deposition and (right panel) Position Resolution

anti-proton of momentum 3.5 GeV/c, at fixed theta 5.2 mrad. Solenoid field gives rise to shift in θ towards beam axis for anti-proton. Further study of the effect due to solenoid magnetic field is being carried out. We have also performed simulation for radiation hard sensor Gallium Nitride GaN. Initial simulation results for energy distribution and position resolution are shown in Fig.8. We have also plans to carry out further simulation with Diamond material.

4. Physics Analysis Tools

The PANDA detector is a complex detector consisting of various sub detectors for tracking and particle identification with almost complete solid angle coverage.The detector consists of a central spectrometer with a solenoid field and a forward spectrometer with dipole field. The tracking in the central part of PANDA is performed by innermost sub detector, which is a micro-vertex detector of Si pixels and strips followed by a central tracker which will be tracker of straw tubes (STT). The simulation and the physics analysis for the PANDA experiment are performed using the Pandaroot software framework [13]. The implementation of the Pandaroot software framework is based on ROOT [14]. The efficient reconstruction of the primary and secondary vertex is crucial for many topics of simulation and data analysis in the PANDA physics program. The vertex position of a set of tracks can be determined by varying the track parameters such that the error is minimized under the condition that the tracks pass through a common vertex point. In addition, the kinematic information of various particles in a particular decay chain can be used as the constraints, for the better determination of track parameters. This leads to an improvement in the momentum and the mass resolution of the measured particles and a larger signal to background ratio in the analysis of the various physics channels. The tools of the vertex and kinematic fitting have been implemented in the Pandaroot software package.

4.1. Algorithm

The algorithms for the vertex and kinematic fitting is based on the minimization of suitably defined χ^2 function using the constraints [15]. The constraints are incorporated by the Lagrange multipliers method. An iterative χ^2 minimization procedure is employed for the vertex fitting. The calculation uses a track representation in a Cartesian frame for the particle tracks with parameters. The trajectory parametrization consists of a three-dimensional position along the trajectory, the particle momentum at this position and the particle energy. The initial track parameters and the covariance matrix are obtained through the conformal mapping based track finder and subsequent track fitting using a Kalman filter procedure in the Pandaroot software. The Cartesian track parameters are obtained by converting the helix representation track parameters and assigning a mass to the final-state particle.

The constraints used for the vertex fitting are expressed in terms of a vector of equations $\mathbf{H}(\alpha, \mathbf{x}) = \mathbf{0}$, where α is a vector of track parameters for all tracks included in the fit and $\mathbf{x}$ is the vertex position. The constraint equations are linearized by using the first order Taylor expansion around a convenient point (α_A; x_A) leading to equation

$$D(\alpha - \alpha_A) + E(x - x_A) + d = 0 \tag{1}$$

where d is the vector of values of the constraint equations at a suitable expansion point, D is the matrix of partial derivatives of the constraint equations at the expansion point with respect to the track parameters and E is the matrix of the partial derivatives of the constraint equations at the expansion point with respect to the vertex coordinates. The constraint equation is included by Lagrange multiplier variable λ in the χ^2 function which is written as:

$$\chi^2 = (\alpha - \alpha_0)^T V_{\alpha 0}^{-1} (\alpha - \alpha_0) + 2\lambda^T (D\delta\alpha + E\delta x + d) \tag{2}$$

where the terms include the contributions from errors in track parameters with respect to initial values α_0 with covariance matrices $V_{\alpha 0}$ and the constraints , respectively. The values of improved

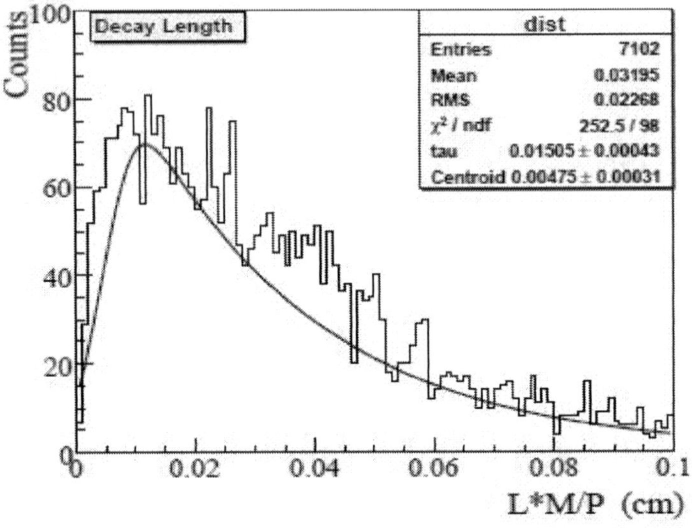

Figure 9. Proper time distribution in terms of decay length for the reconstructed $D^{\pm}$s candidates with the reconstructed mass (M) and 4-momentum (P).

track parameters α and the vertex position x that satisfy the given set of constraints can then be found by minimizing the χ^2 equation with respect to α, x and λ.

4.2. Test simulations

The performance of the vertex fitter has been tested by reconstructing the short lived decay particles $D_s^{\pm}$ mesons using the simulated events from the reaction $p\bar{p} \rightarrow D_s^{\pm}D_s^{*}(2317)^{\mp}$. The decay particle $D_s^{\pm}$ has been reconstructed in the $D_s^{\pm} \rightarrow \phi\pi^{\pm}$, $\phi \rightarrow K^+K^-$ decay modes. The proper lifetime distribution for the reconstructed $D_s^{\pm}$ particles obtained using the 4-momentum obtained from vertex fitting at the vertex point. A fit to the proper time distribution is performed using the function which is the convolution of an exponential and a Gaussian (with the width derived from the fit to the residual of the position distribution) as shown in the Fig. 9. The slope gives a decay length value of $c\tau = 150 \pm 4$ μm which is in good agreement with the PDG value of $c\tau = 147$ μm. The fitting procedure provides the quality of the vertex fit in terms of a normalized χ^2 probability distribution.

In addition to the vertex fit, the possibility to include other kinematic constraints such as mass constraint, total momentum constraint, total energy constraint, four momentum constraints etc. have been developed and tested. Depending on the physics case, one or many constraints can be used for the analysis. For testing the kinematic fit, the simulations have been performed with events generated in the $p\bar{p}$ collisions at the centre of mass energy corresponding to $\psi(2S)$ resonance. After reconstruction of $J\psi$ resonance through the $J/\psi \rightarrow e^+e$ decay channel, a 4-momentum kinematic fit using the energy momentum of the initial $p\bar{p}$ system has been performed. The kinematic fitting leads to improved track parameters for the daughter particles and a better resolution of the mass for the J/ψ resonance as shown in Fig.10.

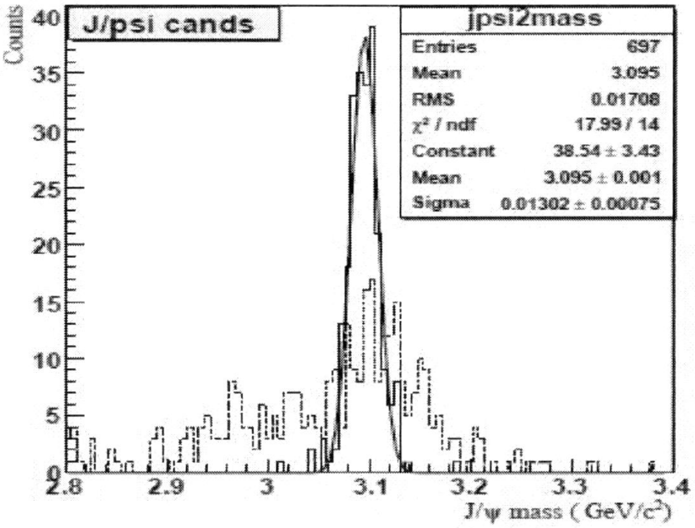

Figure 10. Reconstructed mass of the J/ψ resonance state without (dashed line) and with (solid line) the 4-momentum fit.

5. Summary

The facility for anti proton and ion research(FAIR) will be a unique facility for the investigation of hadron structure and QCD in the charm sector. The Indian hadron physics community has taken a strong initiative to be a part of this exciting physics programme and to contribute to the detector construction, simulation & software development and physics case studies. In collaboration with the research centre GSI, Darmstadt and FZ-Juelich, the Indian group has taken up R&D activities. The initial studies with SiPM show a great promise of this device to be used as photon counter for nuclear physics experiments. For the determination of beam luminosity, double sided silicon micro strip detector as well as radiation hard GaN and Diamond sensors are being considered. The analysis tools of vertex and kinematic fitting have been developed and tested for the studies of PANDA physics programme.

Acknowledgments

We thank the whole India -PANDA collaboration team for their effort and continued interest in this work. The present discussion meeting on the India-PANDA collaboration, which was held at TSH, BARC, Mumbai in November 2011, was sponsored by the BRNS, DAE and DST, India-FAIR coordination centre, Kolkata. We acknowledge the support received from them towards the organization of the meeting.

References

[1] PANDA collaboration- physics performance report, arXiv:0903.3905 and PANDA technical progress report, Gold version, 2005
[2] Geissel H et al, Phys. Rev. Lett. 88, 122301(2002), Phys. Lett.B., 549 (2002) 64 and references therein.
[3] Nekipelov M. et al, Phys. Lett. B, 540 (2002) 207, Rudy Z. et al, Euro. Phys. J A15 (2002) 303
[4] QCD studies with anti-protons at FAIR: Indian participation in PANDA, S.Kailas, B.J.Roy, D.Dutta. V.Jha, R.Varma, Current Science, Vol. 100, No.5 (2011) and references therein.

[5] Proposal for a Scintillation Tile Hodoscope for PANDA, K.Goetzen, H.Orth, G.Schepers, C.Schwarz, A.Wilms (private communication)
[6] SiPM as photon counter for Cherenkov detectors, B.J.Roy, H.Orth, C.Schwarz, A.Wilms, K.Peters, DAE Symp. Nucl. Phys. Vol. 54, 666 (2009)
[7] Study of the spectral sensitivity of G-APDs in the wavelength range from 250 to 800 nm, B.J.Roy et al, , 12th Vienna Conf. on Instrumentation, Feb. 2010, Vienna.
[8] In-beam test of a DIRC Cherenkov radiator with SiPM, B.Kroeck, A.Hayrapetyan, K.Foehl, O.Merle, M.Duren, B.J.Roy, K.Peters, DAE Symp. Nucl. Phys. Vol. 54, 668 (2009)
[9] Development of SiPM based scintillation detector for fast timing application in the PANDA experiment, H.Kumawat, B.J.Roy, V. Jha, U.K.Pal, A.Chatterjee and S.Kailas, DAE symp. Nucl. Phys, Vol.56, 1074 (2011)
[10] D. Renker, PSI (private communication).
[11] N.Anfimov, Dubna (private communication).
[12] Simulations of SiPM based scintillation detector for PANDA, U.K.Pal, B.J.Roy, V.Jha, H.Kumawat, A.Chatterjee and S.Kailas, DAE symp. Nucl. Phys, Vol.56, 1072(2011)
[13] Journal of Physics: Conference Series 119 (2008) 032035.
[14] http://root.cern.ch
[15] P. Avery, Fitting theory write-ups,http://www.phys.ufl.edu/avery/fitting.html(1998).

Search for mesic nuclei at J-PARC

Hiroyuki Fujioka

Department of Physics, Kyoto University, Kitashirakawa-Oiwakecho, Sakyo-ku, Kyoto 606-8502, Japan

E-mail: fujioka@scphys.kyoto-u.ac.jp

Abstract. Experimental programs at J-PARC with regards to $\overline{K}$- and η-nuclear systems are reviewed in this article. Both of them are theoretically expected to exist as bound states, based on the attractive nature of the meson-nucleon interaction. In-medium properties of pseudoscalar mesons may be investigated by looking into these bound states.

1. Introduction

The possibility that a meson and a nucleus may be bound and form a mesic nucleus by the strong interaction is widely discussed. In particular, several experimental observations with regard to antikaon-nuclear bound state (kaonic nuclei) and η-nuclear bound state (η mesic nuclei) have been reported up to now. These mesic nuclei will provide unique information on hadron properties at finite density and hadron-nucleon interaction in the subthreshold region.

In this article, some of experimental programs at J-PARC related to mesic nuclei will be reviewed. The schematic view of the J-PARC hadron experimental facility is shown in Fig. 1. 30 GeV proton beam extracted from the main ring is delivered from the lower left to the production target, from which three secondary beamlines (K1.8, K1.1, and neutral KL beamlines) are separated. The experiments will be performed at either the K1.8 beamline or the K1.8BR beamline branched from K1.8, depending on the required beam momentum. In both the beamlines, secondary particles such as $\pi^{\pm}$ and $K^{\pm}$ are available. The maximum beam momenta are around 2.0 GeV/c and 1.1 GeV/c, respectively.

2. Kaonic nuclei

The mass of the $\Lambda(1405)$ baryon is difficult to explain in a naive quark model, because it is the lightest baryon with $J^P = 1/2^-$, even below the non-strange $N(1535)$ baryon. One probable scenario is that it can be regarded as a $\overline{K}N$ bound state. An attractive interaction between an antikaon and a nucleon enough to form a bound state is reinforced by the $\overline{K}N$ scattering data and the X-ray measurement of kaonic hydrogen by recent experiments (KpX [1], DEAR [2], and SIDDHARTA [3]) which confirmed a repulsive shift of the 1s ground state.

If this is the case, an antikaon can attract more nucleons and antikaon-nuclear bound state may be formed [4, 5]. The $\overline{K}NN$ system is particularly important since it is the simplest system beyond $\Lambda(1405)$. Both experimental and theoretical investigations of this system have been conducted in the last decade.

A variety of calculations of the $\overline{K}NN$ system, using the Faddeev approach or the variational approach, have been performed [6]. The binding energy ranges between $\sim$ 10 MeV and

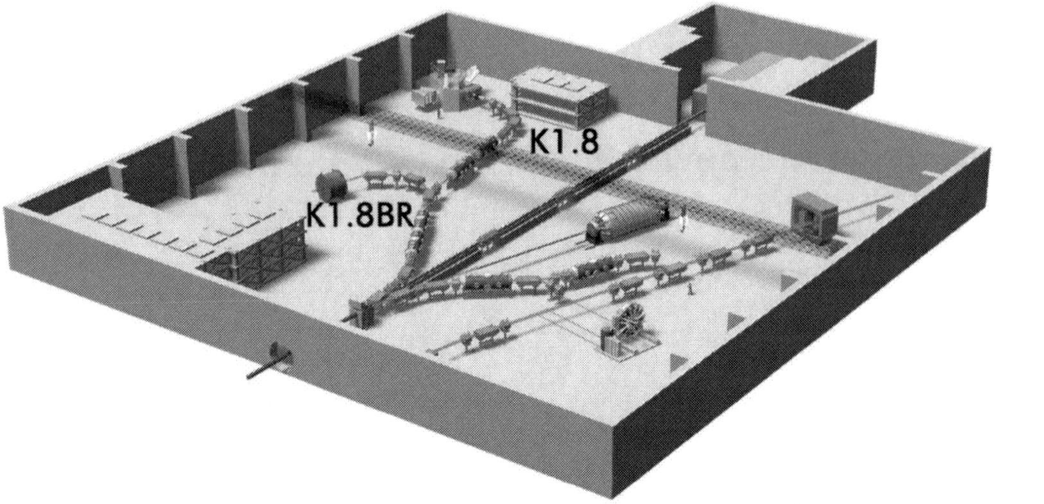

Figure 1. Layout of the J-PARC hadron experimental facility.

$\sim$ 100 MeV, and the decay width is moderately large. The difference mainly comes from the treatment of the $\pi\Sigma$ channel, which couples with the $\overline{K}N$ channel and affects the strength of the $\overline{K}N$ interaction.

Experimental searches for the $\overline{K}NN$ system were carried out by the FINUDA/DAΦNE and the DISTO/SATURNE collaborations. Both the analyses made use of the non-mesonic decay channel of the $\overline{K}NN$ system with the charge +1, i.e. K^-pp, decaying into $\Lambda + p$.

From an experimental point of view, non-mesonic decay of K^-pp states into $\Lambda + p$, instead of mesonic channels, $\Sigma\pi N$ or $\Lambda\pi N$, can be a clean signal in searching for such an exotic state. Many experiments in the past and in the near future focus on this channel, because of the easiness of the detection of charged particles.

The first observation was reported by the FINUDA collaboration[7] at the ϕ-factory DAΦNE, which is an electron-positron collider with $\sqrt{s} = 1.02$ GeV. They detected a pair of Λ and proton from K^- absorption at rest, strongly correlated in the back-to-back direction. Slow and almost monoenergetic K^- beam was produced as the decay product of ϕ into K^+K^-. The invariant mass of the Λp pairs is distributed far below the $K^- + 2p$ threshold (2.370 GeV/c^2), and it is difficult to explain the distribution by the assumption that they originated from a two-nucleon absorption process K^- + "pp" $\rightarrow \Lambda + p$. They interpret Λp pairs may have originated from non-mesonic decay of K^-pp bound states, whose binding energy from the $K^- + 2p$ threshold is $\sim$ 115 MeV and decay width is $\sim$ 67 MeV. However, the distribution may be explained by taking into account final state interaction after the two-nucleon absorption process, as discussed by Magas *et al.* [8] and Pandejee *et al.*[9]

Another indication was reported after a reanalysis of the DISTO experiment, which performed an exclusive measurement of the $p + p \rightarrow p + \Lambda + K^+$ reaction with the incident kinetic energy 2.85 GeV [10]. A broad distinct peak was found in the K^+ missing-mass spectrum and the $p\Lambda$ invariant-mass spectrum, when they select events with large-angle scattered proton and K^+. If the peak corresponds to a K^-pp state, its binding energy and decay width are $\sim$ 105 MeV and $\sim$ 118 MeV, respectively.

At present, the existence of K^-pp bound states observed by the two experiments is not

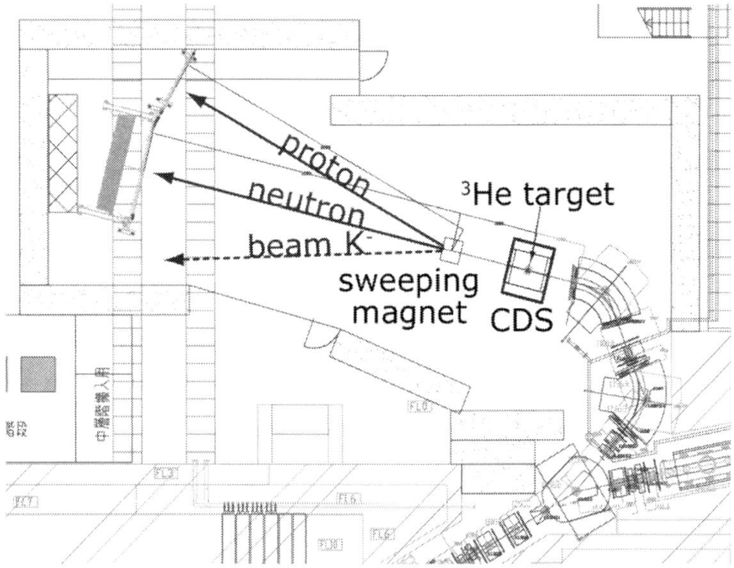

Figure 2. Setup for the J-PARC E15 experiment.

confirmed. Hence, it is important to continue searching for K^-pp states with various reactions. The $p + p \rightarrow p + \Lambda + K^+$ reaction with higher incident energy was measured by the FOPI experiment at GSI in 2009, and the analysis is in progress [11]. The stopped K^- absorption reaction in ^{3}He and ^{4}He will be studied by the AMADEUS experiment at DAΦNE [12]. In J-PARC, two experiments have been approved, the ^{3}He(K^-, n) reaction (J-PARC E15 experiment) and the D(π^+, K^+) reaction (J-PARC E27 experiment).

2.1. *J-PARC E15 experiment*

The J-PARC E15 experiment [13] will be performed at the K1.8BR beamline (Fig. 2). K^- beam at 1.0 GeV/c will be made to impinge on a liquid ^{3}He target. Outgoing neutrons by the (K^-, n) reactions will be detected by a neutron counter, located at 15 m downstream from the target. The K^- beam will be swept away by a sweeping magnet in order not to hit the neutron counter. Furthermore, scattered protons by the (K^-, p) reaction can be detected by installing an additional proton counter next to the neutron counter. The decay particles of K^-pp states into $\Lambda + p$ or $\Sigma^0 + p$ can be detected by a cylindrical detector system (CDS) which consists of a GEM-TPC tracker, a cylindrical drift chamber, and hodoscopes inside a solenoid magnet. Both the missing-mass and invariant-mass techniques will be applied to suppress the background.

The commissioning run of the beamline at 1.0 GeV/c and the CDS has been performed. For example, Λ and K_S are observed by reconstructing the invariant mass of $p\pi^-$ and $\pi^+\pi^-$ pairs, respectively, which were detected by the CDS. The analysis for improving the resolution and the particle identification is under way.

The construction of the neutron counter and the installation of the sweeping magnet will start soon. We expect the first data taking to begin in 2012.

2.2. *J-PARC E27 experiment*

The J-PARC E27 experiment will investigate the D(π^+, K^+) reaction [14]. K^-pp states will be produced by sticking a $\Lambda(1405)$, produced by the elementary reaction $n(\pi^+$, $K^+)$, on the

Journal of Physics: Conference Series **374** (2012) 012015 doi:10.1088/1742-6596/374/1/012015

spectator proton. The experiment will use 1.7 GeV/c π^+ beam at the K1.8 beamline, and the scattered K^+ will be detected by the Superconducting Kaon Spectrometer (SKS).

The main background in the inclusive measurement will come from quasi-free processes ($\pi^+ + $ "N" $\rightarrow Y^{(*)} + K^+$). Since these processes have a spectator nucleon almost at rest, the tagging of two protons by a detector system surrounding the deuterium target will be useful to eliminate their contribution. Thus, two sets of range counter arrays for detection of fast protons will be installed on the left and right sides of the target.

A test experiment with a prototype of the range counter arrays was done during the beam time for the E19 experiment in 2010, which searched for the Θ^+ pentaquark. A part of the range counter arrays was installed near a hydrogen target, and some of scattered $\pi^{\pm}$'s and protons were stopped inside the arrays. The analysis verified the feasibility of clear p/π separation by combining the energy loss in each counter, the time-of-flight between the start counter and the first layer of the array, and the range of the particle.

We plan to take the first data for the E27 experiment in the middle of 2012. The detectors in the beamline and the SKS spectrometer as well as the range counter arrays are almost ready.

3. η mesic nuclei

The ηN interaction is known to be weakly attractive, because the ηN channel strongly couples with the $N(1535)$ resonance, which is above the ηN threshold. However, its scattering length has a large ambiguity with the real part varying between $(0.270\text{–}1.050) fm$ and the imaginary part between $(0.190\text{–}0.399) fm$ [15]. Haider and Liu [16] showed η mesic nuclei with the mass number $A \geq 12$ can be bound when the scattering length is set to $0.28 + 0.29i$ fm or $0.27 + 0.22i$ fm. If the real part of the scattering length is larger, an η meson may be bound in an even lighter nucleus such as ^{3}He and ^{4}He [15, 17].

Recently the COSY-GEM collaboration investigated the ^{27}Al(p, ^{3}He) reaction [18] with the recoilless condition. By detecting back-to-back $\pi^- p$ pairs, which may originate from the decay of η mesic nuclei, they found an indication of η mesic nuclei with the binding energy of ~ 12 MeV. Studies with the (p, ^{3}He) reaction may be continued at J-PARC at the K1.8 beamline together with the SKS or S-2S (under construction) spectrometer.

Itahashi *et al.* propose to study the (π^-, n) reaction on ^{7}Li and ^{12}C target at J-PARC[19]. There are two advantages over a prior experiment at BNL [20]; one is the detection of back-to-back $\pi^- p$ pairs like the COSY-GEM experiment, and the other one is the recoilless condition by adjusting π^- momenta around 0.8–1.0 GeV/c and detecting zero-degree neutrons, while the BNL experiment measured protons at scattering angle 15° from the (π^+, p) reaction, so that the momentum transfer is larger than 200 MeV/c. To achieve these conditions, the use of the E15 experimental setup is desirable. According to Ref. [21], the formation spectrum is sensitive to the in-medium property of $N(1535)$, which affects the η-nucleus optical potential. While the mass difference of $N(1535)$ and N will not change largely in the chiral unitary model, the chiral doublet model leads to a decrease of the mass difference due to the partial restoration of chiral symmetry. The missing-mass spectrum will help to understand the in-medium behavior of $N(1535)$.

A pilot experiment with deuteron target ($\pi^+ d \rightarrow pp\eta$) is also under consideration [22]. The double-scattering reaction (Fig. 3(a)) can be distinguished from the dominant quasi-free process (Fig. 3(b)) by detecting two outgoing protons, since the spectator proton in the latter process has a very small momentum. The differential cross section is sensitive to the low-energy ηN scattering amplitude as the elastic scattering $\eta N \rightarrow \eta N$ will play an important role near the threshold energy [23].

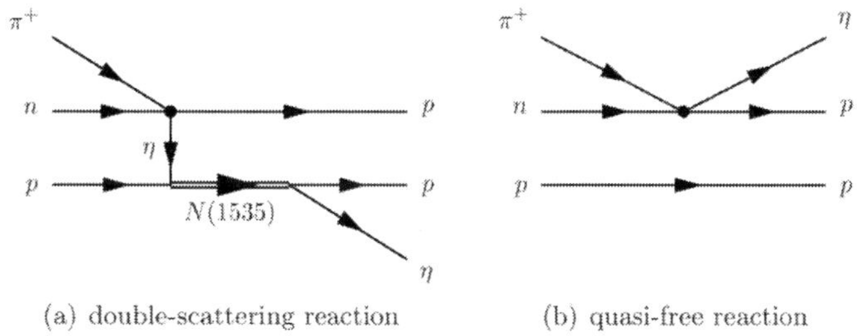

Figure 3. Diagrams for the $\pi^+ d \to pp\eta$ reaction.

4. Summary

Meson-nuclear bound systems, such as kaonic nuclei and η mesic nuclei can be investigated at J-PARC by using intense $\pi^{\pm}$, K^-, and proton beams. In addition to the forward spectrometer for the missing-mass spectroscopy, the detectors for decay particles of mesic nuclei are also important to reduce the background level. The E15 and E27 experiments, both of which will search for K^-pp bound states but with different reactions, will start in 2012. Moreover, search for η mesic nuclei will be also feasible in future.

Acknowledgements

This work was partly supported by the Grants-in-Aid for Scientific Research from MEXT and JSPS (No. 20840047, 22105506).

References

[1] M. Iwasaki *et al.*, Phys. Rev. Lett. **78**, 3067 (1997); T. M. Ito *et al.*, Phys. Rev. C **58**, 2366 (1998).
[2] G. Beer *et al.*, Phys. Rev. Lett. **94**, 212302 (2005).
[3] M. Bazzi *et al.*, Phys. Lett. **B704**, 113 (2011).
[4] S. Wycech, Nucl. Phys. **A450**, 399c (1986).
[5] Y. Akaishi and T. Yamazaki, Phys. Rev. C **65**, 044005 (2002); T. Yamazaki and Y. Akaishi, Phys. Lett. **B535**, 70–76 (2002).
[6] T. Hyodo and D. Jido, arXiv:1104.4474 [nucl-th].
[7] M. Agnello *et al.*, Phys. Rev. Lett. **94**, 212303 (2005).
[8] V. K. Magas, E. Oset, A. Ramos, H. Toki, Phys. Rev. **C74**, 025206 (2006).
[9] G. Pandejee, N. J. Upadhyay, B. K. Jain, Phys. Rev. **C82**, 034608 (2010).
[10] T. Yamazaki, M. Maggiora, P. Kienle, K. Suzuki, A. Amoroso, M. Alexeev, F. Balestra, Y. Bedfer *et al.*, Phys. Rev. Lett. **104**, 132502 (2010).
[11] K. Suzuki *et al.* [FOPI Collaboration], Prog. Theor. Phys. Suppl. **186**, 351-356 (2010).
[12] J. Zmeskal [AMADEUS Collaboration], Int. J. Mod. Phys. **A26**, 414-419 (2011).
[13] M. Iwasaki, T. Nagae *et al.*, J-PARC E15 proposal.
`http://j-parc.jp/NuclPart/pac_0606/pdf/p15-Iwasaki.pdf`
[14] T. Nagae *et al.*, J-PARC E27 proposal.
`http://www.j-parc.jp/NuclPart/pac_0907/pdf/Nagae.pdf`
[15] Q. Haider, L. C. Liu, Phys. Rev. **C66**, 045208 (2002).
[16] Q. Haider, L. C. Liu, Phys. Lett. **B172**, 257-260 (1986); *ibid.* **174**, 465(E) (1986).
[17] A. Khoukaz, Acta Phys. Polon. **B41**, 2271-2280 (2010). [arXiv:1011.0882 [nucl-ex]].
[18] A. Budzanowski *et al.* [COSY-GEM Collaboration], Phys. Rev. **C79**, 012201 (2009).
[19] K. Itahashi *et al.*, J-PARC Letter of Intent.
`http://www.j-parc.jp/NuclPart/pac_0707/pdf/LoI-itahashi.pdf`
[20] R. E. Chrien, S. Bart, P. Pile, R. Sutter, N. Tsoupas, H. O. Funsten, J. M. Finn, C. Lyndon *et al.*, Phys. Rev. Lett. **60**, 2595-2598 (1988).
[21] H. Nagahiro, D. Jido, S. Hirenzaki, Phys. Rev. **C80**, 025205 (2009).
[22] H. Fujioka, Acta Phys. Polon. **B41**, 2261-2270 (2010).
[23] H. Garcilazo, M. T. Pena, Phys. Lett. **B696**, 386-389 (2011).

5th DAE-BRNS Workshop on Hadron Physics (Hadron 2011) IOP Publishing
Journal of Physics: Conference Series **374** (2012) 012016 doi:10.1088/1742-6596/374/1/012016

Heavy Flavour Hadron Spectroscopy: Challenges and Future Prospects

P C Vinodkumar

Department of Physics, Sardar Patel University, Vallabh Vidyanagar - 388120, India

E-mail: p.c.vinodkumar@gmail.com

Abstract. During the last few years, wealth of new experimental results in the heavy flavor hadron sector has become available. The diversity, quantity and accuracy of the data are impressive and include many surprising spectroscopic results. Following the discovery of η_b states and the excited $\eta_c(2S)$ states many excited states of open charm and beauty mesons and plethora of exotic states are reported. Discoveries of many new hadrons at B-factories have shed light on a new class of hadrons beyond the ordinary mesons. Many of these states awaits for its right identification. Though we have the theory (QCD) for the strong interaction, we are still far from extracting the major part of the hadron properties from it. These properties at the hadronic scale obviously play relevant role in many searches for new physics and new phenomenon. For obvious reasons, heavy flavour sector offers unique opportunities in this case. For example, the quarkonium systems are crucially important to improve our understanding of QCD as it falls in the low energy region where the non-perturbative effects dominate. Thus the heavy quarks / quark-antiquark bound states are ideal laboratory where our understanding of non-perturbative QCD and its interplay with perturbative QCD may be tested. A comparative review of different model predictions for example in the case of heavy flavor hadronic systems will be highlighted. The quarkonium studies may be used as a benchmark for our understanding of QCD and for the precise determination of standard model strong interaction parameters such as the constituent quark masses, α_s, the confinement strength (string tension)*etc.*

1. Introduction

Three decades after the discovery of charmonium and bottonium states the hadron spectroscopy within the heavy flavour sector has once again becomes a challenge due to the recent observations of the large number of conventional and nonconventional hadronic states by the different experimental groups at Belle, BaBar, DELPHI, CLEO, CDF, BES-III, SELEX *etc.* These experimental groups are engaged with more focused experiments at the heavy flavour sector discovering new hadronic states with very high precision and better statistics [1]. At the same time theoretical attempts (Lattice QCD, HQET, NRQCD, NRQM, etc.) are matured enough to provide precise quantitative predictions at the hadronic scales [2, 3, 4, 5, 6, 7, 8]. Apart from the excited states of the known charmonium and bottonium resonances including their orbital excitations precisely measured, many open charm and open beauty states and large number of X,Y,Z states in the energy range of 3-6 GeV as well as number of Y_b,Z_b states just above 10.0 GeV are being reported recently. Many of these new discoveries shed lights on new types of hadrons like tetra quark states, hadron molecular states or hybrids [9]. The correct identification of them pose new challenges before theoreticians working in the heavy flavour spectroscopy [10].

2. Recent Observations of new hadronic states and Challenges posed by them

Having played a major role in the understanding of QCD at the hadronic scale, heavy hadron spectroscopy has witnessed a renaissance in the last few years driven by recent experimental reports of various hadronic states by the high energy experimental facilities world over. So far the greatest activity has occurred in the charm sector with energy range 3-5 GeV. In the bound state region of energy less than 4 GeV, major contributions in charm sector came from BES and CLEO. While that above 4 GeV, the charmonium like states have reported from CLEO, Belle and BaBar. Until very recently, the spin singlet state of any quarkonium that was known was $\eta_c(1S)$ and it has provided the hyperfine mass difference between the vector, $J/\Psi(1S)$ and the pseudoscalar $\eta_c(1S)$ states to be equal to 172 MeV [11]. However, one did not know how the hyperfine interaction among the excited states behaves. The great excitement, often referred as the renaissance in hadron spectroscopy, has come from the discovery of a host of new states $X(3872)$, X, Y, Z (in the range 3940), Y(4260) from Belle and BaBar and more recently $X^{'}$ (4160), $X^{''}$ (4324), $X^{'''}$ (4660) etc.,[11, 12, 13]. The challenges posed by these new states include the right identification with the proper J^{PC} values and their decay modes. Some of these may be the orbital excited states of the charmonium, for example, X(3943) as $\eta_c^{''}(3S)$, Z(3929) as the $\chi_{c2}^{'}(2^3P_2)$ and the Y(3943) is speculated to be a hybrid. The Y(4260) observed by BaBar, CLEO and Belle is a vector meson but not likely to be a charmonium vector state. So it is suggested to be a $c\bar{c}g$ hybrid [9]. If so its partner states 0^{-+} and 1^{-+} ought to be also observed nearby this energy. So far it has not convincingly seen from experiments. In the bottonium sector, the Upsilon (1S-6S), χ_b, $\chi_b^{'}$ and now the long awaited η_b [14] are all known and identified. However the higher orbital states are yet to be observed. In general, the bottonium is a much better place to get insight to the quarkonium spectroscopy because the running strong coupling constant is much smaller (α_s 0.2) and the relativistic effects are less important compared to the charmonium case. For the Upsilon (1^{--}) states, all we know is their masses, total widths, and branching fractions for leptonic, radiative, and $\Upsilon(nS) \rightarrow \pi^+\pi^-\Upsilon(n'S)$ decays. A scarce $\Upsilon(3S) \rightarrow \omega\chi_b(2S)$ transition has also been observed, but huge gaps remain.

Another challenging area of spectroscopic interest lie in the open flavour sector. Only the lowest 0^- and 1^- states are listed in the Particle Data Group (PDG) [11]. However, the L3 collaboration [18] reported first measurement of masses of the 1^3P_1 and 1^3P_2 of B_q mesons at $5670 \pm 10 \pm 13$ MeV and $5768 \pm 5 \pm 6$ MeV respectively. Two years ago, DØ and CDF collaborations have reported results on the spectroscopy of orbitally excited beauty mesons [19]. CDF found two states, 1^1P_1 and 1^3P_2, with masses M(1^1P_1)= $5734 \pm 3 \pm 2$ MeV and M(1^3P_2)= $5738 \pm 6 \pm 1$ MeV. DØ also found the same states but with slightly different masses, M(1^1P_1)= $5720.8 \pm 2.5 \pm 5.3$ MeV and M(1^3P_2) - M(1^1P_1)= $25.2 \pm 3.0 \pm 1.1$ MeV. In the strange sector, CDF reported two narrow $B_s(1^3P_1)$ and $B_s(1^3P_2)$ states with masses M($B_s(1^3P_1)$)=5829.4 MeV and M($B_s(1^3P_2)$)=5839 MeV while DØ measured only the $B_s(1^3P_2)$, with mass of $5839.1 \pm 1.4 \pm 1.5$ MeV.

Similar progress has been observed in the open charm sector. The BaBar Collaboration reported the observation of a charm-strange state, the $D_{sJ}^*(2317)$ [20]. It was confirmed by CLEO Collaboration at the Cornell Electron Storage Ring [21] and also by Belle Collaboration at KEK [22]. Besides, BaBar had also pointed out to the existence of another charm-strange meson, the $D_{sJ}(2460)$ [20]. This resonance was measured by CLEO [21] and confirmed by Belle [22]. Belle results are consistent with the spin-parity assignments of $J^P = 0^+$ for the $D_{sJ}^*(2317)$ and $J^P = 1^+$ for the $D_{sJ}(2460)$.

Many heavy flavour baryons are also being observed in recent times at CLEO, BaBar and Belle [1]. Even the positive parity excited states are being observed. However more refined high luminosity measurements are required to identify their J^{PC} values. The progress in this sector is more encouraging as more number of charmed and beauty baryons are predicted long ago by the extension of the Gell-Mann's SU(3) quark model. The B- factories have already reported large number of these baryon states. For example BaBar reported $\Lambda_c(2940)$, $\Omega_{cc}(2770)$, Belle has observed $\Sigma_c(2800)$, $\Sigma_c(2980)$ etc. Before 2006 also, only one bottom baryon (Λ_b) was known, now we have the Σ_b and Ξ_b. These are extremely challenging measurements resolving states at about 6 GeV separated by just 20 MeV also. In short, these high precision measurements really pose challenges before the theorists who are trying to extract the basic QCD properties at these hadronic scales.

3. A Theoretical Attempt Using coulomb Plus Power Potential (CPP$_\nu$)

The investigation of the properties of mesons composed of a heavy quark and antiquark ($c\bar{c}$, $b\bar{c}$, $b\bar{b}$) gives very important insight into heavy quark dynamics and to the understanding of the constituent quark masses. At the hadronic scale, the nonperturbative effects connected with complicated structure of QCD vacuum necessarily play an important role. All this leads to a theoretical uncertainty in the $Q\bar{Q}$ potential at large and intermediate distances. So the success of theoretical model predictions of the hadronic properties with respect to the new experimental results can provide important information about the quark-antiquark interactions. Such information is of great interest, as it is not possible to obtain the $Q\bar{Q}$ potential starting from the basic principle of the quantum chromodynamics (QCD) at the hadronic scale.

Among the various scheme to describe the inter quark interaction inside the hadrons of different flavour compositions, the Cornell potential has become quite successful and it has been deduced from the lattice description of QCD at the hadronic scale. However, for the description of the excited states of the well established quarkonium states and for the description of open charm or open beauty hadrons, it was felt that the string tension in the confinement part of the potential become energy depended. This energy dependent on the string tension can also be viewed in terms of an exponent on the inter-quark separation corresponding to the linear potential of the Cornell type. This has led to the choice of Coulomb plus power potential with exponent ν (CPP_ν). Such a choice of the phenomenological potential for the confinement part would allow us to study the variations in the nature of the inter quark interaction at the formation of the bound states as well as at the different decay processes.

The theoretical predictions of the masses of heavy-light system for ground state as well as excited state are very few [24, 25, 26, 27, 28, 29]. Spectroscopy of heavy flavour mesons ($Q\bar{Q}$, $Q\bar{q}$ systems) have been studied using coulomb plus power potential (CPP$_\nu$) in both relativistic and nonrelativistic formalism with different choices of the potential index ν ($0.1 \leq \nu \leq 2.0$). A comprehensive study based on the CPP$_\nu$ model of the heavy flavour hadrons containing one or more heavy flavour quarks with minimum number of free parameters are being studied by us in recent years [30, 31, 32, 33, 34]. The Hamiltonian of the system is described as

$$H = M_Q + \sqrt{p^2 + m_q^2} + \frac{p^2}{2M_Q} + V(r) + V_{SD}(r) \tag{1}$$

in the case of light-heavy mesonic systems and

$$H = 2M_Q + \frac{p^2}{M_Q} + V(r) + V_{SD}(r) \tag{2}$$

in the case of quarkonia states. Here,

$$V(r) = -\frac{4}{3}\frac{\alpha_s}{r} + Ar^{\nu} \tag{3}$$

and the spin depended part is taken as

$$\begin{aligned} V_{SD}(r) &= V_{SS}(r)\left[S(S+1) - \frac{3}{2}\right] + V_{LS}(r)\left(\vec{L}\cdot\vec{S}\right) + \\ & V_T(r)\left[S(S+1) - \frac{3(\vec{S}\cdot\vec{r})(\vec{S}\cdot\vec{r})}{r^2}\right] \end{aligned} \tag{4}$$

with

$$V_{LS}(r) = \frac{1}{2\, m_Q m_{\bar{Q}}\, r}\left(3\frac{dV_V}{dr} - \frac{dV_S}{dr}\right) \tag{5}$$

$$V_T(r) = \frac{1}{6\, m_Q m_{\bar{Q}}}\left(3\frac{d^2V_V}{dr^2} - \frac{1}{r}\frac{dV_V}{dr}\right) \tag{6}$$

$$V_{SS}(r) = \frac{1}{3\, m_Q m_{\bar{Q}}}\nabla^2 V_V = \frac{16\ \pi\alpha_s}{9\, m_Q m_{\bar{Q}}}\delta^{(3)}(\vec{r}) \tag{7}$$

$$\alpha_s(\mu^2) = \frac{4\pi}{(11 - \frac{2}{3}n_f)ln(\mu^2/\Lambda^2)} \tag{8}$$

Here, we describe mainly our results on the properties of quarkonia only. Our predictions on the S-wave quarkonia are shown in Fig.1 as against the different choices of the potential index ν. The quark mass parameters employed in our calculations are $m_u = m_d = 330MeV$, $m_c = 1.28GeV$ and $m_b = 4.4GeV$. The experimental(PDG average) values are drawn as the horizontal lines. We also find the standard deviation of the predicted S-wave masses with reference to the experimental values and are shown in Fig.2. They show a minimum around $\nu \sim 1.0$ in the case of charmonia and around $\nu \sim 0.8$ in the case of bottonium states. The low lying $P-$wave masses are also shown in Fig.3 against the potential index. The numerical solution of the radial wave function of these states can further be employed to compute the mean square radii, the decay constants $f_{P/V}$, the di-lepton, di-gamma and di-gluon widths as well as the E1 and M1 transitions rates. The di leptonic decay width of the $c\bar{c}$(1S-4S) and $b\bar{b}$(1S-6S) states based in the NRQCD formalism are shown in Fig.4 and Fig.5 respectively against the potential index ν. Corresponding two photon and di gluon decay width are shown in Fig.6 and Fig.7 respectively. The shaded region in all these plots represents the range of ν at which the predicted masses of the state agree with the experimental results. In the heavy-light flavour sector, the model extended to predict the masses of few low-lying states of $Q\bar{q}$ systems (D, D_s, B, B_s), the decay constants $f_{P/V}$, the inclusive semi-leptonic and leptonic branching ratios and the neutral flavour oscillations of $B^0 - \bar{B}^0$ and $B_s^0 - \bar{B}_s^0$ mesons [34].

Unlike in the case of mesons, for baryons, the magnetic moments become an additional property to be studied. Though many of the theoretical attempts successfully predict the masses, there is no consensus among the theoretical predictions of the properties like spin-parity, the form factors, magnetic moments *etc.* Heavy baryons further provide excellent laboratory to understand the dynamics of light quarks in the vicinity of heavy flavour quarks. The present model potential(CPP_ν) has also been extended to study the heavy flavour baryonic properties within a hypercentral scheme [35, 30]. In the hypercentral model, the hyper spherical coordinates are given by the angles

$$\Omega_\rho = (\theta_\rho, \phi_\rho)\ ;\ \Omega_\lambda = (\theta_\lambda, \phi_\lambda) \tag{9}$$

Journal of Physics: Conference Series **374** (2012) 012016 doi:10.1088/1742-6596/374/1/012016

the hyper radius, x and hyper angle ξ as,

$$x = \sqrt{\rho^2 + \lambda^2} \; ; \; \xi = \arctan\left(\frac{\rho}{\lambda}\right) \tag{10}$$

Where ρ and λ are the Jacobi Co-ordinates to describe a three body system of three quarks of masses m_1, m_2 and m_3, given by [36]

$$\vec{\rho} = \frac{1}{\sqrt{2}}(\vec{r}_1 - \vec{r}_2) \; ; \; \vec{\lambda} = \frac{(m_1\vec{r}_1 + m_2\vec{r}_2 - (m_1 + m_2)\vec{r}_3)}{\sqrt{m_1^2 + m_2^2 + (m_1 + m_2)^2}} \tag{11}$$

where

$$m_\rho = \frac{2\, m_1\, m_2}{m_1 + m_2} \; ; \; m_\lambda = \frac{2\, m_3\, (m_1^2 + m_2^2 + m_1 m_2)}{(m_1 + m_2)\,(m_1 + m_2 + m_3)} \tag{12}$$

The model Hamiltonian corresponds to the three body system (baryons) in the hyper central co-ordinates now be expressed as

$$H = \frac{P_x^2}{2\, m} + V(x) \tag{13}$$

where

$$m = \frac{2\, m_\rho\, m_\lambda}{m_\rho + m_\lambda} \tag{14}$$

and

$$V(x) = -\frac{\tau}{x} + \beta x^\nu + \kappa + V_{spin} \tag{15}$$

Here the potential $V(x)$ is not purely a two body interaction but it contains three-body effects also. Within this scheme we have studied the ground state masses, mean square radii, hyperfine mass splitting and the magnetic moments of light flavour (qqq), single heavy flavour (Qqq), double heavy flavour (QQq) and triple heavy flavour (QQQ) baryons. Details and results can be found in [30, 37, 38, 39].

4. Results and Discussions

The predicted results on the $P-$wave masses of $c\bar{c}$ mesons 1^1P_1(3514 - 3542 MeV), 1^3P_1(3514 - 3542 MeV), 1^3P_2(3524 - 3552 MeV) for the potential index between ν = 1.0 to 1.1 are in good agreement with the experimental values of h_c(3526), χ_{c1}(3511), χ_{c2}(3556) [11] while the 1^3P_0(3414) at ν = 0.8 exactly matches with the experimental value of χ_{c0}(3415) [11]. Similar agreement for $b\bar{b}$ states 1^3P_0(9817 - 9909 MeV), 1^3P_1(9831 - 9929 MeV) and 1^3P_2(9838 - 9938 MeV) for the potential index ν = 0.5 to 0.7 are in agreement with the experimental average values of χ_{b0}(9859), χ_{b1}(9893), χ_{b2}(9912) [11]. In the same range of ν, the model predicts the h_b state around (9834 - 9932 MeV), which is in close agreement with the recently reported 9898.25±1.06 MeV by Belle group [40]

The predictions for 2^3P_2(3887 - 3970 MeV), 2^3P_1(3875 - 3958 MeV), 2^3P_0(3835 - 3912 MeV) and 2^1P_1(3877 - 3960 MeV) within the potential index between 1.0 and 1.1 for the $2P-$states of $c\bar{c}$ systems lie close to the experimental states reported by Belle group around 3940 [41]. In the same range of the potential index, $1.0 \leq \nu \leq 1.1$ the results for 3^1S_0(3895 - 3991 MeV) is closer to the experimental charmonium state of X(3938) reported recently by Belle [41] and for 4^1S_0(4180 - 4325 MeV) is close to the Y(4260) state reported by BaBar [42]. The predicted 2^3D_1 states (4130 - 4245 MeV) of $c\bar{c}$ system in the same range of $1.0 \leq \nu \leq 1.1$ is closer to the experimental ψ(4160, $J^P = 1^{--}$) state [43]. The lone known 1^3D_1 (3770) is found to be closer

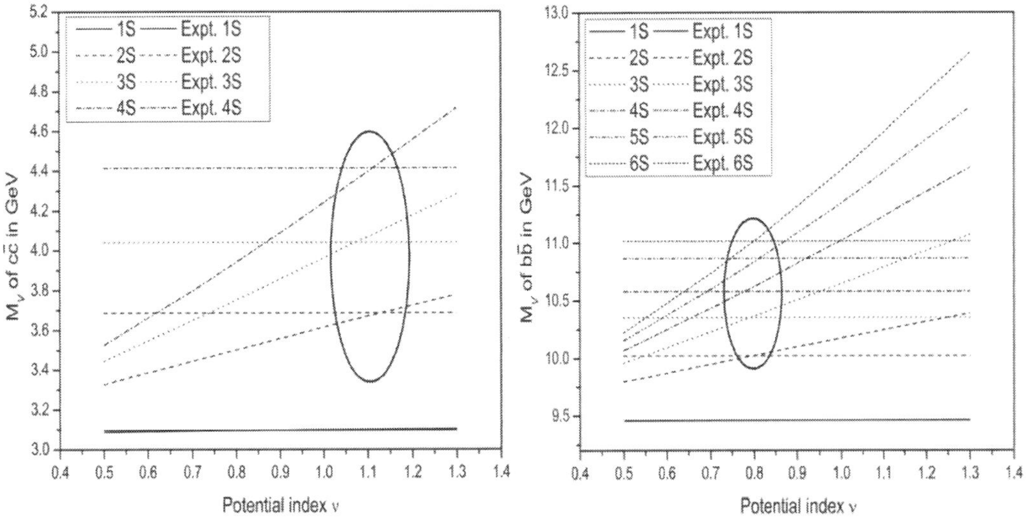

Figure 1. S-wave masses of quarkonia against the potential exponent ν. The horizontal line represents the respective experimental values (PDG). The encircled region shows the intersected region of predictions with the experiment.

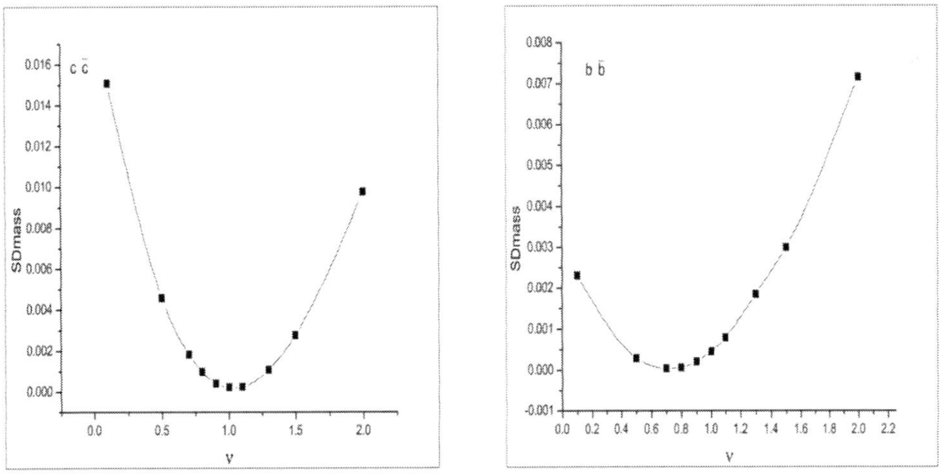

Figure 2. The standard deviation with respect to the experimental values of the predicted masses for each choice of potential exponent ν against the exponent ν

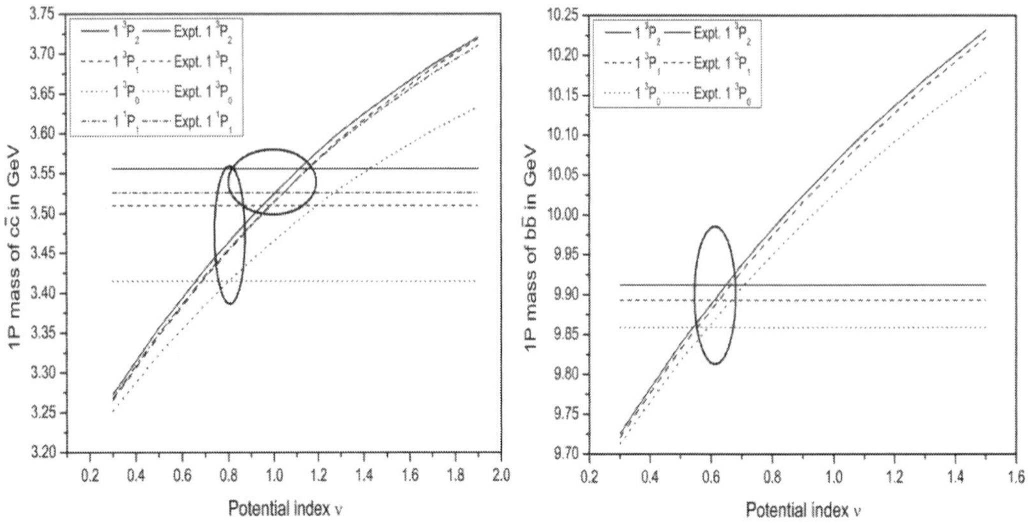

Figure 3. Predicted P-wave masses and their experimental values (horizontal line) of the quarkonia against the potential exponent ν.

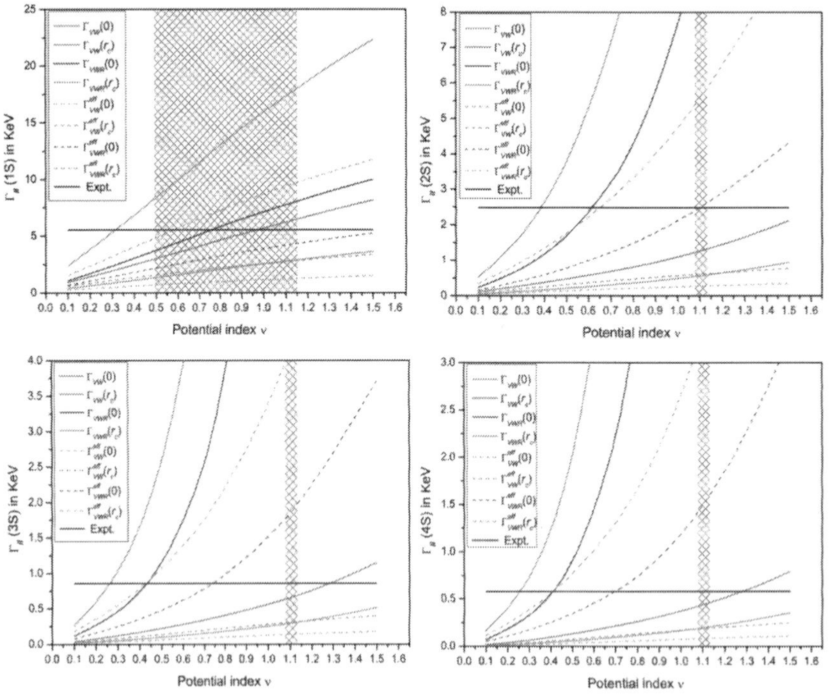

Figure 4. The di-lepton decay widths of $c\bar{c}$ system with potential index ν in NRQCD formalism. The horizontal lines are the respective experimental values.

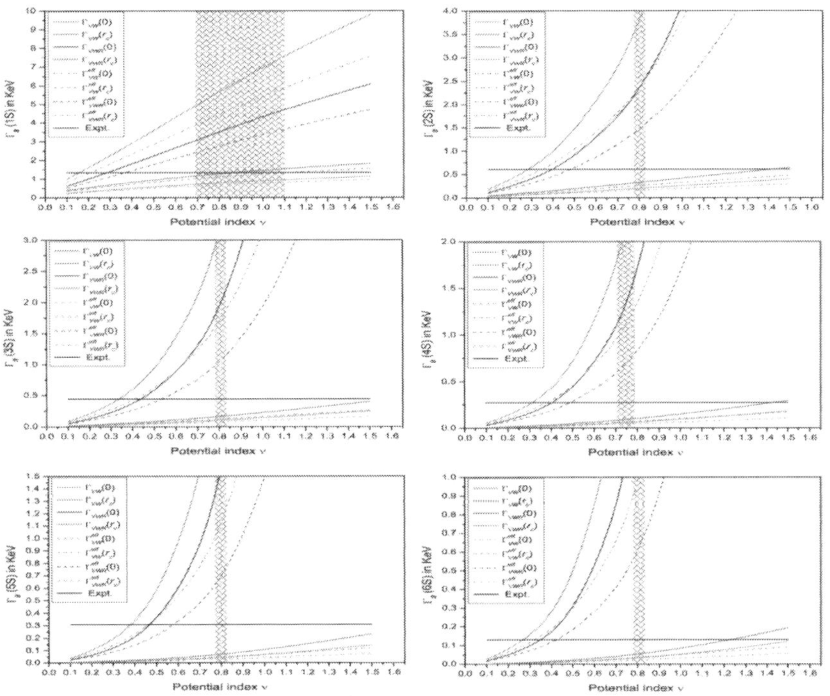

Figure 5. The di-lepton decay widths of $b\bar{b}$ system with potential index ν in NRQCD formalism. The horizontal lines are the respective experimental values.

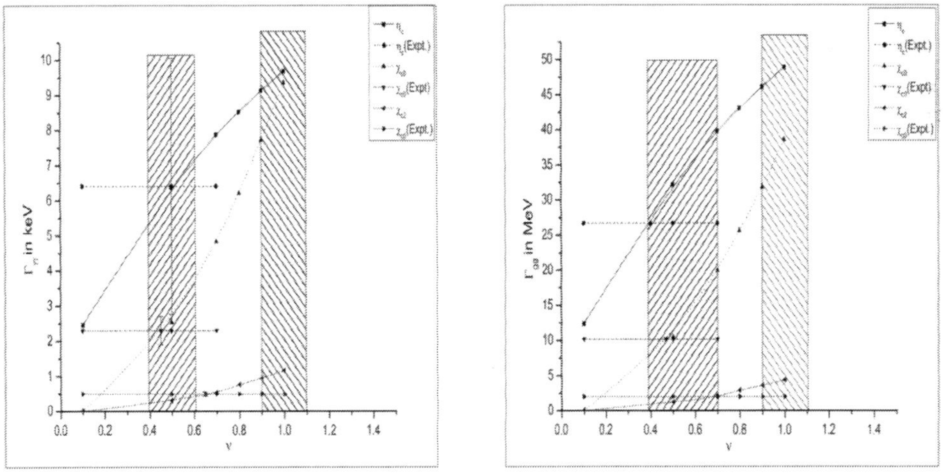

Figure 6. Two photon and two gluon decay width of Charmonia states against potential index ν.

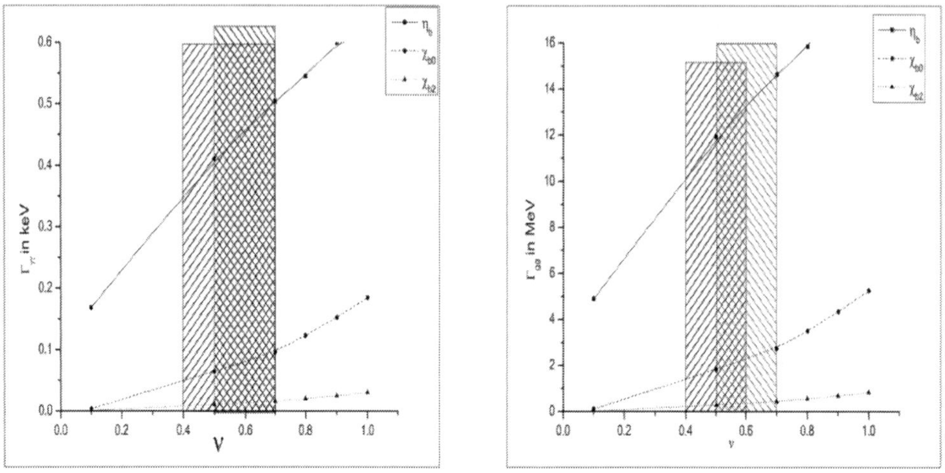

Figure 7. Two photon and two gluon decay width of bottonia states against potential index ν.

to the value of 3796 MeV at $\nu = 0.9$. The D-wave masses obtained here for the potential index $\nu = 0.9$ are close to the lattice predictions [44] for $c\bar{c}$.

The predictions for the $2P-$spin triplet states of $b\bar{b}$ meson, 2^3P_0 (10214 - 10329 MeV), 2^3P_1 (10208 - 10322 MeV) and 2^3P_0(10193 - 102304 MeV), for the potential index ν in the range 0.7 to 0.8 are nearer to the corresponding experimental χ_b states. Its spin singlet state 2^1P_1(10210 - 10234) in the same range of ν is close to the lattice predictions as well as the recently reported values of 10.259±0.64GeV of the $h_b(2P)$by Belle [40]. This shift towards the lower index for the higher angular momentum states probably suggests the orbital energy (n, ℓ) dependence on the string tension A.

The predicted masses of $P-$wave of $D-$mesons provide 1^3P_2(2342 - 2472 MeV) within the range of $1.0 \leq \nu \leq 1.7$, 1^3P_1(2361 - 2464 MeV) for $1.0 \leq \nu \leq 1.5$, 1^3P_0(2312 - 2398 MeV) for $0.9 \leq \nu \leq 1.3$ and 1^1P_1(2269 - 2337 MeV) for $1.0 \leq \nu \leq 1.5$. While the experimental candidate for $J^P = 1^+$ state of $D-$meson (2420 - 2460 MeV) observed by CLEO [45] and Belle [46] and $J^P = 0^+$ state observed in the range 2300 - 2400 MeV by Belle and Focus [47] lie within the predicted range.

In the case of open strange-charm mesons (D_s), the predictions for $P-$states provide 1^3P_2(2416 - 2552 MeV) for $1.0 \leq \nu \leq 2.0$ as against the latest experimental average value (PDG2010) of 2573 MeV, 1^3P_1(2397 - 2484 MeV) for $1.0 \leq \nu \leq 1.5$ as against the PDG(2010) value of 2460 MeV, 1^3P_0(2317 - 2350 MeV) for $0.8 \leq \nu \leq 1.0$ as against the recently reported value of 2317 by BaBar, CLEO and Belle group [1]. The radial excitation of $D_s^*(2112)$ observed by Belle collaboration [48] at 2715 MeV is found to be close to the 3^3S_1 state predicted in this model and at $\nu = 1.0$ as the 2^3S_1 values predicted here lie in the range of 2474 - 2730 MeV for the choice of $1.0 \leq \nu \leq 1.7$, which are below the experimental value in the expected range of the potential index around $\nu = 1.0$. Even higher excited states of $c\bar{s}$ system has been observed by the BaBar collaboration [49] with spin parity $0^+, 1^+$ and 2^+ with mass at 2856±1.5±5.0 which in our case corresponds to the $2P$ state with the predicted mass range of 2^3P_2(2668 - 2842

MeV) for $1.0 \leq \nu \leq 1.3$, 2^3P_1(2651 - 2820 MeV) for $1.0 \leq \nu \leq 1.3$, 2^3P_0(2612 - 2858 MeV) for $1.0 \leq \nu \leq 1.5$ and 2^1P_1(2656 - 2935 MeV) for $1.0 \leq \nu \leq 1.5$. Thus, present study on the open charm and open strange-charm mesons are being identified with the recently discovered $D-$ and D_s meson states. Other predicted high angular momentum states $\ell \geq 2$ of these mesons are expected to be seen in the future experiments at BES-III, BaBar, Belle and CLEO collaborations.

While in the case of open beauty systems (B, B_s), the spectral predictions are in better agreement with the known experimental states and with other theoretical model predictions at potential index lying between $0.7 \leq \nu \leq 1.1$. The predicted mass for the 1^1P_1(5724 MeV) and 1^3P_2 (5431 MeV) at $\nu = 0.7$ of $B-$mesons are very close to the recently observed 1^1P_1($5721 \pm 2.5 \pm 5.3$ MeV) and 1^3P_2 ($5738 \pm 6 \pm 1$ MeV) states by CDF and DØ[19], While in the case of B_s meson the recent CDF observation of 1^3P_1 (5829 MeV) and 1^3P_2 (5839 MeV) lie well within the range of values predicted by 1^3P_2 (5816 - 5850 MeV) and 1^3P_2 (5809 - 5842 MeV) in the potential index between 0.8 to 0.9. Unfortunately there exist only very few experimental data for $B - B_s$ systems [11]. Future, high luminosity $B-$factories are expected to provide more clean and high precision data in the open heavy flavour mesons.

5. Conclusions and Summary

At the end, we summarize that the nonrelativistic Coulomb plus power potential with varying power index using numerical approach to solve the Schrödinger equation is an attempt to understand the nature of the interquark potential and their parameters that provide us the spectroscopic properties as well as the decay properties of the $Q\bar{Q}$ system with the potential index between 0.7 and 1.1. It also provides us the importance of the quark mass parameters and the energy dependence on the potential strength for the study of the spectral properties. The radial wave functions obtained as the solution from the study are not only important for the determination of hyperfine and fine splitting of their mass spectra but also essential inputs for evaluating decay constants, decay rates, NRQCD parameters and production cross sections for quarkonium states.The spectroscopic mass difference due to the hyperfine/fine splitting are found to be sensitive to the choice of quark mass parameters. A closer look at the different properties of the heavy flavour mesons studied using phenomenological models reveals strong correlation between the model quark mass($m_{Q/q}$) parameter and the confinement strength (A).

The study on the spectroscopy of heavy flavour mesons clearly indicate the dependence on the energy scales for the nature of inter-quark potential. For instance below 3.0 GeV energy scale the confinement part of the interquark potential seems to be above $\nu = 1.0$ while for the energy scales beyond 3.0 GeV, the confinement part of the potential seemed to be flattened with $\nu < 1.0$. The deviation from the linear behavior ($\nu = 1$) indicates the relative importance of nonperturbative behavior of QCD below and above 3.0 GeV scale. It indicates 3 GeV scale as a transition energy between the perturbative and nonperturbative domains.

The spectroscopic parameters of the CPP$_\nu$ model, are also being employed to compute the neutral the (B^0, B_s^0) oscillation parameters quite satisfactorily [30]. The predictions for Δm_q and other parameters are very close to values observed from the recent experiments.

The most challenging problem at present is the description of the recent observation of states such as X(1835), observed at BES, the observations of X(3940), Y(3940), Z_b(10.61,10.65), Y_b(10.88) at Belle, the observations of Y states in the (4.1 to 4.6). Many extremely interesting questions in hadron spectroscopy remain unanswered at present. However, there is every hope that the upcoming facilities, PANDA at GSI, JPARC at KEK, and the 12 GeV upgrade at

Jefferson Laboratory (JLab), will rise to meet the challenges and pose new challenges to the theorists and phenomenologist to have serious attention in heavy flavour spectroscopy.

The Large Hadron Collider (LHC), with 14 TeV center of mass energy proton-proton collisions, will offer an opportunity for the study of QCD at unprecedented energy scales. The CMS facility at the LHC will be able to detect hadron production at an unprecedented center-of-mass energy of 14 TeV. The CMS physics program is mostly devoted to searches related to Higgs and new physics beyond standard model. Heavy flavor physics is also a field where many interesting observations are expected, both in terms of hadron production and decays, especially in the low-luminosity scenario. Heavy flavor processes are also interesting because they can constrain indirectly transitions that involve scales much higher than m_b, through loop propagation of new particles. The CMS Experiment will also study the J/ψ reconstruction and decay of B_c meson.

The ATLAS facilities at LHC will cover central proton-proton collisions and plans to study heavy flavor hadrons including quarkonium states. The expected statistics will permit high precision measurements of production polarization of the Λ_b and J/ψ. ATLAS is well instrumented for B Physics, having been designed with precision vertexing and tracking, good muon identification, high resolution calorimetry, and a flexible dedicated B trigger. A rich B Physics program is planned, including CP violation studies (especially for the Λ_b and B_s systems, which are not accessible to the B factories), rare decays sensitive to new physics (including $b \to s\ell^+\ell^-$ and $b \to d\ell^+\ell^-$). The ATLAS Collaborations planned high statistics studies of heavy flavor hadrons, including quarkonia, will throw light on the properties of bound states, the spin dependence of quark confinement, the nature of the strong interaction potential, factorization in heavy quark effective theory, the source of CP Violation, and perhaps the resolution of some puzzles in heavy flavour physics [60]. The LHC will be the first opportunity for significant statistics on excited states in the B_c family. These measurements can constrain models of the strong potential and cast light on the inter-dependance between the electroweak and strong forces.

Acknowledgement: Major part of the results reported here are done through the financial supports from the University Grant commission, Government of India under Major research projects F. 40-457/2011(SR), F. 32-31/2006(SR) and Department of Science and Technology (DST) Govt. of India under the major research project SR/S2/HEP-20/2006. I also acknowledge the efforts put up by all my research students.

References

[1] Aubert B *et al.* (BABAR Collaboration), Phys. Rev. Lett. **98**, 122011 (2007); Phys. Rev. Lett. **98** 012001 (2007); Phys. Rev. Lett. **97** 232001 (2006); Phys. Rev. **D 77** 012002 (2008); [BaBar Collaboration], Phys. Rev. Lett. **90**, 242001 (2003); Besson D *et al.* [CLEO Collaboration], Phys. Rev. **D 68**, 032002 (2003);Abe K *et al.* [Belle Collaboration], Phys. Rev. Lett. **92**, 012002 (2004); Mizuk R *et al.* (Belle Collaboration), Phys. Rev. Lett. **94**, 122002 (2005); Feindt M *et al.* (DELPHI Collaboration), Report no. CERN-PRE/95-139 (2007); Edwards K W *et al.*(CLEO Collaboration), Phys. Rev. Lett. **74**, 3331 (1995); Artuso M *et al.* (CLEO Collaboration), Phys. Rev. Lett. **86**, 4479 (2001); Gorelov I V (CDF Collaboration), J. Phys. Conference Series. **69**, 012009 (2007).

[2] Davies C T H *et al.*, Fermilab, HPQCD, MILC, UKQCD collaborations, Phys. Rev. Lett. **92**, 022001 (2004).

[3] Davies C, Frascati Physics Series Vol. XLVI, 43 (2007).

[4] Bañuls, Pich A and Scimemi I, Phys. Rev. **D61** 94009 (2000).

[5] Bodwin G T, Braaten E and Lepage G P, Phys. Rev. **D51**, 1125 (1995) [Erratum-ibid. **D55**, 5853 (1997)] [arXiv:hep-ph/9407339].

[6] Isgur N, Karl G, Phys. Rev.**D18**, 4187 (1997).

[7] S. Godfrey *et al.*, Phys. Rev. **D32**, 189 (1985).

[8] Eichten E *et al.*, Phys. Rev. **D17**, 3090 (1978).

[9] N. Brambila *et al.*, European Phy. J. **c 71**, 1534 (2011).

[10] P C Vinodkumar, Proc. National Seminar on New Frontiers in Nuclear, Hadron and Mesoscopic Physics, Edited by V K B Kota and Arun Pratap, Allied Pub. Pvt. Ltd. Ahmedabad. 51-63 (2010).
[11] Amsler C *et al.* (Particle Data Group), Phys. Lett. **B**, 1 (2008).
[12] D. Besson *et al.*(CLEO Collaboration), Phys. Rev. **D68**, 032002 (2003).
[13] Antimo P, Nucl. Phys. **B156**, 105 (2006).
[14] Aubert B *et al.*(BaBar Collaboration), Phys. Rev. Lett. **101**, 071801 (2008).
[15] C Y Wong, Phys. Rev. **C 69** 055202 (2004).
[16] E Braaten and M Kusunoki, Phys. Rev. **D 69** 074005 (2004).
[17] E S Swanson, Phys. Lett. **B 588** 189 (2004).
[18] Acciarri M *et al.* (L3 Collobration), Phys. Lett. **B465**, 323 (1999).
[19] Paulini M (CDF and DO Collobrations) arXiv:hep-ex/0702047.
[20] BaBar Collaboration, Aubert B *et al.*, Phys. Rev. Lett. **90**, 242001 (2003).
[21] CLEO Collaboration, Besson D *et al.*, Phys. Rev. D **68**, 032002 (2003).
[22] Belle Collaboration, Mikani Y *et al.*, Phys. Rev. Lett. **92**, 012002 (2004).
[23] Godfrey S and Kokoski R, Phys. Rev. D **43**, 1679 (1991). Ebert D , Galkin V O, and Faustov R N, Phys. Rev. D **57**, 5663 (1998). Pierro M Di and Eichten E, Phys. Rev. D **64**, 114004 (2001). Lucha W and Schöberl F, Mod. Phys. Lett A **18**, 2837 (2003). Godfrey S, J. Phys. Conf. Ser. **9**, 59 (2005).
[24] Godfrey S and Kokoski R, Phys. Rev. **D43**, 1679 (1991).
[25] M. Di Pierro *et al.*, Phys. Rev. **D64**, 114004 (2001).
[26] Ebert D, Faustov R N and Galkin V O, Phys. Rev. **D57**, 014027 (1998).
[27] Bardeen W A, Eichen E J and Hill C T, Phys. Rev. **D68**, 054024 (2003).
[28] Colangelo P, Fazio F De and Ferrandes R, Nucl. Phys. (Proc. Suppl.)**D163**, 177 (2007).
[29] Falk A F and Mehen T, Phys. Rev. **D53**, 231 (1996).
[30] Bhavin Patel and P C Vinodkumar, J. Phys. G: Nucl. Part. Phys. **36** 115003 (2009);
[31] Bhavin Patel, Ajay Rai and P C Vinodkumar, J. Phys. G: Nucl. Part. Phys. **35** 065001 (2008).
[32] Ajay Majethiya, Bhavin Patel and P C Vinodkumar, Eur. Phys. J. **A 38** 307 (2008).
[33] Bhavin Patel and P C Vinodkumar, J. Phys. G: Nucl. Part. Phys. **36** 035003 (2009).
[34] Bhavin Patel, Ph.D. Thesis [Spectroscopy and decay properties of heavy flavour hadrons.] Sardar Patel University, (2009).
[35] Giannini M M, Santopinto E and Vassallo A, Eur. Phys. J. **A12**, 447 (2001).
[36] R. Bijker, F. Iachello and A. Leviatan, Annals of Physics **284**, 89-133 (2000).
[37] Patel B, Majethiya A and Vinodkumar P C, Pramana - J. Phys. **72**, 679 (2009).
[38] Kaushal Thakkar, Bhavin Patel, Ajay Majethiya and P C Vinodkumar,Pramana - J. Phys. **77**, 1053 (2011).
[39] Ajay Majethiya, Ph.D. Thesis [Properties of Heavy Flavour Baryons using Quark Models.] Sardar Patel University, (2011).
[40] Belle Collaboration [arXiv:1103.3419/hep-ex]
[41] Abe K *et al.* [Belle Collaboration], Phys. Rev. Lett. **96**. 082003 (2006).
[42] Aubert B *et al.* [BaBar Collaboration], Phys. Rev. Lett. **95**, 142001 (2005).
[43] Branes T, Godfrey S and Swanson E S, Phys.Rev. **D72**, 054026 (2005).
[44] Bali G S, Schilling K and Wachter A, Phys. Rev. **D56**, 2566 (1997).
[45] Anderson S *et al.* [CLEO Collaboration], Nucl. Phys. **A663**, 647 (2000).
[46] Abe K*et al.* [Belle Collaboration], Phys. Rev. **D69**, 112002 (2004).
[47] Link J M *et al.* [FOCUS Collaboration],Phys. Lett. **B586**, 11 (2004).
[48] Abe K *et al.* [Belle Collaboration], Belle report BELLE-CONF-0643, hep-ex/0608031.
[49] Aubert B *et al.* [BaBar Collaboration], Phys. Rev. Lett. **97**, 222001 (2006).
[50] Abe F *et al.*, CDF Collaboration, Phys. Rev. **D58**, 112004 (1998).
[51] Gouz I P, Kiselev V V, Likhoded A K, Romanovsky V I and Yushchenko O P, Phys. Atom. Nucl. **67**, 1559 (2004) [Yad. Fiz. **67**, 1581 (2004)] [arXiv:hep-ph/0211432].
[52] Aaltonen T *et al.*(CDF Collaboration),Phys. Rev. Lett. **99**, 202001 (2007) [arXiv:hep-ex/0706.3868].
[53] Abazov V *et al.* (DØCollaboration),Phys. Rev. Lett. **99**, 052001 (2007) [arXiv:hep-ex/0706.1690].
[54] Faessler A *et al.*, Phys. Rev. **D73**, 094013 (2006).
[55] M. A. Gomshi Nobarya, and R. Sepahvanda, Nucl. Phys. **B 741**, 34-41 (2006); [arXiv:hep-ph/508115v3].
[56] Mattson M *et al.*(SELEX Collaboration), Phys. Rev. Lett.**89**, 112001 (2002).
[57] Rosner J L, J. Phy. G : Nucl Part. Phy. **34**, S127 (2007).
[58] Seth K K, Frascati Physics Series Vol. XLVI, 407 (2007).
[59] Li W, Frascati Physics Series Vol. XLVI, 371 (2007).
[60] Seidel S, Frascati Physics Series Vol. XLVI, 397 (2007).

Journal of Physics: Conference Series **374** (2012) 012017 doi:10.1088/1742-6596/374/1/012017

Properties of heavy flavoured mesons using NRQCD formalism

Ajay Kumar Rai

Applied Physics Department, Sardar Vallabh National Institute of Technology, Surat, Gujarat, India, 395 007

E-mail: raiajayk@gmail.com,akr@ashd.svnit.ac.in

Abstract. The mass spectrum of the $c\bar{c}$ meson is reviewed in non-relativistic phenomenological quark anti-quark potential of the type $V(r) = -\frac{\alpha_c}{r} + Ar^{\nu}$, with ν varying from 0.5 to 2. The spin hyperfine and spin-orbit interactions are employed to obtain the masses of the pseudoscalar and vector mesons. The decay constants with QCD corrections are computed in this model. The di-gamma and di-leptonic decays of $c\bar{c}$ meson are investigated using some of the known potential models by incorporating radiative corrections up to the lowest order, bound state effects and contribution from quark/antiquark propagator. The decay rates of $c\bar{c}$ meson are studied in the NRQCD formalism, and also with the relativistic corrections of order v^4. These decay widths are also computed using the pNRQCD for comparison.

1. Introduction

Heavy flavoured mesons have a rich spectroscopy with many narrow states lying under the threshold of open flavour production. The investigation of the various properties of these mesons gives very important insight into heavy quark dynamics. Recently, there have been renewed interest in the spectroscopy of the heavy flavoured hadrons due to number of experimental facilities (CLEO at the Cornell Electron Storage Ring , DELPHI, Belle at KEK, BaBar, LHCb at the CERN, Panda at the GSI etc) which have been continuously providing and expected to provide more accurate and new information about the hadrons from low flavour to heavy flavour sector [1].

Among many theoretical attempts or approaches to explain the hadron properties based on its quark structure very few were successful in predicting the hadronic properties starting from mass spectra to decay widths. For the mass predictions, the non-relativistic potential models with Buchmüller and Tye [2], Martin [3, 4, 5], Log [6, 7], Cornell [8] etc., were successful at the heavy flavour sectors. There exist relativistic approaches for the study of the different hadronic properties [9, 10]. Some potential models have also predicted the masses and various decays of the heavy-heavy mesons which are in fair agreement with the experimental results [11, 12, 13, 14, 15, 16, 17, 18]. A comprehensive review of developments in heavy quarkonium physics is available in ref [19].

The new role of the heavy flavour studies as the testing ground for the non-perturbative aspects of QCD, demands extension of earlier phenomenological potential model studies on quarkonium masses to their predictions of decay widths with the non-perturbative approaches like NRQCD [20]. It is in terms of their short distance and long distance coefficients. The long distance

Journal of Physics: Conference Series **374** (2012) 012017 doi:10.1088/1742-6596/374/1/012017

coefficients are obtained through phenomenological potential model description of the mesons. The mass spectrum of $c\bar{c}$, meson in the potential scheme of coulomb plus power potential (CPP_ν) with the power index (ν), varying from 0.5 to 2. Spin hyperfine interaction is introduced to get the S-wave masses of the pseudoscalar and vector mesons. We present details of the non-relativistic treatment of the heavy quarks along with the computed results in section-2. The decay constants $f_{P,V}$ of these mesons incorporating QCD corrections up to $O(\alpha_s)$ are also presented in this section. In section-3, we present the details of the computations of the di-gamma decays of pseudoscalar state and the leptonic decay widths of the vector states of the $c\bar{c}$ quarkonium in the frame work of the NRQCD formalism as well as other treatments incorporating different correction terms to the respective decay widths. Though the NRQCD formalism takes advantage of the fact that heavy quark mass is much larger than the other energy scales such as the binding energy scale, Λ_{QCD} and $|\vec{p}|$, the energy fluctuations of the heavy quarks of the order of the light energy scale are implemented in pNRQCD [21, 22, 23]. Thus we also compute the decay widths in the pNRQCD formalism for comparison. Conclusions are drawn in section-4.

2. Non-relativistic Treatment for $Q\bar{Q}$ systems

For the study of heavy-heavy flavoured bound state system such as $c\bar{c}$ we consider a non-relativistic Hamiltonian given by [16, 17, 18]

$$H = M + \frac{p^2}{2m} + V(r) \tag{1}$$

where $M = m_Q + m_{\bar{Q}}$ is the total mass of quark and antiquark, m is reduced mass, p is the relative momentum of each quark and $V(r)$ is the quark antiquark potential. Recently, we have considered a general power potential with colour coulomb term of the form

$$V(r) = \frac{-\alpha_c}{r} + Ar^\nu \tag{2}$$

as the static quark-antiquark interaction potential (CPP_ν) [24, 25]. Here, for the study of mesons, $\alpha_c = \frac{4}{3}\alpha_s$, α_s being the strong running coupling constant, A is the potential parameter and ν is a general power. For the present study of heavy-heavy flavour mesons, we employ the exponential trial wave function of the hydrogenic type to generate the Schrödinger mass spectra. Within the Ritz variational scheme using the trial radial wave function we obtain the expectation values of the Hamiltonian as $\langle H\rangle = E(\mu, \nu)$, Which gives the spin average mass of the ground state. For excited states the trial wave function is multiplied by an appropriate orthogonal polynomial function such that the excited trial wave function gets ortho-normalized (For more details see Refs. [26, 27, 28, 29, 30]). Parameters used in calculations are $\alpha_c(c\bar{c}) = 0.40$, $m_c = 1.31\ GeV$ and A(taken from ref[26]). The spin-spin interaction among the constituent quarks provides the mass splitting of $J = 0^{-+}$ and 1^{--} states. Accordingly, the spin-spin interactions are taken as [31]

$$V_{S_Q\cdot S_{\bar{Q}}}(r) = \frac{8}{9}\frac{\alpha_s}{m_Q m_{\bar{Q}}}\vec{S}_Q\cdot\vec{S}_{\bar{Q}}\,4\pi\delta(r) \tag{3}$$

The decay constants of mesons are important parameters in the study of leptonic or non-leptonic weak decay processes. The decay constants of pseudoscalar (f_P) and vector (f_V) mesons are obtained by incorporating a first order QCD correction factor, we compute them using the relation,

$$f^2_{P/V} = \frac{12\left|\psi_{P/V}(0)\right|^2}{M_{P/V}}\left(1 - \frac{\alpha_s}{\pi}\left[2 - \frac{m_Q - m_{\bar{Q}}}{m_Q + m_{\bar{Q}}}\,ln\frac{m_Q}{m_{\bar{Q}}}\right]\right) \tag{4}$$

Table 1. Mass Spectra (in GeV) of $c\bar{c}$ meson (S states).

State	Potential Index ν										[1]	[33]
	0.5	0.7	0.9	1.0	1.1	1.3	1.5	1.7	1.9	2.0		
1^1S_0	3.000	2.980	2.960	2.950	2.942	2.926	2.912	2.899	2.887	2.882	2.980	2.979
1^3S_1	3.092	3.100	3.109	3.112	3.116	3.123	3.129	3.134	3.141	3.144	3.097	3.097
2^1S_0	3.352	3.427	3.495	3.522	3.549	3.597	3.636	3.668	3.696	3.708	3.637	3.623
2^3S_1	3.375	3.464	3.547	3.583	3.619	3.683	3.739	3.788	3.832	3.852	3.686	3.673
3^1S_0	3.541	3.697	3.846	3.912	3.979	4.102	4.212	4.309	4.396	4.436		3.991
3^3S_1	3.553	3.717	3.878	3.95	4.024	4.161	4.285	4.395	4.500	4.547	4.039	4.022
4^1S_0	3.684	3.912	4.140	4.246	4.354	4.558	4.748	4.920	5.082	5.158		4.250
4^3S_1	3.689	3.922	4.157	4.267	4.379	4.592	4.792	4.975	5.149	5.230	4.263	4.270
5^1S_0	3.799	4.093	4.395	4.539	4.687	4.973	5.243	5.497	5.739	5.853		4.446
5^3S_1	3.803	4.101	4.408	4.555	4.707	5.001	5.281	5.544	5.797	5.916	4.421	4.463
6^1S_0	3.899	4.254	4.628	4.809	4.997	5.366	5.722	6.060	6.387	6.543		4.595
6^3S_1	3.903	4.260	4.638	4.823	5.014	5.390	5.754	6.101	6.439	6.600		4.608

Table 2. Pseudoscalar and Vector decay constants(f_P and f_V) in MeV of $c\bar{c}$ States

ν	1S		2S		3S		1S		2S		3S	
	f_P	f_{Pcor}	f_P	f_{Pcor}	f_P	f_{Pcor}	f_V	f_{Vcor}	f_V	f_{Vcor}	f_V	f_{Vcor}
0.5	430	348	204	165	140	113	437	353	205	166	141	114
0.7	495	401	256	207	184	149	506	409	258	208	184	149
0.9	549	444	304	246	224	181	564	456	306	248	225	182
1.0	572	463	325	263	243	196	589	476	328	265	244	197
1.1	593	480	345	279	261	211	613	495	349	282	262	212
1.3	627	507	383	309	295	239	655	527	387	313	297	240
1.5	663	536	416	336	325	263	692	559	422	341	328	266
1.7	691	559	445	360	349	282	724	585	453	366	282	285
1.9	716	579	472	382	379	306	753	609	481	389	383	310
2.0	728	589	484	391	390	316	767	620	493	399	395	320
[34]	413	327	288	228	241	191	420	304	290	210	242	175

Where quantity in bracket is the QCD correction factor given by [32]. The computed f_P and f_V for $c\bar{c}$ system using Eqn(4) and results are tabulated in Table(2).

3. Decay rates of quarkonia

Along with the mass spectrum, successful predictions of various decay widths of heavy flavoured systems have remained as testing ground for the success of phenomenological models. As an attempt to improve the theoretical predictions involving the phenomenological description of the meson, using the redial wave functions and other model parameters of the different potential models we study the decay of $^1S_0$ quarkonium into gamma pairs and the decay of $^3S_1$ into leptons pairs based on NRQCD formalism [20]. It is expected that the NRQCD formalism has all the corrective contributions for the right predictions of the decay rates. NRQCD factorization expressions for the decay rates of quarkonium is given by [35]

$$\Gamma(^1S_0 \to \gamma\gamma) = \frac{F_{\gamma\gamma}(^1S_0)}{m^2}X + \frac{G_{\gamma\gamma}(^1S_0)}{m^4}Y \tag{5}$$

$$\Gamma(^3S_1 \to e^+e^-) = \frac{F_{ee}(^3S_1)}{m^2}X + \frac{G_{ee}(^3S_1)}{m^4}Y \tag{6}$$

Table 3. Decay rates of $0^{-+} \to \gamma\ \gamma$ and the relevant correction terms of η_c meson.

Models	Γ_0 (keV)	Γ_B (keV)	Γ_C (keV)	Γ_R (keV)	Γ_T (keV)	Γ_{NRQCD} (keV)	Γ_{wrc} (keV)	Γ_{pNRQCD} (keV)	Γ_{Others} (keV)
ERHM	7.460	0.645	-0.860	-2.855	4.391	4.005	7.956	8.450	
BT	10.870	0.146	-0.195	-4.206	6.615	6.554	7.406	11.291	7.4
PL(Martin)	13.406	-5.578	7.438	-6.196	9.069	8.432	5.007	8.542	±1.4[1]
Log	10.937	-0.147	0.196	-4.349	6.636	6.309	7.210	10.965	8.5[33]
Cornell	19.512	-9.167	12.222	-6.581	16.066	13.787	4.792	12.957	7.5 [36]
CPP_ν= 0.5	8.173	2.070	-2.761	-2.635	4.847	9.622	10.535	11.581	9.02
0.7	10.918	2.625	-3.500	-3.521	6.522	13.323	10.179	15.231	±0.8[20]
0.9	13.465	3.093	-4.124	-4.342	8.092	16.990	9.910	18.602	9.32[34]
1.0	14.649	3.277	-4.370	-4.724	8.833	18.700	9.804	20.093	9.684[25]
1.1	15.812	3.461	-5.615	-5.099	9.560	20.425	9.706	21.558	
1.3	17.987	3.767	-5.0221	-5.800	10.931	23.710	9.552	24.245	
1.5	19.971	4.005	-5.340	-6.440	12.196	26.768	9.431	26.677	
1.7	22.788	4.190	-5.586	-7.026	13.363	29.615	9.338	28.838	
1.9	23.502	4.347	-5.796	-7.578	14.474	32.353	9.265	30.836	
2.0	24.297	4.418	-5.890	-7.835	14.990	33.661	9.234	31.784	

ERHM [14, 15], BT [2], PL (Martin) [3], Log [6], Cornell [8]

The long distance coefficients X and Y are the NRQCD matrix elements for the decay. The vacuum saturation approximations allowed the matrix elements of the four fermion operators to be expressed in terms of renormalized wave function parameters as [20]

$$X = \left\langle {}^{2S+1}L_J|O_1(h)|{}^{2S+1}L_J\right\rangle \simeq \frac{2}{3\pi}\left|R_{cw}(0)\right|^2\left[1+O(v^4)\right] \tag{7}$$

$$Y = \left\langle {}^{2S+1}L_J|P_1(h)|{}^{2S+1}L_J\right\rangle \simeq -\frac{2}{3\pi}\left|R_{cw}\,\nabla^2\,R_{cw}\right|\left[1+O(v^4)\right] \tag{8}$$

The short distance coefficients F's and G's of the order of α_s^2 and α_s^3 are given by [35]

$$F_{\gamma\gamma}({}^1S_0) = 2\pi Q^4\alpha^2\left[1+\left(\frac{\pi^2}{4}-5\right)C_F\frac{\alpha_s}{\pi}\right],\quad G_{\gamma\gamma}({}^1S_0) = -\frac{8\pi Q^4}{3}\alpha^2 \tag{9}$$

$$F_{ee}({}^3S_1) = \frac{2\pi Q^2\alpha^2}{3}\left\{1-4C_F\frac{\alpha_s(m)}{\pi}+\left[-117.46+0.82n_f+\frac{140\pi^2}{27}ln\left(\frac{2m}{\mu_A}\right)\right]\left(\frac{\alpha_s}{\pi}\right)^2\right\} \tag{10}$$

$$G_{ee}({}^3S_1) = -\frac{8\pi Q^2}{9}\alpha^2 \tag{11}$$

Where $C_F = (N_c^2-1)/(2N_c) = 4/3$, and $C_A = N_c = 3$ is the colour factor. .

3.1. Decay rates with relativistic corrections

We compute the decay widths of quarkonia in NRQCD formalism incorporating the relativistic correction through $O(v^4)$ as [35] $\Gamma_{wrc}({}^1S_0 \to \gamma\gamma)$

$$= 2\left[\left(\frac{1}{m^2}\right)F_{\gamma\gamma}({}^1S_0)+\left(\frac{p^2}{m^4}\right)G_{\gamma\gamma}({}^1S_0)+\left(\frac{p^4}{m^6}\right)H^1_{\gamma\gamma}({}^1S_0)+\left(\frac{p^4}{m^6}\right)H^2_{\gamma\gamma}({}^1S_0)\right] \tag{12}$$

Table 4. Decay rates of $1^{--} \to l^+ \, l^-$ and the relevant correction terms of J/ψ meson.

Systems	Models	Γ_{VW} (keV)	Γ_{cor} (keV)	Γ_{rad} (keV)	Γ_{T} (keV)	Γ_{NRQCD} (keV)	Γ_{wrc} (keV)	Γ_{pNRQCD} (keV)	Γ_{EXP}[1] (keV)
	ERHM	5.595	-0.289	-3.381	1.925	2.542	6.040	4.360	
J/ψ	BT	8.152	-0.240	-4.982	2.930	2.539	5.623	5.810	5.55
	PL (Martin)	10.055	-1.089	-7.341	3.803	3.312	3.801	3.473	±14
	Log	8.203	-0.171	-5.152	2.879	1.956	5.474	5.515	± 0.02
	Cornell	14.634	-1.837	-7.701	8.769	7.925	3.638	7.457	6.60[33]
	CPP_ν= 0.5	6.130	-3.122	-0.624	2.384	4.212	7.927	6.693	7.37[34]
	0.7	8.189	-4.171	-0.845	3.173	6.199	7.811	8.990	
	0.9	10.153	-5.171	-1.065	3.917	8.246	7.485	11.198	
	1.0	11.053	-5.629	-1.165	4.259	9.353	7.412	12.210	
	1.1	11.946	-6.084	-1.268	4.594	10.430	7.345	13.242	
	1.3	13.621	-6.937	-1.463	5.221	12.558	7.241	15.150	
	1.5	15.165	-7.723	-1.645	5.797	14.636	7.162	16.938	
	1.7	16.582	-8.445	-1.813	6.324	16.643	7.103	18.572	
	1.9	17.920	-9.126	-1.982	6.812	18.659	7.060	20.163	
	2.0	18.549	-9.447	-2.061	7.041	19.634	7.043	20.908	

ERHM [14, 15], BT [2], PL (Martin) [3], Log [6], Cornell [8]

where the short-distance coefficients at leading order in α_s^0 are

$$F_{\gamma\gamma}(^1S_0) = 2\pi Q^4\alpha^2, \quad G_{\gamma\gamma}(^1S_0) = -\frac{8\pi}{3}Q^4\alpha^2, \quad H^1_{\gamma\gamma}(^1S_0) + H^2_{\gamma\gamma}(^1S_0) = \frac{136\pi}{45}Q^4\alpha^2 \tag{13}$$

We see that, in NRQCD through relativistic corrections of $O(v^4)$, the decay width for a $^3S_1$ state in to a e^+e^- pair is is given by $\Gamma_{wrc}(^3S_1 \to e^+e^-)$

$$= 2\left[\left(\frac{1}{m^2}\right)F_{ee}(^3S_1) + \left(\frac{p^2}{m^4}\right)G_{ee}(^3S_1) + \left(\frac{p^4}{m^6}\right)H^1_{ee}(^3S_1) + \left(\frac{p^4}{m^6}\right)H^2_{ee}(^3S_1)\right] \tag{14}$$

where the short-distance coefficients at leading order in α_s^0 are

$$F_{ee}(^3S_1) = \frac{2\pi}{3}Q^2\alpha^2, \quad G_{ee}(^3S_1) = -\frac{8\pi}{9}Q^2\alpha^2, \quad H^1_{ee}(^3S_1) + H^2_{ee}(^3S_1) = \frac{58\pi}{54}Q^2\alpha^2 \tag{15}$$

where p and m are the relative momentum and mass of the quarks. The relative momentum in our computations using CPP_ν models are obtained from the Fermi momentum parameter $\bar{\mu}$. For comparison, we also compute these decay widths using pNRQCD formalism [23] as well as based on a systematically improved gauge invariant description of all hard processes involving the heavy quark anti-quark systems of pseudoscalar and vector mesons given by [37].

Accordingly, the two photon decay width of the pseudoscalar meson is given by [17]

$$\Gamma_T(0^{-+}{\to}2\gamma) = \Gamma_0 + \Gamma_B + \Gamma_C + \Gamma_R \tag{16}$$

Here Γ_0 is the conventional Van Royen-Weisskopf term for the $0^{-+} \to \gamma\gamma$ decays [38], while Γ_B, Γ_c and Γ_R are the corrections due to bound state effects, correction coming from the quark propagator and the radiative corrections respectively for this decay. Where the different terms of Eqn.(17) are computed as [37]

$$\Gamma_0 = \frac{12\alpha_e^2 e_Q^4}{M_P^2} R^2(0), \ \Gamma_B = -2\frac{\varepsilon_B}{M_P}\Gamma_0, \ \Gamma_c = \frac{16}{3M_P^2}\frac{\nabla^2 R(0)}{R(0)}\Gamma_0, \ \Gamma_R = \frac{\alpha_s}{\pi}\left(\frac{\pi^2-20}{3}\right)\Gamma_0 \quad (17)$$

Here the quark antiquark binding energy ε_B is defined in terms of the m_1 and m_2 of the quark and antiquark masses as $\varepsilon_B = m_1 + m_2 - M_P$. Similarly, the leptonic decay width of the vector meson with correction terms is computed as

$$\Gamma_T(1^{--} \to l^+ l^-) = \Gamma_{VW} + \Gamma_{cor} + \Gamma_{rad} \quad (18)$$

where

$$\Gamma_{VW} = \frac{4\alpha_e^2 e_Q^2}{M_V^2} R^2(0), \ \Gamma_{rad} = \frac{-16}{3\pi}\alpha_s \Gamma_{VW} \ \ and \ \ \Gamma_{cor} = \frac{4}{3M_V^2}\frac{\nabla^2 R(0)}{R(0)}\Gamma_{VW} \quad (19)$$

Γ_{rad}, Γ_{cor} are the radiative correction and the correction term due to the quark propagator within the vector meson. It is obvious to note that the computations of the decay rates and the various correction terms described here require the right description of the meson state through its radial wave function at the origin, R(0) and its mass M along with other model parameters like α_s and quark masses. Generally, due to lack of exact solutions for colour dynamics, R(0) and M can be considered as free parameters of the theory [37]. However, we found appropriate to employ phenomenological model predictions of the mesonic mass and corresponding wave function to calculate the decay rates.

4. Discussion and conclusion

Here we have studied the mass spectrum (S-Wave), decay constants with and without QCD corrections and decay rates in CPP_ν and other potential models. The potential model parameters and the masses of the charmed quark obtained from the respective potential model predictions have been employed to study their decay properties in the conventional Van Royen-Weisskopf formula with various corrections, in NRQCD with and without relativistic corrections and also in pNRQCD formalism.

It is interesting to note that the ERHM[15] predictions of the di-gamma decay widths of η_c and leptonic decay width of J/ψ are in good agreement with the respective experimental results with out any corrections. The decay properties studied here shown similar behaviour with NRQCD and pNRQCD formulism. While it is found that the relativistic correction of $O(v^4)$ contribute considerably to the decay widths.

The Γ_{NRQCD}, widths for $\eta_c \to \gamma\gamma$ computed using the parameters of the phenomenological models, BT, PL and Log are in excellent agreement with the respective experimental values (see Tables 3). Their predictions along with CPP_ν using the conventional formula Γ_0 with corrections (Γ_T) are also close to the experimental values at $\nu = 0.7$. With the relativistic corrections ERHM, BT and Log potential results are close to experimental as well as other theoretical values but in case of pNRQCD only ERHM and PL results are in order rest are overestimating. In case di-leptonic decay of J/ψ in conventional formula, NRQCD and CPP_ν at 1.3 and 0.7 are close to experimental values. But in case of relativistic corrections and pNRQCD formalism BT and Log predictions are better. We have calculated the decay rates in various potential models using their input parameters like quark masses and running coupling constant and predicted meson mass and wave function at origin. All these parameters and predictions are different in different potential models. Therefore the decay rates calculated in these models are not in mutual agreement with each other.

The present study of the decay rates of quarkonia clearly indicates the relative importance of QCD related corrections on the phenomenological potential models. The success of CPP_ν in the determination of the Swave masses and decay rates of $c\bar{c}$ system provide future scopes to study various transition rate and excited states of these mesonic system. With the masses and wave functions of the heavy flavour mesons at hand, it would be rather simple to compute various transition rates such as $E1$ and $M1$ in these mesons.

4.1. Acknowledgments

Author would like to thanks Prof. P. C. Vinodkumar of Sardar Patel University, Vallabh Vidyanagar, Anand, Gujarat, for the valuable discussion.

5. References

[1] Particle Data Group, K. Nakamura et. al. 2010 *Jnl. Phys G.* **37** *075021*
[2] Buchmuller and Tye 1981 *Phys.Rev* **D24** *132*
[3] A. Martin 1980 *Phys. Lett.* **93 B** *338*
[4] A. Martin 1979 *Phys. Lett.* **82 B** *272*
[5] J. L. Richardson 1979 *Phys. Lett.* **82 B** *272*
[6] C. Quigg and J. L. Rosner 1977 *Phys. Lett. 71 B 153*
[7] C. Quigg and J. L. Rosner 1979 *Phys. Rept.* **56** *167*
[8] E. Eichten et. al. 1978 *Phys. Rev.* **D 17** *3090*
[9] Altarelli G, Cabibbo N, Corbo G, Maiani L and Martinelli G 1982 *Nucl. Phys.* **B 208.** *365*
[10] D. Ebert R.N. Faustov and V. O. Galkin 2003 *Phys. Rev.* **D 67** *014027*
[11] S. N. Gupta, J. M. Johnson and W. W. Repko 1996 *Phys. Rev.* **D 54** *2075*
[12] S. Godfrey 1986 *Phys. Rev.* **D 33** *1391*
[13] Hwang D S, Kim C S and Wuk Namgung 1996 *Phys. Rev.* **D 53** *4951*
[14] J N Pandya and P C Vinodkumar 2001 *Pramana J. Phys* **57** *821*
[15] P. C. Vinodkumar, J. N. Pandya, V. M. Bannur and S. B. Khadkikar 1999 *Eur. Phys. J.* **A 4** *83*
[16] Ajay Kumar Rai, R H Parmar and P C Vinodkumar 2002 *Jnl. Phys G.* **28**, *2275*
[17] Ajay Kumar Rai, J N Pandya and P C Vinodkumar 2005 *Jnl. Phys G.* **31** *1453*
[18] Ajay Kumar Rai and P C Vinodkumar 2006 *Pramana J. Phys.* **66** *953*
[19] N. Barmbilla, et al. 2004 *CERN Yellow Report, CERN-2005-005, Geneva: arXive:hep-ph/0412158*
[20] Bodwin G T, Braaten Erich and Lepage G P 1995 *Phys. Rev.* **D 51**, *1125;* **55** *(E) 1997*
[21] A. Pineda and J. Soto 1998 *Nucl. Phys.* **B**, *Proc. Suppl,* **64** *428*
[22] Nora Brambilla et. al 2003 *Phys. Rev.* **D 67** *034018*
[23] Nora Brambilla, Antonia Pineda and Joan Soto, Antonio Vairo 2005 *Rev. Mod. Phys.* **77** *1423*
[24] Naynesh Devlani and Ajay Kumar Rai and P C Vinodkumar 2011 *Phys. Rev.* **D 84** *074030*
[25] A. Parmar Bhavin Patel and P C Vinodkumar 2010 *Nucl. Phys.* **A 848** *299*
[26] Ajay Kumar Rai, B. Patel and P.C. Vinodkumar 2008 *Phys. Rev.* **C 78** *055202*
[27] Ajay Kumar Rai, J.N. Pandya and P.C. Vinodkumar 2008 *Eur. Phys. J.* **A 38** *77*
[28] Ajay Kumar Rai and P.C. Vinodkumar 2010 *AIP Conf. Proc.* **1257** *316*
[29] Ajay Kumar Rai 2011 *AIP Conf. Proc.* **1343** *415*
[30] J.N. Pandya, Ajay Kumar Rai and P.C. Vinodkumar 2007 *hep-ph/0701026*
[31] Gershtein S S, Kiselev V V, Likhoded A K and Tkabladze A V 1995 *Phys. Rev.* **D 51**, *3613*
[32] E. Braaten and S. Fleming 1995 *Phys. Rev.* **D 52** *181*
[33] Bai-Quing Li and Kuang-Ta Chao 2009 *Phys. Rev.* **D 79** *094004*
[34] Bhavin Patel and P C Vinodkumar 2009 *J. Phys.* **G 36** *035003*
[35] Bodwin G T and Petrelli A 2002 *Phys. Rev.* **D 66** *094011*
[36] Braaten Erich and Lee J 2003 *Phys. Rev.* **D 67** *054007*
[37] Hafsakhan and Pervez Hoodbhoy 1996 *Phys. Rev.* **D53**, *2534*
[38] R. Van Royen and V. F. Weisskopf 1967 *Nuovo Cimento* **50**, *617*

Energy of partial A-body problem for ${}^{9}_{\Lambda}$Be and ${}^{10}_{\Lambda\Lambda}$Be

Sonika

Theory Division, Saha Institute of Nuclear Physics, 1/AF Bidhan Nagar, Kolkata - 700 064, INDIA

E-mail: sonikag7@gmail.com

Abstract. The energy of the degenerate doublet $(3^+/2, 5^+/2)$ of ${}^{9}_{\Lambda}$Be, treating it as a partial nine-body system in the $\Lambda\alpha\alpha$ cluster model, has been calculated in the variational Monte Carlo framework. A simplified treatment, with the central two-body Urbana type ΛN and the three-body dispersive and two-pion exchange ΛNN forces along with the central two- and three-body correlations, is found to be adequate in explaining the energy of observed γ-ray transition from the excited degenerate doublet to the ground state. This partial nine-body model of ${}^{9}_{\Lambda}$Be has been extended for the case of partial ten-body problem ${}^{10}_{\Lambda\Lambda}$Be in the $\Lambda\Lambda + \alpha\alpha$ model where nucleonic degrees of freedom of αs are taken into consideration ignoring antisymmetrization between two αs. The central two-body ΛN and $\Lambda\Lambda$ and the three-body dispersive and two-pion exchange ΛNN forces, constrained by the Λp scattering data and the observed ground state energies of ${}^{5}_{\Lambda}$He and ${}^{6}_{\Lambda\Lambda}$He, are employed. The product type trial wavefunction predicts binding energy for the ground state considerably less than for the event reported by Danysz *et al* but however, consistent with the value deduced assuming a γ-ray of 3.04 MeV must have escaped undetected in the decay of the product ${}^{9}_{\Lambda}\mathrm{Be}^* \longrightarrow {}^{9}_{\Lambda}\mathrm{Be} + \gamma$ of the emulsion event ${}^{10}_{\Lambda\Lambda}\mathrm{Be} \longrightarrow \pi^- + p + {}^{9}_{\Lambda}\mathrm{Be}^*$ and for the excited 2^+ state closer to the value measured in the Demachi-Yanagi event. The results of the present work are consistent with the earlier three- and four-body cluster model approaches where αs are assumed to be structureless entities.

1. Introduction

Recently, Shoeb [1] and coworkers [2, 3, 4] have analysed the ground state binding energies of s- and p-shell hypernuclei and excited states of p-shell hypernuclei in the α cluster model using the variational Monte Carlo (VMC) method. In these analyses [1, 2, 3], a phenomenological dispersive three-body $\Lambda\alpha\alpha$ force of Yukawa shape was proposed in analogy with the one suggested in explaining the spectra of ^{12}C using the three-body cluster $\alpha\alpha\alpha$ potential [5, 6]. The phenomenological dispersive three-body $\Lambda\alpha\alpha$ force [1, 2, 3] along with the appropriate $\Lambda\alpha$, $\alpha\alpha$ and $\Lambda\Lambda$ potentials explains the ground and excited state energies of ${}^{9}_{\Lambda}$Be and ${}^{10}_{\Lambda\Lambda}$Be. The need to incorporate a repulsive dispersive three-body force has been pointed out not only in the α cluster model analysis [7] but investigated in detail in the earlier microscopic study [8]. In contrast to analyses [1, 2, 3, 4], Hiyama *et al* [9] have found the modified odd/even state ΛN potential important in explaining the binding energy and excited state data of $S = -1$ and -2 hypernuclei. Although these potentials have been microscopically calculated but were tailored to adjust the Λ-cluster potentials so that ground-state or ground- and excited-state energies of hypernuclei containing clusters are reproduced thus rendering these as phenomenological ones. In the calculation of the properties of ${}^{9}_{\Lambda}$Be in the $\Lambda\alpha\alpha$ model, using the Faddeev method, Cravo *et al* [10] needs neither a dispersive force nor modified odd/even state ΛN potential. In the last few years, other cluster model analyses [11, 12] have also been performed for the binding energy of s-shell hypernuclei. Here it is noteworthy to point out that oldest calculation for the excited state of ${}^{9}_{\Lambda}$Be is by Ali *et al* [13].

In all the above-mentioned analyses, α clusters are treated as rigid entities devoid of structure and thus the effects of the role played by the dynamical correlations among the baryons are not explicitly manifested in the energy calculation. Therefore, such studies are deficient in accommodating the realities of the internal structure and seem, therefore, far from satisfactory. Earlier, Shoeb *et al* [14] have calculated the ground state energy of ${}^{9}_{\Lambda}$Be, treating it as a partial nine-body system in the $\Lambda + 2\alpha$ model. This work presupposes the existence of the αs structure. The two-body NN correlations within αs were explicitly incorporated. However, the effect of NN correlations, where each α contributes a nucleon, is simulated through $\alpha\alpha$ correlation. Thus, antisymmetrization between two αs has been ignored. However, soft repulsive core in the $\alpha\alpha$ potential [6] simulates the effect of NN antisymmetrization in the wavefunction. The three-body ΛNN correlations were included in the trial wavefunction but ΛN space-exchange correlations were ignored. From this study it was concluded that dispersive three-body ΛNN force and space-exchange ΛN potential can not be determined uniquely. The presence of one masks the determination of other. However, B_{Λ} of ${}^{9}_{\Lambda}$Be is satisfactorily explained.

The energy of the ground and excited states of ${}^{10}_{\Lambda\Lambda}$Be has been extensively analysed in the $\Lambda\Lambda\alpha\alpha$ cluster model in [1, 2, 9, 15, 16] using a variety of methods: VMC framework, Gaussian-basis coupled-rearrangement channel method and Faddeev-Yakubovsky method. The input $\Lambda\Lambda$, $\Lambda\alpha$, and $\alpha\alpha$ potentials with soft repulsive core and reasonable shapes, constrained by the data relevant to the each interacting pair, have been used. The analyses [2, 9] predict the ground state $\Lambda\Lambda$ binding of ${}^{10}_{\Lambda\Lambda}$Be, $B_{\Lambda\Lambda}$ about 15% less than the currently accepted experimental [17] value 17.6 ± 0.4 MeV. Thus strengthening the speculation that in the measurement of $B_{\Lambda\Lambda}$, a γ-ray [18] of 3.04 MeV must have escaped undetected from the decay products of ${}^{10}_{\Lambda\Lambda}$Be in the emulsion. Further, the cluster model calculations of Shoeb [1] and Hiyama *et al* [9] for the energy of the 2^{+} excited state of ${}^{10}_{\Lambda\Lambda}$Be incidentally agree with the $\Lambda\Lambda$ binding energy $B_{\Lambda\Lambda} = 12.33^{+0.35}_{-0.21}$ MeV found in the E373 experiment, named as Demachi-Yanagi event [19]. Thus Demachi-Yanagi event, based on the calculation of $\Lambda\Lambda\alpha\alpha$ cluster model, is identified with the 2^{+} excited state of ${}^{10}_{\Lambda\Lambda}$Be.

The satisfactory application of a partial nine-body problem [14] within the $\Lambda + 2\alpha$ model in explaining the B_{Λ} of ${}^{9}_{\Lambda}$Be, though for a single set of potential parameters, motivated us to apply it to analyse the experimental [18] binding energy $B_{\Lambda} = 3.67$ MeV of the degenerate doublet $(3^{+}/2, 5^{+}/2)$ of ${}^{9}_{\Lambda}$Be. We have also extended the partial nine-body model for the case of ${}^{10}_{\Lambda\Lambda}$Be, treating it as a partial ten-body problem in the $\Lambda\Lambda + \alpha\alpha$ model and to investigate how far it succeeds in explaining the energy of the ground and 2^{+} excited states. We may point out that antisymmetrization which has been ignored among the nucleons of two well separated αs is being simulated through the soft repulsive core in the $\alpha\alpha$ potential [6, 20]. Following the spirit of the earlier work [14], for the simplicity of the calculation, we have chosen simple baryon-baryon potentials along with the corresponding simple correlation functions.

The paper is organized as follows: In the next section, we briefly describe the Hamiltonians of ${}^{9}_{\Lambda}$Be and of hypernucleus ${}^{10}_{\Lambda\Lambda}$Be in the α cluster model along with the trial wavefunctions and the energy calculations. The expressions for the ΛN, NN, $\alpha\alpha$ and $\Lambda\Lambda$ and three-body ΛNN potentials are given in our earlier work [21, 22]. In section 3, the results and discussion are presented. The conclusions of our study are given in the last section.

2. Hamiltonians of the p-shell hypernuclei in the α cluster model, Trial Wavefunctions and Energy Calculations

2.1. Hamiltonians of the p-shell hypernuclei

The 2^+ state of ${}^{10}_{\Lambda\Lambda}$Be is assumed to be built on the first excited state $J_C = 2^+$ of the core nucleus ^{8}Be. $J_C = 2^+$ is a coupled state of $L_C = 2$, and $S_C = 0$. The two Λ particles of $s_\Lambda = 1/2$ coupled to spin singlet function ${\chi^0}_0$ when combined to the $J_C = 2^+$ of ^{8}Be core in ${}^{10}_{\Lambda\Lambda}$Be gives rise the $J = 2^+$ spin state. The Hamiltonian of the $A(=10)$ baryons system ${}^{10}_{\Lambda\Lambda}$Be in the $2\alpha + 2\Lambda$ is given as:

$$\begin{aligned} H_H^{10} &= \sum_{i=1}^{A-2} K_N(i) + \sum_{i<j}^{4} V_{NN}(r_{ij}) + \sum_{i<j\, i=5}^{A-2} V_{NN}(r_{ij}) + V_{\alpha\alpha}^{(l)}(r_{\alpha_1\alpha_2}) + \sum_{k=1}^{2}\Big(K_{\Lambda_k} + \sum_{i=1}^{4} V_{\Lambda N}(r_{\Lambda_k i}) \\ &+ \sum_{i=5}^{A-2} V_{\Lambda N}(r_{\Lambda_k i}) + \sum_{i<j}^{A-2} V_{ij\Lambda}(r_{i\Lambda_k}, r_{j\Lambda_k})\Big) + V_{\Lambda\Lambda}(r_{\Lambda_1\Lambda_2}), \end{aligned} \tag{1}$$

where labels 1 to 8 specify the nucleons and symbols α_1 and α_2 are identified with the α particles. V_{xy} denotes the potential for a pair of particles xy ($=NN$, ΛN, $\Lambda\Lambda$, $\alpha\alpha$) and in the case of $\alpha\alpha$ pair, $V_{\alpha\alpha}^{(l)}$ is potential in the relative angular momentum l (=0 for the ground and 2 for the excited state). The three-body potential $V_{ij\Lambda}$ is the sum of the dispersive force $V_{\Lambda NN}^D$ and the two-pion exchange (TPE) three-body force $V_{\Lambda NN}^{2\pi}$. The contribution of $\langle V_{\Lambda NN}^D\rangle$ to the energy, from triads ΛNN, where each α contributes a nucleon, is substantial as shown in [8, 14], neglecting it not only over binds the ${}^9_\Lambda$Be but also ${}^{10}_{\Lambda\Lambda}$Be. The expressions for the ΛN, NN, $\alpha\alpha$ and $\Lambda\Lambda$ and for three-body ΛNN potentials used here are given in [21, 22]. The suppression of Λ index in equation (1) leads to the Hamiltonian of the partial nine-body problem ${}^9_\Lambda$Be.

2.2. Trial Wavefunctions

The variational wavefunction for ${}^{10}_{\Lambda\Lambda}$Be in the state (J, J_z) as usual is constructed from the product of central two-body correlation functions f_{xy} (for $xy = \alpha\alpha$, relative angular momentum state $l = 0$ or 2), three-body correlations $f_{\Lambda NN}$ and the ls coupled function $(y_{lm}(\Omega_{\alpha_1\alpha_2})\otimes{\chi^0}_0)_{JJ_z}$, a appropriate combination of ${\chi^0}_0$, spin singlet function of the two Λ particles and $y_{lm}(\Omega_{\alpha_1\alpha_2})$, spherical harmonic for the motion of two αs in relative angular momentum state l:

$$\begin{aligned} \Psi_H^{(10)}(J,J_z) = \prod_{k=1}^{2}\Bigg(\Big[\prod_{i=1}^{4} f_{\Lambda N}(r_{\Lambda_k i})\Big]\Big[\prod_{i=5}^{A-2} f_{\Lambda N}(r_{\Lambda_k i})\Big]\Big[\prod_{i<j\, i=1}^{4} f_{\Lambda NN}(r_{i\Lambda_k}, r_{j\Lambda_k})\Big] \\ \times\Big[\prod_{i<j\, i=5}^{A-2} f_{\Lambda NN}(r_{i\Lambda_k}, r_{j\Lambda_k})\Big]\Bigg)\Big[\prod_{i<j}^{4} f_{NN}(r_{ij})\Big]\Big[\prod_{i<j\, i=5}^{A-2} f_{NN}(r_{ij})\Big] \\ \times f_{\alpha\alpha}^{(l)}(r_{\alpha_1\alpha_2}) f_{\Lambda\Lambda}(r_{\Lambda_1\Lambda_2})(y_{lm}(\Omega_{\alpha_1\alpha_2})\otimes{\chi^0}_0)_{JJ_z}. \end{aligned} \tag{2}$$

The central two-body spin-independent correlation functions $f_{\Lambda N}$, f_{NN}, $f_{\alpha\alpha}^{(l)}(r)$ and $f_{\Lambda\Lambda}$, similar to earlier analyses [1, 2, 14, 15] are obtained from the procedure developed by the Urbana group and three-body correlations $f_{\Lambda NN}$ have the analytical forms as used in our earlier work [3]. The $f_{\Lambda NN}$ correlations arising from the triads ΛNN, where a participating nucleon comes from each α, have been ignored as these make a negligibly small contributions owing to large $\alpha\alpha$ separation. The wavefunction $\Psi_H^{(10)}$ depends on a total of seventeen variational parameters $\kappa_{\Lambda N}$, $c_{\Lambda N}$, $a_{\Lambda N}$, $R_{\Lambda N}$, $s_{\Lambda N}$, κ_{NN}, c_{NN}, a_{NN}, R_{NN}, $\kappa_{\alpha\alpha}$, $c_{\alpha\alpha}$, $a_{\alpha\alpha}$, $R_{\alpha\alpha}$, $\kappa_{\Lambda\Lambda}$, $c_{\Lambda\Lambda}$, $a_{\Lambda\Lambda}$ and $R_{\Lambda\Lambda}$ for ground state of ${}^{10}_{\Lambda\Lambda}$Be and exactly the same number of variational parameters for the excited state. The trial wavefunction for ${}^9_\Lambda$Be involving 13 variational parameters, is obtained from wavefunction of ${}^{10}_{\Lambda\Lambda}$Be after suppressing the Λ index and restricting baryon number to 9.

2.3. Energy calculations

The energy $-B_{\Lambda\Lambda}(J, J_z)$ for a hypernucleus of baryon number A is the difference of energy of hypernucleus in the state $\Psi_H^{(A)}(J, J_z)$ and of the nuclear core in the state $\Psi_C^{(A-2)}(J_C, M_C)$ and is written as:

$$-B_{\Lambda\Lambda}(J, J_z) = \frac{\langle\Psi_H^{(A)}(J, J_z)|H_H^A|\Psi_H^{(A)}(J, J_z)\rangle}{\langle\Psi_H^{(A)}(J, J_z)|\Psi_H^{(A)}(J, J_z)\rangle} - \frac{\langle\Psi_C^{(A-2)}(J_C, M_C)|H_C^{A-2}|\Psi_C^{(A-2)}(J_C, M_C)\rangle}{\langle\Psi_C^{(A-2)}(J_C, M_C)|\Psi_C^{(A-2)}(J_C, M_C)\rangle}. \quad (3)$$

The variational parameters entering in the wavefunction are varied to optimize the energy using the standard optimizing routine. The parameters $\kappa_{\Lambda N}$, κ_{NN}, $\kappa_{\alpha\alpha}$ and $\kappa_{\Lambda\Lambda}$ related to the separation energy of the pair xy are those on which the energy depends sensitively. The estimates for the energy were made for 100,000 points. The two terms in equation (3) were separately calculated. The second term in equation (3) is $-31.20(-30.34)$ MeV for ^{4}He and $-62.3(-60.58)$ MeV for ^{8}Be core for Malfliet-Tjon [23] (Volkov) [24] NN potential. Similarly, expression for the energy for partial nine-body problem $^9_\Lambda$Be is obtained.

3. Results and Discussion

3.1. Ground State and Degenerate Doublet $(3^+/2, 5^+/2)$ of $^9_\Lambda$Be

The experimental data for B_Λ of the systems $^5_\Lambda$He and $^9_\Lambda$Be have been taken from [18, 25]. Prior to the analysis of the energy of the degenerate doublet $(3^+/2,\ 5^+/2)$ of $^9_\Lambda$Be we need to recalculate the ground state energies of $^5_\Lambda$He and $^9_\Lambda$Be. The parameters of the two-body ΛN force are $\bar{V}$ = 6.15, 6.10 and 6.20 MeV and ϵ = 0.25 and the strength W_d of dispersive force is adjusted for the sets of combinations of C_p and $\hat{c}$ to reproduce B_Λ of $^5_\Lambda$He for a chosen NN potential. The results for ground states of $^5_\Lambda$He and $^9_\Lambda$Be and for degenerate doublet $(3^+/2, 5^+/2)$ of $^9_\Lambda$Be are listed in our earlier work [21, 26]. Here we will only discuss the final results, listed in table 1, for the difference between excited- and ground-state energies for all the combinations of the three-body ΛNN potentials for Malfliet-Tjon (MT) and Volkov NN potentials.

We note that the difference $\Delta E_\Lambda[=E(^9_\Lambda\text{Be}^*) - E(^9_\Lambda\text{Be})]$, between excited and ground states energies for the strength parameter $\bar{V}$ = 6.15, 6.10 and 6.20 MeV for MT and Volkov NN potential, is in good agreement with the observed emitted γ-ray [18] provided small spin-orbit splitting is ignored.

The Volkov NN potential induces weaker correlations compared to MT NN potential. However, the results for ΔE_Λ, listed in the fifth column of table 1 are not very different from those obtained for the MT potential. Thus both the central NN potentials with soft repulsive core give equally good fit to the energy spacing between the two states of $^9_\Lambda$Be.

3.2. Ground and Excited 2^+ States of $^{10}_{\Lambda\Lambda}$Be

Before we proceed to discuss the results for the calculation of the energy of the system $^{10}_{\Lambda\Lambda}$Be, we need to constrain some of the potential parameters from a fit to the ground state energies of $^5_\Lambda$He and $^6_{\Lambda\Lambda}$He. The results for ground state energies of $^5_\Lambda$He and $^6_{\Lambda\Lambda}$He have been discussed in our earlier work [21, 22, 26]. From the previous subsection, we have found that the results for the energy spacing of the excited degenerate doublet $(3^+/2, 5^+/2)$ from the ground state of $^9_\Lambda$Be for the MT and Volkov NN potentials are essentially the same. Therefore, we have carried out calculations for the energy of $^{10}_{\Lambda\Lambda}$Be using the MT NN potential alone. We have also noticed from last subsection that the energy difference of the ground and excited states of $^9_\Lambda$Be for the strength parameter $\bar{V}$ = 6.15, 6.10 and 6.20 MeV is not very different, therefore, we have restricted the calculations of the energy of $^{10}_{\Lambda\Lambda}$Be for $\bar{V}$ = 6.15 MeV alone. For the parameters $\bar{V}$ = 6.15 MeV and ϵ = 0.25 of the two-body ΛN Urbana-type force, the strength W_d of dispersive force is adjusted for the sets of combinations of C_p and $\hat{c}$ from a fit to B_Λ of $^5_\Lambda$He for the NN

Table 1. The energy spacing ΔE_Λ (in MeV) between excited and ground states of ${}^{9}_{\Lambda}$Be for the MT and Volkov NN potentials. All the potential strengths are in MeV and $\hat{c}$ in fm^{-2}. Experimental $\Delta E_\Lambda = 3.04$ MeV.

$\bar{V}$	W_d	$C_p(\hat{c})$	Theoretical ΔE_Λ	
			Malfliet-Tjon [23]	Volkov [24]
6.15	0.012	0 (0)	3.18	3.17
	0.009	1 (1)	3.24	3.19
	0.010	1 (2)	3.00	3.11
	0.0115	1 (3)	3.10	3.19
	0.006	2 (1)	3.06	3.21
	0.009	2 (2)	2.99	3.16
	0.016	2 (3)	3.08	3.17
6.10	0.008	0 (0)	3.25	3.10
	0.005	1 (1)	2.98	3.09
	0.006	1 (2)	3.22	3.13
	0.0073	1 (3)	3.27	3.08
	0.002	2 (1)	3.27	3.00
	0.005	2 (2)	3.23	3.16
	0.0115	2 (3)	2.96	3.12
6.20	0.017	0 (0)	3.10	3.16
	0.013	1 (1)	3.23	3.24
	0.013	1 (2)	3.08	3.19
	0.0153	1 (3)	3.06	3.11
	0.010	2 (1)	2.99	3.21
	0.013	2 (2)	3.16	3.23
	0.020	2 (3)	3.17	3.18

potential, as described in detail in [21, 26]. The sets of ΛN and ΛNN potential parameters, so obtained, along with $\Lambda\Lambda$ potential are used to analyse the ${}^{6}_{\Lambda\Lambda}$He. The parameters sets that give a good account of the binding energies of ${}^{5}_{\Lambda}$He and ${}^{6}_{\Lambda\Lambda}$He, are employed to carry out the energy calculations for the ground and excited states of ${}^{10}_{\Lambda\Lambda}$Be. The experimental binding energy data for the systems ${}^{6}_{\Lambda\Lambda}$He and ${}^{10}_{\Lambda\Lambda}$Be are taken from [17, 19, 27].

Here, we may remark that for the TPE ΛNN force, for the ease of computation, the calculations were restricted for two extreme values of cut-off radii $\hat{c} = 1$ and 3 fm^{-2} as $\hat{c} = 2$ fm^{-2} is expected not to give results different from the other two chosen values, such a situation has been observed in the last subsection for ${}^{9}_{\Lambda}$Be.

The energies of ground and excited 2^+ states of ${}^{10}_{\Lambda\Lambda}$Be are discussed in our earlier work [22, 26]. Here we will only quote the final results, listed in table 2, for the difference between excited- and ground-state energies for all the combinations of the three-body ΛNN potentials for MT NN potential.

The difference $\Delta E_{\Lambda\Lambda}[=E({}^{10}_{\Lambda\Lambda}\mathrm{Be}^*)-E({}^{10}_{\Lambda\Lambda}\mathrm{Be})]$, between excited and ground states energy for the combinations of the potential parameters sets under consideration, is listed in the column three of table 2. We note from the table that its value 3.1—3.3 MeV is consistent with the cluster model calculation of Shoeb [1] but about 12% larger than the one found by Hiyama *et al* [9]. However, theoretically calculated $\Delta E_{\Lambda\Lambda}$ lies in between the two extreme experimental values 2.17 ± 0.4 MeV and 5.27 ± 0.4 MeV but closer to the lower limit.

Table 2. The energy spacing $\Delta E_{\Lambda\Lambda}$ (in MeV) along with the statistical error between 2^+ excited and ground states of ${}^{10}_{\Lambda\Lambda}$Be. Other quantities are the same as in the preceding table. Experimental $\Delta E_{\Lambda\Lambda} = 5.27 \pm 0.4$ MeV from Danysz *et al* [17] (2.17 ± 0.4 MeV deduced from the assumption of a missing γ-ray).

W_d	$C_p(\hat{c})$	Theoretical $\Delta E_{\Lambda\Lambda}$
0.012	0 (0)	3.24 ± 0.08
0.009	1 (1)	3.11 ± 0.08
0.0115	1 (3)	3.24 ± 0.15
0.006	2 (1)	3.32 ± 0.11
0.016	2 (3)	3.13 ± 0.17

4. Conclusions

We conclude that we have applied, to our knowledge, for the first time, the variational Monte Carlo method to analyse the degenerate doublet $(3^+/2, 5^+/2)$ of ${}^{9}_{\Lambda}$Be, treating it as a partial nine-body system in the $\Lambda\alpha\alpha$ cluster model and to analyse the energy of the ground and excited 2^+ states of ${}^{10}_{\Lambda\Lambda}$Be, treating it as a partial ten-body system in the $\Lambda\Lambda\alpha\alpha$ cluster model using the simple NN and ΛN potentials and corresponding correlation functions. The use of Urbana ΛN potential consistent with the Λp-scattering data along with the dispersive ΛNN or dispersive plus two-pion exchange ΛNN forces not only explains B_Λ of ${}^{5}_{\Lambda}$He and ${}^{9}_{\Lambda}$Be but also explains the emission of γ-ray of energy 3.04 MeV from the degenerate doublet to the ground state of ${}^{9}_{\Lambda}$Be, ignoring very small spin-orbit force. The $\Lambda\Lambda$ potential constrained by $B_{\Lambda\Lambda}$ of ${}^{6}_{\Lambda\Lambda}$He from NAGARA event gives energy of 2^+ excited state of ${}^{10}_{\Lambda\Lambda}$Be close to the Demachi-Yanagi event thus confirming the spin and parity assignment 2^+ of the event suggested from the α cluster model analyses. However, the ground state energy does not agree with the currently accepted experimental value. The partial A-body model gives results for ${}^{9}_{\Lambda}$Be and ${}^{10}_{\Lambda\Lambda}$Be consistent with the earlier α cluster model approach. Therefore, our microscopic calculations strongly suggest that afresh measurement be made for the ground state energy of ${}^{10}_{\Lambda\Lambda}$Be.

Acknowledgments

I am thankful to the Organizing committee for providing me an opportunity to present my work. I am indebted to the Head, Theory Division, Saha Institute of Nuclear Physics and the Chairman, Department of Physics, Aligarh Muslim University, for providing necessary facilities for carrying out the work presented in this manuscript. I also would like to thank to Professor Mohammad Shoeb, for his continuous help in doing this work.

References

[1] Shoeb Mohammad 2006 *Phys. Rev.* C **74** 064316
[2] Shoeb Mohammad, Mamo Alemiye and Fessahatsion Amanuel 2007 *Pramana* **68** 943
[3] Shoeb Mohammad and Sonika 2009 *Int. J. Mod. Phys.* E **18** 1404
[4] Shoeb Mohammad and Sonika 2009 *J. Phys. G: Nucl. Part. Phys.* **36** 045104
[5] Maeda S and Schmid E W 1984 *Few-Body Problem in Physics* **II** 379
[6] Fedorov D V and Jensen A S 1996 *Phys. Lett.* B **389** 631
[7] Filikhin I N, Gal A and Suslov V M 2004 *Nucl. Phys.* A **743** 194
[8] Bodmer A R, Usmani Q N and Carlson J 1984 *Phys. Rev.* C **29** 684
[9] Hiyama E, Kamimura M, Motoba T, Yamada T and Yamamoto Y 2002 *Phys. Rev.* C **66** 024007 (and references therein)
[10] Cravo E, Fonseca A C and Koike Y 2002 *Phys. Rev.* C **66** 014001
[11] Myint K S, Shinmura S and Akaishi Y 2003 *Eur. Phys. J.* A **16** 21
[12] Filikhin I N, Gal A and Suslov V M 2003 *Phys. Rev.* C **68** 024002

[13] Ali S, Murphy J W and Bodmer A R 1965 *Phys. Rev. Lett.* **15** 534
[14] Shoeb Mohammad, Usmani Q N and Bodmer A R 1998 *Pramana* **51** 421 (and references therein)
[15] Bodmer A R and Usmani Q N 1987 *Nucl. Phys.* A **468** 653
[16] Filikhin I N and Gal A 2002 *Nucl. Phys.* A **707** 491 (and references therein)
[17] Danysz M *et al* 1963 *Nucl. Phys.* **49** 121
Dalitz R H, Davis D H, Fowler P H, Montwill A, Pniewski J and Zakrzewski J A 1989 *Proc. Roy. Soc. London* A **426** 1
[18] Akikawa H *et al* 2002 *Phys. Rev. Lett.* **88** 082501
[19] Ichikawa A 2001 Ph. D. thesis, Kyoto University
[20] Ali S and Bodmer A R 1966 *Nucl. Phys.* **80** 99
[21] Shoeb Mohammad and Sonika 2009 *Phys. Rev.* C **79** 054321
[22] Shoeb Mohammad and Sonika 2009 *Phys Rev.* C **80** 024317
[23] Malfliet R A and Tjon J A 1969 *Nucl. Phys.* A **127** 161
[24] Volkov A B 1965 *Nucl. Phys.* A **74** 33
[25] Juric M *et al* 1973 *Nucl. Phys.* B **52** 1
[26] Sonika 2010 Ph. D. thesis (unpublished), Aligarh Muslim University
[27] Takahashi H *et al* 2001 *Phys. Rev. Lett.* **87** 212502

The Capabilities of the upgraded MIPP experiment with respect to Hypernuclear Physics

Rajendran Raja
For the MIPP collaboration
Fermilab, P.O. Box 500, Batavia, Illinois 60510
E-mail: raja@fnal.gov

Abstract. We describe the state of analysis of the MIPP experiment, its plans to upgrade the experiment and the impact such an upgraded experiment will have on hypernuclear physics.

1. Introduction

The upgraded MIPP experiment is designed to measure the properties of strong interaction spectra from beams $\pi^{\pm}, K^{\pm}$ and $p^{\pm}$ for momenta ranging from 1 GeV/c to 120 GeV/c. The layout of the apparatus in the data taken so far can be seen in Figure 1. The centerpiece of the experiment is the time projection chamber, which is followed by the time of flight counter, a multi-cell Cerenkov detector and the RICH detector. The TPC can identify charged particles with momenta less than 1 GeV/c using dE/dx, the time of flight will identify particles below approximately 2 GeV/c, the multi-cell Cerenkov detector is operational from 2.5 GeV/c to 14 GeV/c and the RICH detector can identify particles up to 120 GeV/c. Following this is an EM and hadronic calorimeter capable of detecting forward going neutrons and photons.

The experiment has been busy analyzing its data taken on various nuclei and beam conditions. The table 2 shows the data taken by MIPP I to date.

We have almost complete acceptance in the forward hemisphere in the lab using the TPC. The reconstruction capabilities of the TPC can be seen in Figure 3.

The particle identification capabilities of the TPC can be seen in Figure 4.

The time of flight system provides further measurement of the particles with momenta less than 2 GeV/c. Figure 5 shows the time of flight data where a kaon peak is clearly visible.

2. Results from data taken so far

MIPP has published forward going neutron spectra off various nuclei and compared them with Monte Carlo predictions [1]

It can be seen clearly that the Monte Carlo predictions differ significantly from each other and the data.

Figure 7 shows (in a paper in preparation, [2]) preliminary results of particle production from the full NuMI target as a function of momenta.

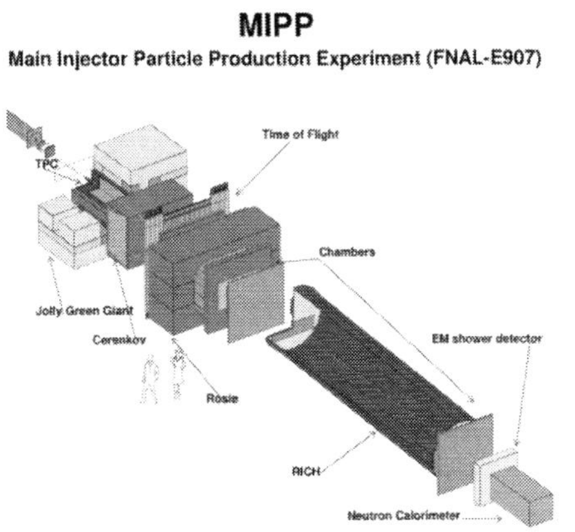

Figure 1. Layout of the experiment - Run I. With the MIPP upgrade we propose to add the plastic ball detector as a recoil detector surrounding the target.

Data Summary 27 February 2006			Acquired Data by Target and Beam Energy Number of events, x 10^6									
Target			E									Total
Z	Element	Trigger Mix	5	20	35	40	55	60	65	85	120	
0	Empty	Normal		*0.10*	*0.14*			*0.52*			*0.25*	*1.01*
	K Mass	No Int.				*5.48*	*0.50*	*7.39*	*0.96*			*14.33*
	Empty LH	Normal		*0.30*				*0.61*		*0.31*		7.08
1	LH	Normal	0.21	1.94				1.98		1.73		
4	Be	*p* only									1.08	1.75
		Normal			0.10			0.56				
6	C	Mixed						0.21				1.33
	C 2%	Mixed		0.39				0.26			0.47	
	NuMI	*p* only									1.78	1.78
13	Al	Normal			*0.10*							*0.10*
83	Bi	*p* only									1.05	2.83
		Normal			0.52			1.26				
92	U	Normal						1.18				1.18
Total			0.21	2.73	0.86	5.48	0.50	13.97	0.96	2.04	4.63	31.38

Figure 2. Data taken by MIPP I to date

3. The MIPP Upgrade proposal

The MIPP upgrade proposal consists primarily of an electronics upgrade of the whole detector so that the detector reads out at 3 kiloHZ as compared to 20 Hz in Run I. All the new electronics modules have been prototyped. The TPC has the largest data rate and we display one of its elctronics modules in Figure 8. The document [3] contains the details of the upgrade proposal. The MIPP upgrade proposal will acquire high quality data of particle production with high statistics and much improved particle identification. The available beam momenta and the calculated beam rates can be seen in table 1 and range from 1 GeV/c to 120 GeV/c. As such the upgrade can straddle the resonance region and the scaling region of strong interaction physics using 6 beam species. The MIPP upgrade intends to acquire data for a variety of interactions. These include

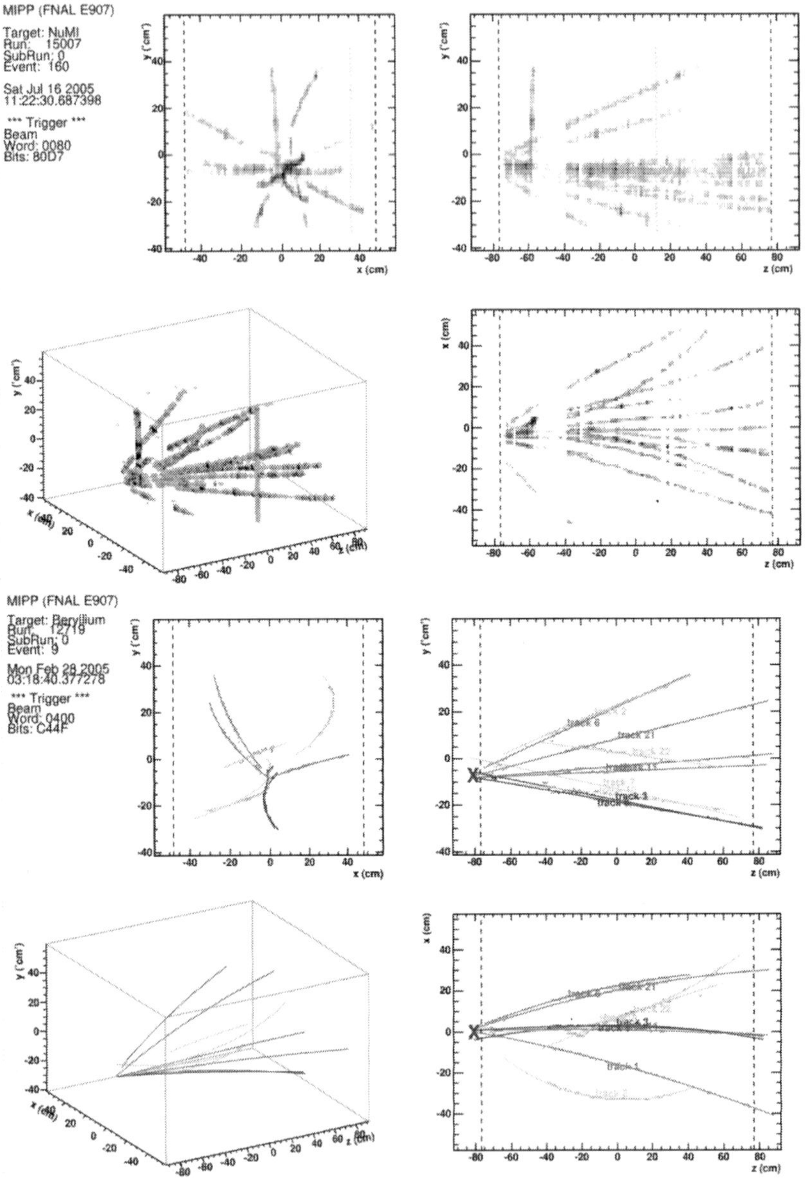

Figure 3. RAW and Reconstructed TPC tracks from two different events.

- A systematic study of missing baryon resonances between 1 GeV and 3 GeV.
- An investigation of fundamental scaling laws in inclusive reactions.
- A thorough investigation of forward neutron spectra.
- A systematic acquisition of particle production data and comparison and tuning of existing Monte Carlo generators.
- A measurement of particle production data from full scale targets used in neutrino experiments and using the data to improve the target design.

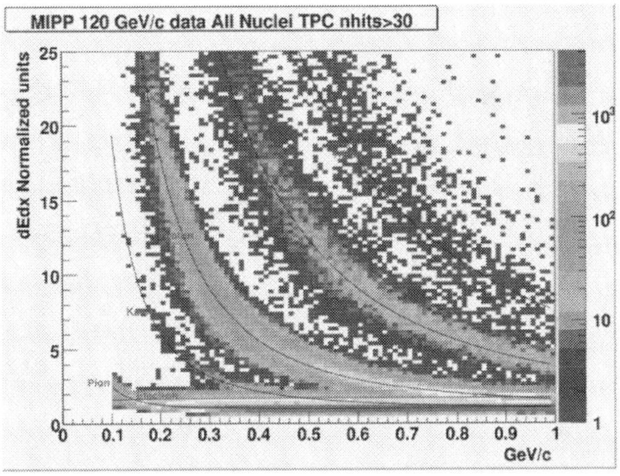

Figure 4. TPC dE/dx data from Run I. Electrons, pions, kaons and protons are distinguishable. Deuterons and tritons can also be seen

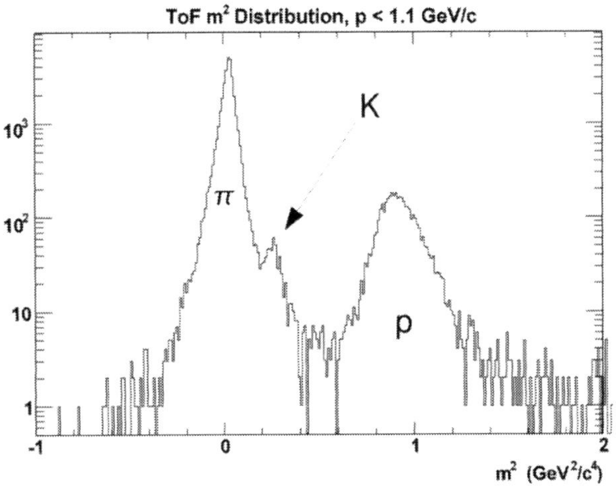

Figure 5. ToF data with momenta less than 1.1 GeV/c. A kaon peak is clearly visible.

- A systematic study of hypernuclei
- Measurement of cross section off nitrogen targets which is essential to understanding the cosmic ray showers and atmospheric neutrino production for experiments such as INO.

This is made possible since MIPP upgrade will be able to collect 5 million events per day assuming a beam spill rate of one 4 second spill every 2 minutes,

4. Hypernuclear Physics with MIPP Upgrade

Singly strange hypernuclei (with one Λ or one Σ) have production cross sections in the 100 to 250 nbarn/steradian at 1 to 2 GeV/c proton beams on nuclear targets as can be seen for Figure 9. Doubly strange hypernuclei (with two Λ) are 10 to 50 nbarn/sr with a kaon or hyperon beam on

Journal of Physics: Conference Series **374** (2012) 012019 doi:10.1088/1742-6596/374/1/012019

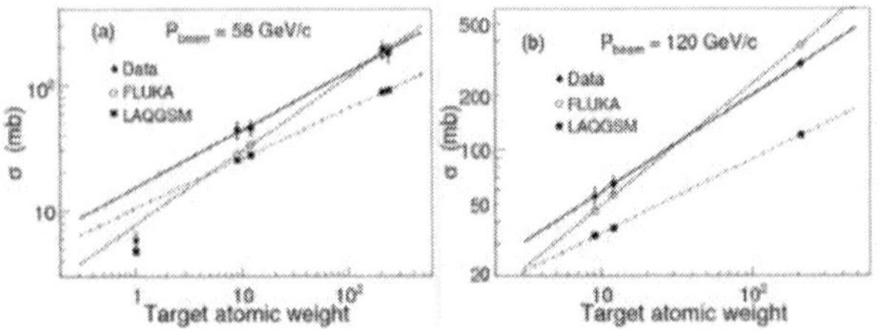

Figure 6. Comparison of the A-dependence of MIPP cross sections with those from Monte Carlos. The lines are fits to the data. The cross sections are for producing neutrons with momentum greater than the threshold and within an angular range of 20.4 mrad. The errors are combined statistical and systematic uncertainties. Note that the hydrogen data point is not included in the fit. The cross sections are not corrected for geometric acceptance.

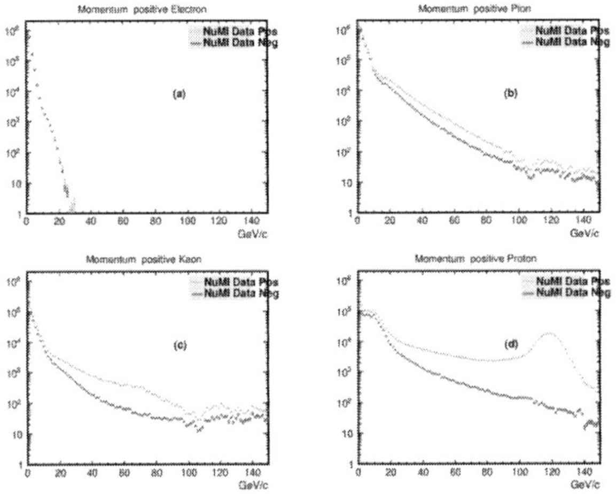

Figure 7. Analysis of NuMI target spectra (data) as a function of particle type (preliminary)

Figure 8. Electronics prototype of TPC upgrade board with ALTO/PASA chips

nuclear target and are more difficult to measure. The diagrams for doubly strange hypernuclei can be found in Figure 10. Doubly strange hypernuclei with one Ξ are not discovered yet, but 7 candidate events are known to exist. The binding energies for nuclei as a function of atomic weight can be found in Figure 11. The binding energies decrease with atomic weight indicating that the hypernuclei are less tightly bound as atomic weight increases. This behavior is also reflected in the lifetimes plot in Figure 12 where the lifetime of the bound lambda is seen to decrease with increasing atomic weight.

Using MIPP upgrade, we have complete forward acceptance using the TPC and full particle

momentum GeV/c	p	K^+	π^+	momentum GeV/c	$\bar{p}$	K^-	π^-
1	5752	0	64798	1	7907	0	30425
2	23373	194	459718	2	26863	142	236494
3	53431	3060	1069523	3	51424	2221	598742
5	153220	32763	2400799	5	103996	23164	1550810
10	663916	223210	5006708	10	195767	142777	3862225
15	1618120	443557	7141481	15	221602	245868	5248463
20	3113387	655426	9290219	20	212171	306685	5841030
30	8158054	1043430	12770579	30	160329	340144	5837467
40	16664431	1294189	13944272	40	101617	288728	5156862
50	29288928	1338452	12788523	50	53056	196400	4114582
60	45985629	1191744	10094311	60	22092	108032	2905091
70	65227010	919279	6834097	70	6987	47093	1762060

Table 1. Calculated Beam rates for 2×10^{11} protons on primary target for MIPP upgrade

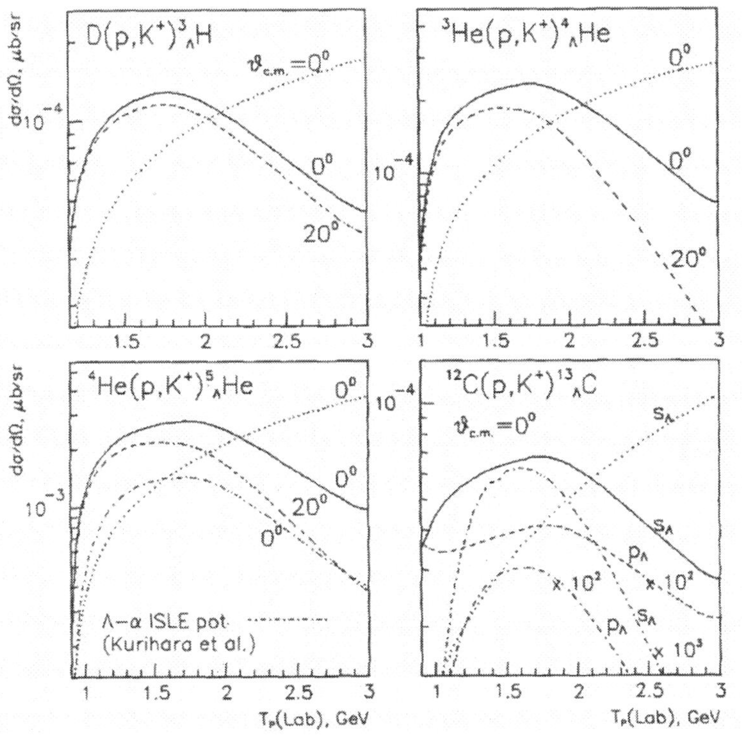

Figure 9. Differential cross sections of single hypernucleus production on various nuclei with a proton beam and a final state kaon [4]

identification (using the TPC and the time of flight) for particles of momentum less than 1.5 GeV/c. In the backward hemisphere in the lab frame, we have the plastic ball that catches recoil particles.

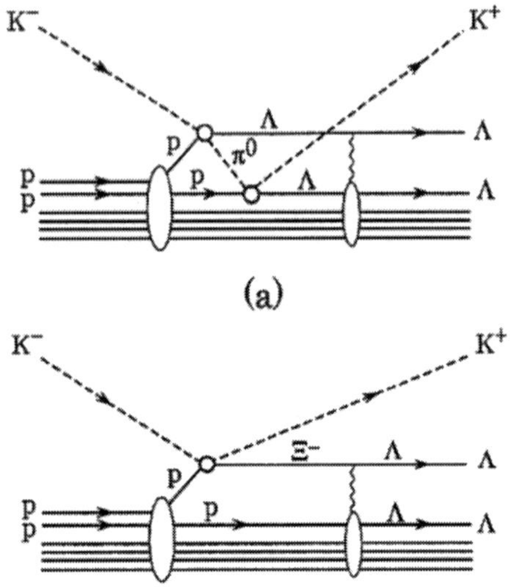

Figure 10. Mechanisms for producing double hypernuclei using a K^- beam. K^+ beams can study the background to this process [5]

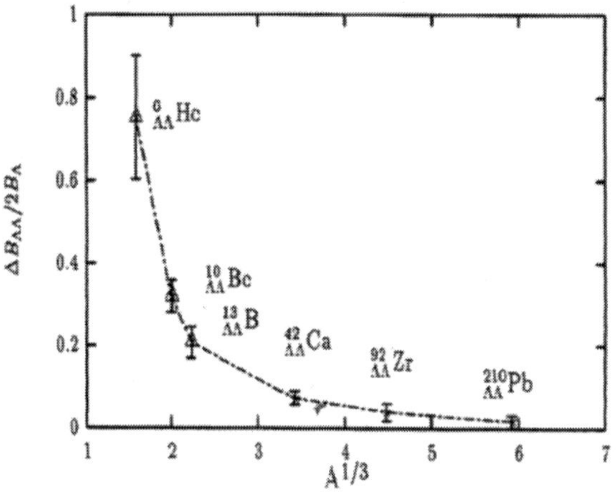

Figure 11. Binding energies of hypernuclei as a function of atomic weight.

We investigate the channels that are likely to yield hypernuclei.
The reaction involving a π^+ beam can be denoted by

$$\pi^+ + p \to \pi^+ + K^+ + \Lambda \tag{1}$$

In the above equation, π^+ beam remains a π^+ beam and fragments a proton target to form a K^+ and a Λ. Such reactions are called diffractive dissociations of the target and these maintain a large cross section as the beam energy increases. The K^+ in the final state is detected by the TPC and the time of flight systems and the Λ is absorbed in the nucleus to make the

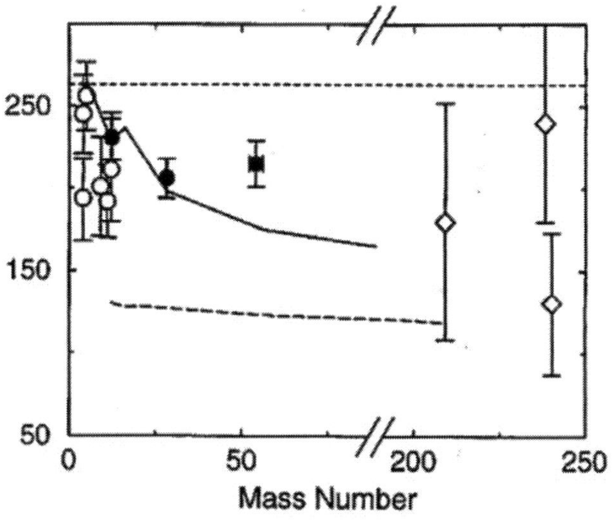

Figure 12. Lifetimes of hypernuclei as a function of atomic weight A. As A increases, lifetimes get shorter, implying less tightly bound hypernuclei

hypernucleus. There may be other particles in the final state, which the particle identification of MIPP will ensure contain a net strangeness of zero. The counterpart of the above reaction with a neutron target can be written

$$\pi^+ + n \rightarrow \pi^+ + K_S^0 + \Lambda \tag{2}$$

The K_s^0 in the final state can be detected by the TPC. Similarly with a proton beam

$$p + p \rightarrow p + K^+ + \Lambda \tag{3}$$

$$p + n \rightarrow p + K_S^0 + \Lambda \tag{4}$$

These form the proton counter part. Using a K$^-$ beam, one can look for the reaction

$$K^- + p \rightarrow \pi^+ + \Lambda \tag{5}$$

$$K^- + n \rightarrow \pi^0 + \Lambda \tag{6}$$

One can test the backgrounds to this reaction using K$^+$ beams which should not produce hypernuclei.

5. Conclusions

The high acceptance and particle identification of MIPP upgrade promises to produce a high quality hypernucleus experiment. This work was done at Fermilab operated by Fermi Research Alliance, LLC under Contract No. De-AC02-07CH11359 with the United States Department of Energy.

References

[1] T. S. Nigmanov et. al, *Phys. Rev.* D **83** 012002 (2011).
[2] "Measurement of Particle Production on the NuMI target with a 102 GeV/c proton beam in the MIPP Experiment", MIPP collaboration, paper in preparation.
[3] A copy of the MIPP upgrade proposal may be found at `https://mipp-docdb.fnal.gov:440/cgi-bin/ShowDocument?docid=1046`
[4] V. N. Fetisova, *Nucl. Phys.* A **639** 177, (1998).
[5] H. C. Bhang et al., *Nucl. Phys.* A **639**,269 (1998).
[6] S. Ajimura et al., *Nucl. Phys.* A **639** 93 (1998).

Nuclear medium effects in structure functions

M. Sajjad Athar
Department of Physics, Aligarh Muslim University, Aligarh-202 002, India
E-mail: sajathar@gmail.com

I. Ruiz Simo
Departamento de Física Teórica and IFIC, Centro Mixto Universidad de Valencia-CSIC, Institutos de Investigación de Paterna, E-46071 Valencia, Spain
Departamento de Física Atómica Molecular y Nuclear, Universidad de Granada, E-18071 Granada, Spain

H. Haider
Department of Physics, Aligarh Muslim University, Aligarh-202 002, India

M. J. Vicente Vacas
Departamento de Física Teórica and IFIC, Centro Mixto Universidad de Valencia-CSIC, Institutos de Investigación de Paterna, E-46071 Valencia, Spain

Abstract. We discuss the nuclear-medium effects in the weak structure functions $F_2(x, Q^2)$ and $F_3(x, Q^2)$ in the charged current neutrino and antineutrino induced deep inelastic reactions in some nuclei which are being used in the present or proposed neutrino oscillation experiments. Our study of the medium effects has been done using relativistic nuclear spectral functions which incorporates Fermi motion, binding energy and nucleon correlations. We also consider the pion and rho meson cloud contributions calculated from a microscopic model for meson-nucleus self-energies. Using these structure functions, the results for the differential cross section have been obtained and compared with some of the available experimental data.

1. Introduction

At the present time many ongoing and proposed neutrino oscillation experiments are being performed to determine precisely the parameters of neutrino mixing matrix, emphasis has also been laid towards better understanding of the low and intermediate energy neutrino interactions and specifically attention has been brought to study both theoretically and experimentally the strong dynamics of the nucleon and nucleus that affect these interactions. Statistically plentiful data are available from T2K, MiniBooNE, MINOS, SciBooNE, etc. and more results are expected from the Main INjector Experiment ν-A(MINERνA) being performed at Fermi Lab. MINERνA is studying the medium effects in neutrino-nucleus cross-section measurements using various nuclear targets like helium, carbon, oxygen, iron and lead using muon neutrinos in the energy region of 1-20 GeV and the aim is to study effect of nuclear medium in the calculation of the event rates.

In the few GeV energy region the contribution to the cross section comes from the quasielastic, inelastic as well as the deep inelastic processes. Several theoretical calculations have been performed for the quasielastic and inelastic one pion production processes. In the deep inelastic region, both experimentally as well as theoretically, limited efforts have been made to understand the medium effects for the weak interaction induced processes. In the case of experimental measurements performed for $F_{2,3}^{Weak}$ the error bars are large and need better precision. In general, nuclear modifications in the weak structure functions $F_{2,3}^{Weak}(x,Q^2)$ may be different from the nuclear modification in F_2^{EM}. Particularly neutrino data is important in the determination of valence quark distributions in the nucleon. This is due to the fact that the parity-violating (F_3^{Weak}) structure function directly probes into the valence distributions. Precise determination of parton distribution functions (PDFs) is also necessary for the new physics at the colliders. But for the determination of PDFs, nuclear medium effects should be properly accounted for. Nuclear-medium effects may be responsible for the anomaly observed in the NuTeV experiment for the weak mixing angle $\sin^2\theta_W$ [1]. The weak structure functions $F_2^N(x,Q^2)$ and $F_3^N(x,Q^2)$ have also been measured using neutrino (antineutrino) beams [2, 3, 4, 5, 6, 7, 8, 9] with some nuclear targets. CERN Dortmund Heidelberg Saclay Warsaw CDHSW [3] and NuTeV [9] collaborations have studied differential scattering cross sections in iron.

CERN Hybrid Oscillation Research apparatUS CHORUS collaboration [10, 11] has performed high-statistics measurement of the differential (anti)neutrino cross-sections using mainly lead target at various neutrino and antineutrino energies as a function of Bjorken scaling variables x and y. Analysis is being performed at the Neutrino Oscillation MAgnetic Detector (NOMAD) [12] for the weak structure functions and cross section measurements with carbon target using neutrino beam. The Neutrino Scattering On Glass (NuSOnG) experiment [13] has been proposed at Fermilab to study the structure functions in the deep inelastic region using neutrino scattering on carbon target. The Oscillation Project with Emulsion-tRacking Apparatus (OPERA) experiment [14] is a long baseline experiment using lead emulsion target and the main purpose of the experiment is the observation of ν_μ to ν_τ oscillations in the direct appearance mode and the charged current deep inelastic scattering process is expected to be dominant. Therefore, several experimental activities are going on to study neutrino as well as antineutrino event rates in the deep inelastic region.

On the theoretical side, for the weak interaction induced processes in the deep inelastic region, the studies can be divided into two parts. One in which nuclear medium effects have been phenomenologically described in terms of a few parameters which are determined by fitting the experimental data of charged leptons and (anti)neutrino deep inelastic scattering from various nuclear targets. The various phenomenological studies differ in the choice of the data sets(Lepton-Nucleus+Drell Yan data, Lepton-Nucleus+Drell Yan+$\nu(\bar{\nu})$-Nucleus data, and $\nu(\bar{\nu})$-Nucleus data), experimental cuts in their analysis, parametrization of the parton distributions, etc. for example see the discussions given in Refs. [15, 16, 17]. Another approach is to understand the dynamics of the nucleon in the nuclear medium. This approach has been used by few authors [18, 19, 20] and more work is needed in this direction.

In the present work, we study nuclear medium effects on the structure functions $F_2(x,Q^2)$ and $F_3(x,Q^2)$ in some nuclei like carbon, iron and lead. We shall treat iron and lead to be a nonisoscalar nuclear targets. This study has been performed using a relativistic nuclear spectral function [21] to describe the momentum distribution of nucleons in the nucleus within a field-theoretical approach where nucleon propagators are written in terms of this spectral function, and nuclear many-body theory is used to calculate it for an interacting Fermi sea in nuclear matter. A local-density approximation is then applied to translate these results to finite nuclei [19, 20, 22]. We have assumed the Callan-Gross relationship for nuclear structure functions ${F_1}^A(x)$ and ${F_2}^A(x)$. The contributions of the pion and rho meson clouds are taken into account in a many-body field-theoretical approach which is based on Refs. [22, 23]. We have taken into account target mass correction (TMC) following Ref. [24] which has significant effect at low Q^2, moderate, and high Bjorken x. To take into account the shadowing effect, which is important at low Q^2 and low x, and modulates the contribution of pion and rho cloud contributions, we have followed the works of Kulagin and Petti [18, 25]. Using structure functions F_2^A and F_3^A as an input we have calculated differential scattering cross section for charged current neutrino and antineutrino induced reactions. The Next-to-leading Order(NLO) evolution has been incorporated using the prescription of Moch et al [26]. In the next section we present the formalism in brief and in Sec. III we present the results and discussions. In Sec. IV we conclude our findings.

2. Formalism

The expression of the cross section, for deep inelastic scattering (DIS) of neutrino with a nucleon target induced by charged current reaction

$$\nu_l(k) + N(p) \rightarrow l^-(k') + X(p'), \; l = \mu, \tag{1}$$

is written as

$$\begin{aligned} \sigma &= \frac{1}{v_{rel}} \frac{2m_\nu}{2E_\nu(\mathbf{k})} \frac{2M}{2E(\mathbf{p})} \int \frac{d^3k'}{(2\pi)^3} \frac{2m_l}{2E_l(\mathbf{k}')} \times \prod_{i=1}^{N} \int \frac{d^3p_i'}{(2\pi)^3} \prod_{l\epsilon f} \left(\frac{2M_l'}{2E_l'}\right) \prod_{j\epsilon b} \left(\frac{1}{2\omega_j'}\right) \bar{\sum} \sum |T|^2 (2\pi)^4 \\ &\times \delta^4 \left(p + k - k' - \sum_{i=1}^{N} p_i'\right) \end{aligned} \tag{2}$$

where f stands for fermions and b for bosons in the final state X. The index i is split into l and j for fermions and bosons respectively,

T is the invariant matrix element for the above reaction and is, written as

$$-iT = \left(\frac{iG}{\sqrt{2}}\right) \bar{u}_l(k')\gamma^\alpha(1-\gamma_5)u_\nu(k) \left(\frac{m_W^2}{q^2 - m_W^2}\right) \langle X|J_\alpha|N\rangle . \tag{3}$$

After performing the phase space integration in Eq.(2), the double differential scattering cross section evaluated for a nucleon target in its rest frame is expressed as,

$$\frac{d^2\sigma_{\nu,\bar{\nu}}^N}{d\Omega' dE'} = \frac{{G_F}^2}{(2\pi)^2} \frac{|\mathbf{k}'|}{|\mathbf{k}|} \left(\frac{m_W^2}{q^2 - m_W^2}\right)^2 L_{\nu,\bar{\nu}}^{\alpha\beta} W_{\alpha\beta}^N , \tag{4}$$

where G_F is the Fermi coupling constant, m_W is the mass of the W boson, k is the incoming neutrino four momentum and k' is the outgoing four momentum of the lepton, $q = k - k'$ is the four momentum transfer and Ω', E' refer to the outgoing lepton.

The lepton tensor for antineutrino(neutrino) scattering $L^{\alpha\beta}$ is given by

$$L^{\alpha\beta} = k^\alpha k'^\beta + k^\beta k'^\alpha - k.k' g^{\alpha\beta} \pm i\epsilon^{\alpha\beta\rho\sigma} k_\rho k'_\sigma , \tag{5}$$

and the hadronic tensor $W_{\alpha\beta}^N$ is defined as

$$W_{\alpha\beta}^N = \frac{1}{2\pi} \bar{\sum_{s_N}} \sum_X \sum_{s_i} \prod_{i=1}^{n} \int \frac{d^3p_i'}{(2\pi)^3} \prod_{l\epsilon f} \left(\frac{2M_l'}{2E_l'}\right) \prod_{j\epsilon b} \left(\frac{1}{2\omega_j'}\right) \langle X|J_\alpha|N\rangle \langle X|J_\beta|N\rangle^* (2\pi)^4 \delta^4(p + q - \sum_{i=1}^{n} p_i'), \tag{6}$$

where q is the momentum of the virtual W, s_N the spin of the nucleon and s_i the spin of the fermions in X. N is a nucleon, X is a jet of n hadrons consisting of fermions(f) and bosons(b) in the final state labeled by l and j in the following. In the case of antineutrino $\langle X|J_\alpha|N\rangle$ is replaced by $\langle X|J^\dagger_\alpha|N\rangle$. The most general form of the hadronic tensor $W^N_{\alpha\beta}$ is expressed as

$$W^N_{\alpha\beta} = \left(\frac{q_\alpha q_\beta}{q^2} - g_{\alpha\beta}\right) W_1^{\nu(\bar\nu)} + \frac{1}{M^2}\left(p_\alpha - \frac{p.q}{q^2}q_\alpha\right)\left(p_\beta - \frac{p.q}{q^2}q_\beta\right) W_2^{\nu(\bar\nu)} - \frac{i}{2M^2}\epsilon_{\alpha\beta\rho\sigma}p^\rho q^\sigma W_3^{\nu(\bar\nu)} + \frac{1}{M^2}q_\alpha q_\beta W_4^{\nu(\bar\nu)} + \frac{1}{M^2}(p_\alpha q_\beta + q_\alpha p_\beta)W_5^{\nu(\bar\nu)} + \frac{i}{M^2}(p_\alpha q_\beta - q_\alpha p_\beta)W_6^{\nu(\bar\nu)}, \tag{7}$$

where M is the nucleon mass and W_i^N are the structure functions, which depend on the scalars q^2 and $p.q$. The terms depending on W_4, W_5 and W_6 in Eq. (7) do not contribute to the cross section in Eq. (4) in the limit of lepton mass $m_l \to 0$.

In terms of the Bjorken variables x and y defined as

$$x = \frac{Q^2}{2M\nu}, \quad y = \frac{\nu}{E_\nu}, \quad Q^2 = -q^2, \quad \nu = \frac{p.q}{M} \tag{8}$$

we can write the expression for the differential scattering cross section (in the limit of lepton mass $m_l \to 0$) as

$$\frac{d^2\sigma^{\nu(\bar\nu)}}{dxdy} = \frac{G_F{}^2 M E_\nu}{\pi}\left\{xy^2 F_1^{\nu(\bar\nu)}(x,Q^2) + \left(1 - y - \frac{xyM}{2E_\nu}\right)F_2^{\nu(\bar\nu)}(x,Q^2) \pm xy(1-y/2)F_3^{\nu(\bar\nu)}(x,Q^2)\right\}, \tag{9}$$

where the + (−) sign stands for the neutrino (antineutrino) cross section, and the $F_i^{\nu,\bar\nu}(x,Q^2)$ are dimensionless structure functions defined as

$$F_1^{\nu(\bar\nu)}(x,Q^2) = MW_1^{\nu(\bar\nu)}(\nu,Q^2), \quad F_2^{\nu(\bar\nu)}(x,Q^2) = \nu W_2^{\nu(\bar\nu)}(\nu,Q^2), \quad F_3^{\nu(\bar\nu)}(x,Q^2) = \nu W_3^{\nu(\bar\nu)}(\nu,Q^2) \tag{10}$$

F_1 and F_2 are related by the Callan-Gross relation [27] leaving only two unknown structure functions $F_2^{\nu(\bar\nu)}(x,Q^2)$ and $F_3^{\nu(\bar\nu)}(x,Q^2)$. The nucleon structure functions are determined in terms of parton distribution functions for quarks and anti-quarks. The parton distribution functions for the nucleons have been taken from CTEQ6.6 [28] collaboration for our numerical calculations.

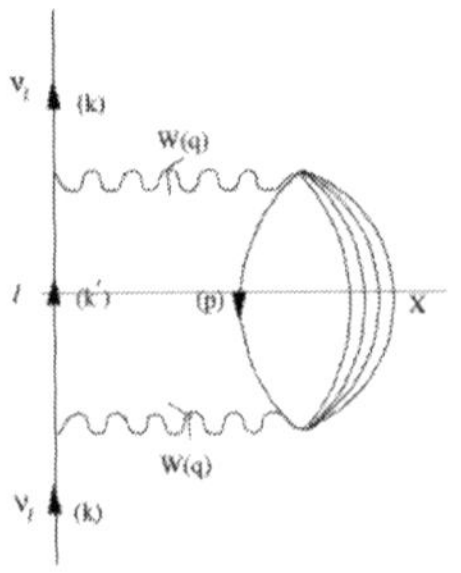

Figure 1. Self-energy diagram of the neutrino in the nuclear medium associated with the process of deep inelastic neutrino-nucleon scattering.

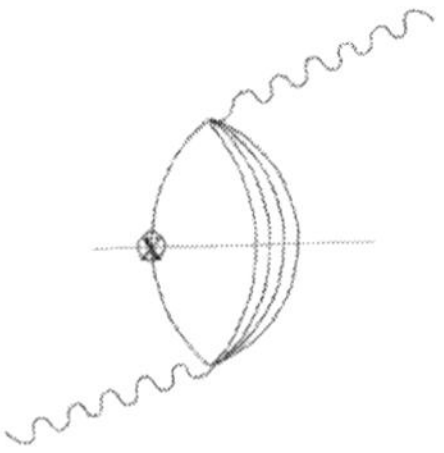

Figure 2. Self-energy diagram of the W-boson in the nuclear medium.

3. Neutrino nucleus scattering

In a nuclear medium the expression for the cross section given in Eq.(4) is modified as:

$$\frac{d^2\sigma^A_{\nu,\bar\nu}}{d\Omega' dE'} = \frac{G_F^2}{(2\pi)^2}\frac{|\mathbf{k}'|}{|\mathbf{k}|}\left(\frac{m_W^2}{q^2-m_W^2}\right)^2 L^{\alpha\beta}_{\nu,\bar\nu}\, W^A_{\alpha\beta}\,, \tag{11}$$

where $W^A_{\alpha\beta}$ is the nuclear hadronic tensor defined in terms of nuclear hadronic structure functions $W_{iA}(x,Q^2)$ through Eq.(7).

In our formalism the neutrino nuclear cross sections are obtained in terms of neutrino self energy $\Sigma(k)$ in the nuclear medium which also defines the dimensionless nuclear structure functions $F_i^A(x,Q^2)$. This is when compared with the equation for the differential scattering cross section for neutrino scattering with a nucleon with momentum $p=(E,\mathbf{p})$ in the rest frame of the nucleus

$$\frac{d^2\sigma^N_{\nu,\bar\nu}}{d\Omega' dE'} = \frac{{G_F}^2}{(2\pi)^2}\frac{|\mathbf{k}'|}{|\mathbf{k}|}\frac{M}{E(\mathbf{p})}\left(\frac{m_W^2}{q^2-m_W^2}\right)^2 L^{\alpha\beta}_{\nu,\bar\nu}\, W^N_{\alpha\beta}\,, \tag{12}$$

gives the nuclear hadronic tensor $W^A_{\alpha\beta}$. With proper choice of tensor components α,β, the dimensionless nuclear structure functions $F_i^A(x,Q^2)$ are obtained[22].

The probability per unit time for the neutrino to collide with nucleons when traveling through nuclear matter may be expressed as:

$$\Gamma(k) = -\frac{2m_\nu}{E_\nu(k)}\,\mathrm{Im}\,\Sigma(k)\,, \tag{13}$$

and the cross section for neutrino scattering from an element of volume d^3r and surface dS in the nucleus is given by

$$d\sigma \;=\; \Gamma dt dS = \Gamma\frac{dt}{dl}\,dl dS = \frac{\Gamma}{v}\,d^3r = \Gamma\,\frac{E_\nu(k)}{|k|}\,d^3r = -\frac{2m_\nu}{|k|}\,\mathrm{Im}\,\Sigma\,d^3r\,. \tag{14}$$

The neutrino self-energy in nuclear matter corresponding to Fig.1 is given by,

$$\Sigma(k) = \quad (-i)\frac{G_F}{\sqrt{2}}\frac{4}{m_\nu}\int\frac{d^4k'}{(2\pi)^4}\frac{1}{k'^2-m_l^2+i\epsilon}\left(\frac{m_W}{q^2-m_W^2}\right)^2 L_{\alpha\beta}\;\Pi^{\alpha\beta}(q)\,, \tag{15}$$

where $L_{\alpha\beta}$ is given by Eq.(5) and $\Pi^{\alpha\beta}(q)$ is the W self-energy in the nuclear medium and is written with the help of Fig.2 as:

$$-i\Pi^{\alpha\beta}(q) = \quad (-)\int\frac{d^4p}{(2\pi)^4}iG(p)\sum_X\sum_{s_p,s_i}\prod_{i=1}^{n}\int\frac{d^4p_i'}{(2\pi)^4}\prod_l iG_l(p_l')\prod_j iD_j(p_j')\left(\frac{-G_F m_W^2}{\sqrt{2}}\right)$$
$$\times\langle X|J^\alpha|N\rangle\langle X|J^\beta|N\rangle^*(2\pi)^4\delta^4(q+p-\Sigma_{i=1}^n p_i')\,. \tag{16}$$

In the above expression $G_l(p_l')$ and $D_j(p_j')$ are respectively the nucleon and meson relativistic propagators in the final state which are taken as the standard free relativistic propagators. $G(p)$ is the nucleon propagator with mass M and energy $E(\mathbf{p})$ in the initial state, which is calculated for a relativistic nucleon in the interacting Fermi sea.

The relativistic Dirac propagator G(p) for a free nucleon, is written in terms of the contribution from the positive and negative energy components of the nucleon described by the Dirac spinors $u(\mathbf{p})$ and $v(\mathbf{p})$ using their appropriate normalizations is written as

$$G^0(p) = \frac{\not p + M}{p^2-M^2+i\epsilon} = \frac{M}{E(p)}\left\{\frac{\sum_r u_r(\mathbf{p})\bar u_r(\mathbf{p})}{p^0-E(\mathbf{p})+i\epsilon} + \frac{\sum_r v_r(-\mathbf{p})\bar v_r(-\mathbf{p})}{p^0+E(\mathbf{p})-i\epsilon}\right\} \tag{17}$$

The nucleon propagator G(p) is then calculated by making a perturbative expansion of $G(p)$ in terms of $G^0(p)$ given in Eq.(17) by retaining the positive energy contributions only(the negative energy components are suppressed). This perturbative expansion is summed in ladder approximation to give [22]:

$$G(p) = \frac{M}{E(\mathbf{p})}\sum_r \frac{u_r(\mathbf{p})\bar{u}_r(\mathbf{p})}{p^0 - E(\mathbf{p}) - \bar{u}_r(\mathbf{p})\sum^N(p^0,\mathbf{p})u_r(\mathbf{p})\frac{M}{\mathbf{E(p)}}} \tag{18}$$

where $\Sigma^N(p^0,p)$ is the nucleon self energy in nuclear matter taken from Ref. [21]. The relativistic nucleon propagator G(p) in a nuclear medium is then expressed as [22]:

$$G(p) = \frac{M}{E(\mathbf{p})}\sum_r u_r(\mathbf{p})\bar{u}_r(\mathbf{p})\left[\int_{-\infty}^{\mu} d\omega\frac{S_h(\omega,p)}{p^0-\omega-i\eta} + \int_{\mu}^{\infty} d\omega\frac{S_p(\omega,p)}{p^0-\omega+i\eta}\right], \tag{19}$$

where $S_h(\omega,p)$ and $S_p(\omega,p)$ being the hole and particle spectral functions respectively, which are given by [22, 21]:

$$S_h(p^0,\mathbf{p}) = \frac{1}{\pi}\frac{\frac{M}{E(\mathbf{p})}Im\Sigma^N(p^0,p)}{(p^0-E(\mathbf{p})-\frac{M}{E(\mathbf{p})}Re\Sigma^N(p^0,p))^2+(\frac{M}{E(\mathbf{p})}Im\Sigma^N(p^0,p))^2} \tag{20}$$

for $p^0 \leq \mu$

$$S_p(p^0,\mathbf{p}) = -\frac{1}{\pi}\frac{\frac{M}{E(\mathbf{p})}Im\Sigma^N(p^0,p)}{(p^0-E(\mathbf{p})-\frac{M}{E(\mathbf{p})}Re\Sigma^N(p^0,p))^2+(\frac{M}{E(\mathbf{p})}Im\Sigma^N(p^0,p))^2} \tag{21}$$

for $p^0 > \mu$.

The normalization of this spectral function is obtained by imposing the baryon number conservation which is written as:

$$2V\int\frac{d^3p}{(2\pi)^3}\int_{-\infty}^{\mu} S_h(\omega,p)d\omega = B = 1. \tag{22}$$

In the local density approximation, we do not have a box of constant density, and the reaction takes place at a point $\mathbf{r}$, lying inside a volume element d^3r with local density $\rho_p(\mathbf{r})$ and $\rho_n(\mathbf{r})$ corresponding to the proton and neutron. Therefore, the upper limit in the integration over nucleon momentum in Eq. (22) is the local Fermi momentum $k_{F_{p,n}}(\mathbf{r})$ of the nucleon given by:

$$k_{F_p}(\mathbf{r}) = \left[3\pi^2\rho_p(\mathbf{r})\right]^{1/3}, k_{F_n}(\mathbf{r}) = \left[3\pi^2\rho_n(\mathbf{r})\right]^{1/3}. \tag{23}$$

This makes the spectral function $S_h(\omega,p)$ density dependent i.e. $S_h(\omega,p,k_F(\mathbf{r}))$ and the normalization condition given in Eq. (22) is modified to

$$2\int\frac{d^3p}{(2\pi)^3}\int_{-\infty}^{\mu} S_h(\omega,p,k_{F_{p,n}}(\mathbf{r}))d\omega = \rho_{p,n}(\mathbf{r}) \tag{24}$$

For a symmetric nuclear matter of density $\rho(\mathbf{r})$, there is a unique Fermi momentum given by $k_F(\mathbf{r}) = \left[\frac{3}{2}\pi^2\rho(\mathbf{r})\right]^{1/3}$ for which we obtain

$$4\int\frac{d^3p}{(2\pi)^3}\int_{-\infty}^{\mu} S_h(\omega,p,k_F(\mathbf{r}))d\omega = \rho(\mathbf{r}), \tag{25}$$

where $\rho(r)$ is the baryon density for the nucleus.

This leads to the normalization condition given by

$$4\int d^3r \int \frac{d^3p}{(2\pi)^3}\int_{-\infty}^{\mu} S_h(\omega, p, \rho(r))\, d\omega = A \tag{26}$$

In the antineutrino case the expressions obtained are very similar. $L^{\alpha\beta}$ appears, as in Eq. (5), with a minus sign in front and in the W self-energy, Eq. (16), we have $\langle X|J^\dagger_\alpha|N\rangle$, instead of $\langle X|J_\alpha|N\rangle$.

Using Eq.(14) in Eq.(15), we get the expression for the total scattering cross section in the local density approximation as

$$\sigma = \frac{4\sqrt{2}G_F}{|\mathbf{k}|}\mathrm{Im}\int d^3r \int \frac{d^4k'}{(2\pi)^4}\frac{1}{k'^2 - m_\mu^2 + i\epsilon}\left(\frac{m_W}{q^2 - m_W^2}\right)^2 L_{\alpha\beta}\Pi^{\alpha\beta}(q)\,. \tag{27}$$

The imaginary part of the neutrino self energy is evaluated by means of the Cutkosky rules by cutting the Feynman diagram shown in Fig. 2 along the dotted line which puts the particles corresponding to the cut propagators on the mass shell by replacing the fermion and meson propagators by their imaginary parts as

$$\Sigma(k) \rightarrow 2iIm\Sigma(k), \quad D(p_j') \rightarrow 2i\theta(p_{0j})\; ImD(p_j') \tag{28}$$
$$G(p_l') \rightarrow 2i\theta(p_{0l}')\; ImG(p_l'), \quad \frac{1}{k'^2 - m_l^2 + i\epsilon} \rightarrow 2\pi\delta(k'^2 - m_l^2).$$

After performing the k_0', p_0' and p_0 integrations for all momenta in Eq.(27) and using Eqs.(19) and (28), we get the differential scattering cross section which is written in the local density approximation as:

$$\frac{d^2\sigma^A_{\nu,\bar\nu}}{d\Omega' dE'} = \frac{{G_F}^2}{(2\pi)^2}\frac{|\mathbf{k}'|}{|\mathbf{k}|}\left(\frac{m_W^2}{q^2 - m_W^2}\right)^2 L^{\alpha\beta}_{\nu,\bar\nu}\, W^A_{\alpha\beta}\,, \tag{29}$$

where

$$W^A_{\alpha\beta} = 4\int d^3r \int \frac{d^3p}{(2\pi)^3}\int_{-\infty}^{\mu} dp^0 \frac{M}{E(\mathbf{p})} S_h(p^0, p, \rho(r)) W^N_{\alpha\beta}(p, q),\,. \tag{30}$$

with $W^N_{\alpha\beta}(p,q)$ is given by Eq.(6).

Finally $W_i^A(x, Q^2)$ (i=1-3) are redefined in terms of the dimensionless structure functions $F_i^A(x, Q^2)$ through

$$M_A W_1^A(\nu, Q^2) = F_1^A(x, Q^2), \quad \nu W_2^A(\nu, Q^2) = F_2^A(x, Q^2), \quad \nu W_3^A(\nu, Q^2) = F_3^A(x, Q^2). \tag{31}$$

We have also assumed the Callan-Gross relationship for nuclear structure functions ${F_1}^A(x)$ and ${F_2}^A(x)$, therefore, we are left with only two independent structure functions viz. ${F_2}^A(x)$ and ${F_3}^A(x)$. Taking xy and xx component on both sides of Eq. (30) and q along the z axis we can evaluate structure functions ${F_2}^A(x)$ and ${F_3}^A(x)$.

$$F_2^A(x_A, Q^2) = 4\int d^3r \int \frac{d^3p}{(2\pi)^3}\frac{M}{E(\mathbf{p})}\int_{-\infty}^{\mu} dp^0\, S_h(p^0, \mathbf{p}, \rho(\mathbf{r}))\frac{x}{x_N}\left(1 + \frac{2x_N p_x^2}{M\nu_N}\right) F_2^N(x_N, Q^2) \tag{32}$$

$$F_3^A(x_A, Q^2) = 4\int d^3r \int \frac{d^3p}{(2\pi)^3}\frac{M}{E(\mathbf{p})}\int_{-\infty}^{\mu} dp^0 S_h(p^0, \mathbf{p}, \rho(\mathbf{r}))\frac{p^0\gamma - p_z}{(p^0 - p_z\gamma)\gamma} F_3^N(x_N, Q^2) \tag{33}$$

where

$$\gamma = \frac{q_z}{q^0} = \left(1 + \frac{4M^2x^2}{Q^2}\right)^{1/2}, x_N = \frac{Q^2}{2(p^0q^0 - p_zq_z)}. \tag{34}$$

Eqs.32 and 33 are used for our base calculations where Pauli blocking and Fermi motion has been taken into account. Thereafter, we have added other medium effects namely shadowing, anti-shadowing, pion, and rho cloud contributions. The expression for the target mass correction (TMC) is taken from Ref. [24] and the shadowing effect has been incorportaed using the model of Kulagin and Petti [18, 25]. We shall give the expressions for the pion and rho cloud contributions in the next subsection.

In ${F_3}^A(x)$ only valence quarks contribute due to which there is no pion and rho cloud contributions. Target mass correction has also been considered in our base calculations. Numerical calculations in our paper with all medium effects are called as full model calculations. All of the above calculations have been done for isoscalar targets. For non-isoscalar targets like iron and lead where proton and neutron numbers are different few modifications are needed to be done in Eqs.30, 32, and 33.

$$\begin{aligned} W_{\alpha\beta}^{\nu(\bar\nu)A} &= 2\int d^3r \int \frac{d^3p}{(2\pi)^3}\frac{M}{E(\mathbf{p})}\int_{-\infty}^{\mu_p} dp^0\; S_h^{proton}(p^0,\mathbf{p},k_{F,p}) W_{\alpha\beta}^{\nu(\bar\nu)p} \\ &+ 2\int d^3r \int \frac{d^3p}{(2\pi)^3}\frac{M}{E(\mathbf{p})}\int_{-\infty}^{\mu_n} dp^0\; S_h^{neutron}(p^0,\mathbf{p},k_{F,n}) W_{\alpha\beta}^{\nu(\bar\nu)n} \end{aligned} \tag{35}$$

where the factor 2 in front of the integral accounts for the two degrees of freedom of the spin of the proton/neutron. There are two different spectral functions, each one of them normalized to the number of protons or neutrons in the nuclear target and are functions of Fermi momentum of protons and neutrons respectively which are given by $k_{F,p} = (3\pi^2\rho_p)^{1/3}$ and $k_{F,n} = (3\pi^2\rho_n)^{1/3}$. For the proton and neutron densities in iron and lead, we have used two-parameter Fermi density distribution and harmonic oscillator density for carbon. The density parameters are taken from Refs. [29, 30].

The expressions for $F_2^A(x)$ and $F_3^A(x)$ are obtained as:

$$\begin{aligned} F_2^A(x_A,Q^2) &= 2\int d^3r \int \frac{d^3p}{(2\pi)^3}\frac{M}{E(\mathbf{p})}\left[\int_{-\infty}^{\mu_p} dp^0\; S_h^{proton}(p^0,\mathbf{p},k_{F,p}) F_2^{proton}(x_N,Q^2)\right. \\ &+ \left.\int_{-\infty}^{\mu_n} dp^0\; S_h^{neutron}(p^0,\mathbf{p},k_{F,n}) F_2^{neutron}(x_N,Q^2)\right]\frac{x}{x_N}\left(1+\frac{2x_N p_x^2}{M\nu_N}\right) \end{aligned} \tag{36}$$

$$\begin{aligned} F_3^A(x_A,Q^2) &= 2\int d^3r \int \frac{d^3p}{(2\pi)^3}\frac{M}{E(\mathbf{p})}\left[\int_{-\infty}^{\mu_p} dp^0\; S_h^{proton}(p^0,\mathbf{p},k_{F,p}) F_3^{proton}(x_N,Q^2)\right. \\ &+ \left.\int_{-\infty}^{\mu_n} dp^0\; S_h^{neutron}(p^0,\mathbf{p},k_{F,n}) F_3^{neutron}(x_N,Q^2)\right]\frac{p^0\gamma - p_z}{(p^0 - p_z\gamma)\gamma} \end{aligned} \tag{37}$$

Using expressions for structure functions in nucleus for isoscalar as well as non-isoscalar we can calculate differential scattering cross section for charge current in nucleus which is written as:

$$\begin{aligned} \frac{d^2\sigma_{CC}^{\nu(\bar\nu)A}}{dx_A\, dy_A} &= \frac{G_F^2 M_A E_\nu}{\pi}\left(\frac{m_W^2}{Q^2+m_W^2}\right)^2 \left(y_A^2 x_A F_1^{\nu(\bar\nu)A}\right. \\ &+ \left.\left\{1-y_A-\frac{M_A x_A y_A}{2E_\nu}\right\} F_2^{\nu(\bar\nu)A} \pm x_A y_A\left(1-\frac{y_A}{2}\right) F_3^{\nu(\bar\nu)A}\right) \end{aligned} \tag{38}$$

m_l is the mass of lepton, E_ν is the incident $\nu/\bar\nu$ energy, M_A is the mass of nucleus, in F_3^A, +sign(-sign) is for $\nu/\bar\nu$, $x_A = \frac{Q^2}{2M_A\nu}$ is the Bjorken variable, $y_A = \frac{\nu}{E_\nu}$, ν and $Q^2 = -q^2$.

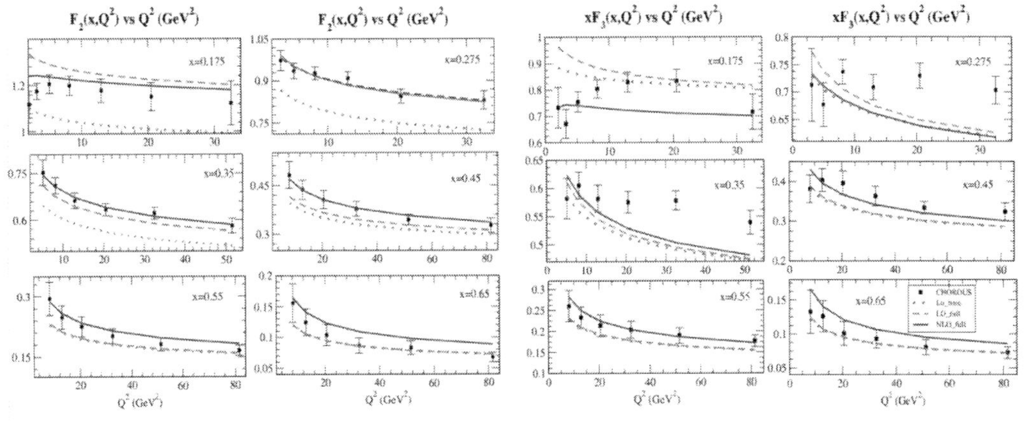

Figure 3. (Color online) $F_2(x, Q^2)$ vs Q^2(left panel) and $xF_3(x, Q^2)$ vs Q^2(right panel) in ^{208}Pb calculated using our base equation with TMC at LO, the full model at LO and NLO.

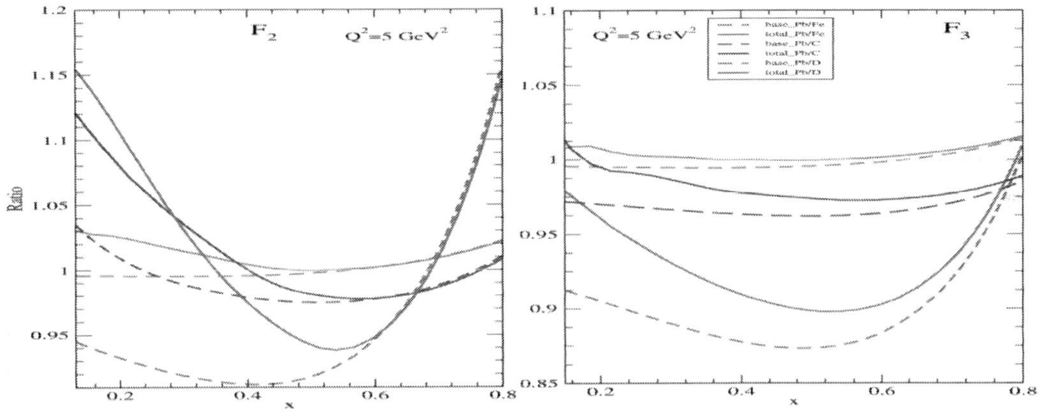

Figure 4. (Color online) Ratio $R(x, Q^2)$=$\frac{2F_2^{Pb}}{208F_2^D}, \frac{12F_2^{Pb}}{208F_2^C}$, and $\frac{56F_2^{Pb}}{208F_2^{Fe}}$(left panel) and $\frac{2F_3^{Pb}}{208F_3^D}, \frac{12F_3^{Pb}}{208F_3^C}$, and $\frac{56F_3^{Pb}}{208F_3^{Fe}}$(right panel) with full calculation for $Q^2 = 5$ GeV^2 using our base equation with TMC (dashed line) and the full model (solid line) at NLO.

3.1. π and ρ mesons contribution to the nuclear structure function

The pion and rho meson cloud contributions to the F_2 structure function have been implemented following the many body field theoretical approach of Refs. [22, 23].

The pion structure function $F_{2A,\pi}(x)$ is written as [22];

$$F^A_{2,\pi}(x) = -6 \int d^3r \int \frac{d^4p}{(2\pi)^4} \, \theta(p^0) \, \delta ImD(p) \, \frac{x}{x_\pi} \, 2M \, F_{2\pi}(x_\pi) \, \theta(x_\pi - x) \, \theta(1 - x_\pi) \tag{39}$$

where $D(p)$ is the pion propagator in the nuclear medium which is given in terms of the pion self energy Π_π:

$$D(p) = [p_0^2 - \vec{p}^{\,2} - m_\pi^2 - \Pi_\pi(p^0, p)]^{-1}, \tag{40}$$

where

$$\Pi_\pi = \frac{f^2/m_\pi^2 F^2(p)\vec{p}^{\,2}\Pi^*}{1 - f^2/m_\pi^2 V_L' \Pi^*}. \tag{41}$$

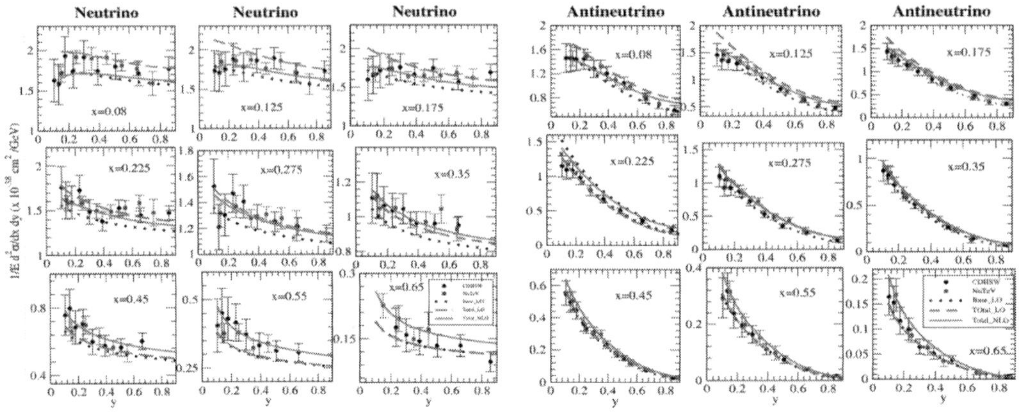

Figure 5. (Color online)$\frac{1}{E}\frac{d^2\sigma}{dxdy}$ vs y(left panel is for neutrino and right panel is for antineutrino) at different x for $\nu_\mu(E_{\nu_\mu} = 65$ GeV) induced reaction in ^{56}Fe. Dotted line is the results using Eq.38 with TMC. Solid (Dashed line) line is full calculation at NLO (LO). The experimental points are from CDHSW [3] and NuTeV [9].

Here, $F(p) = (\Lambda^2 - m_\pi^2)/(\Lambda^2 + \vec{p}^2)$ is the πNN form factor and Λ=1 GeV, $f = 1.01$, V_L' is the longitudinal part of the spin-isospin nucleon-nucleon interaction and Π^* is the irreducible pion self energy that contains the contribution of particle-hole and delta-hole excitations. In Eq.(39), $\delta ImD(p)$ is given by

$$\delta ImD(p) \equiv ImD(p) - \rho\,\frac{\partial ImD(p)}{\partial\rho}\,|_{\rho=0} \tag{42}$$

and

$$\frac{x}{x_\pi} = \frac{-p^0 + p^z}{M} \tag{43}$$

Assuming SU(3) symmetry, the pion structure function at LO can be written in terms of pionic PDFs [31, 32] as

$$F_{2\pi}(x_\pi) = x_\pi\left(2v_\pi(x_\pi) + 6\bar{q}_\pi(x_\pi)\right) \tag{44}$$

where $v_\pi(x_\pi)$ is the valence distribution and $\bar{q}_\pi(x_\pi)$ is the light SU(3)-symmetric sea distribution.

Similarly, the contribution of the ρ-meson cloud to the structure function is written as [22]

$$F_{2,\rho}^A(x) = -12\int d^3r \int \frac{d^4p}{(2\pi)^4}\theta(p^0)\delta ImD_\rho(p)\frac{x}{x_\rho}\,2MF_{2\rho}(x_\rho)\theta(x_\rho - x)\theta(1 - x_\rho) \tag{45}$$

where $D_\rho(p)$ is the ρ-meson propagator and $F_{2\rho}(x_\rho)$ is the ρ-meson structure function, which we have taken equal to the pion structure function $F_{2\pi}$ using the valence and sea pionic PDFs from Ref. [31]. Λ_ρ in ρNN form factor $F(p) = (\Lambda_\rho^2 - m_\rho^2)/(\Lambda_\rho^2 + \vec{p}^2)$ has also been taken as 1 GeV.

4. Results and Discussions

$F_2^A(x, Q^2)$ and $F_3^A(x, Q^2)$ structure functions have been calculated using Eqs.(32) and (33) for the carbon target and Eqs.(36) and (37) have been used for the nonisoscalar nuclear targets like iron and lead, with target mass correction (TMC) and CTEQ6.6 parton distribution functions (PDFs) at LO [28]. We call this as our base (Base) result. Thereafter, we include pion and rho cloud contributions in F_2^A and the shadowing corrections in F_2^A and F_3^A, which we call as our full calculation (Total). The effect of shadowing is about 5-7% at x=0.1, Q^2=1-5 GeV^2 and 1-2% at x=0.2, Q^2=1-5 GeV^2 which dies out with the increase in x and Q^2. In the case of F_2^A there are pion and rho cloud contributions. The pion contribution is very dominant in comparison to the rho contribution. Pion contribution is significant in the region of $0.1 < x < 0.4$. Thus, we find that the shadowing correction seem to be negligible as compared to the other nuclear effects. It is the meson cloud contribution which is dominant at low and intermediate x for F_2. In Fig.3 we have presented the results for lead nuclear target and also compared our numerical results with the CHORUS data [11]. We find that numerical results at NLO are in good agreement with the CHORUS data [11]. For the results of other nuclear targets like carbon and iron see Ref.[20].

We have shown the results for $\frac{2F_i^{Pb}}{208F_i^{D}}$, $\frac{12F_i^{Pb}}{208F_i^{C}}$, and $\frac{56F_i^{Pb}}{208F_i^{Fe}}$ $(i = 2, 3)$ at $Q^2 = 5$ GeV^2 in Fig.4 using CTEQ6.6 PDFs at NLO [28]. These results are presented for our base as well as full calculations. The deuteron structure functions have been calculated using the same formulae as for ^{12}C in Ref. [20], but performing the convolution with the deuteron wave function squared instead of the spectral function. We have used the parametrization given in Ref. [33] for the deuteron wave function of the Paris N-N potential. We observe that the EMC effect is more prominent for ratios of structure functions between a heavy nucleus such as ^{208}Pb and a much lighter one like ^{12}C than for ratios between a heavy nucleus (^{208}Pb) and a medium-size one like ^{56}Fe.

The differential scattering cross section for charged current neutrino and antineutrino induced reactions has been presented in Figs.5 and 6 in iron(isoscalar) and lead(nonisoscalar) respectively. In these figures the results for iron have been presented for isoscalar nuclear target as the experimental results are corrected for the isoscalar target [3, 9]. However, if we do nonisoscalar calculations then there would be 2-3% change in the results(not shown here). The medium effects in carbon and iron are small and for details see the discussion given in Ref. [20]. In the case of lead nucleus, for neutrino induced process when the differential scattering cross section is obtained using the full calculation, the change from the results obtained by using the base calculation is around 12-18% for 0.1<x<0.2, that decreases with increase in x and becomes 2-5% for 0.4<x<0.5. For antineutrino induced reaction this difference is around 22-28% for 0.1<x<0.2 which becomes 8-15% for 0.4<x<0.5.

5. Conclusions

In this paper we have studied structure functions $F_2^A(x)$ and $F_3^A(x)$ in nuclei. For parton distribution functions(PDFs) in nucleon we have used CTEQ6.6 parametrization. Many-body theory is used to describe the spectral function of the nucleon in the nuclear-medium for all Q^2. The local density approximation has been used to apply the results for the finite nuclei. The use of the spectral function is to incorporate Fermi motion and binding effects. Target mass correction has been considered. We have taken the effects of mesonic degrees of freedom, shadowing, anti-shadowing in the calculation of F_2^A and shadowing and anti-shadowing effects in the calculation of F_3^A. We have found that the mesonic cloud (mainly pion) gives an important contribution to the cross section. Using these structure functions we obtained differential scattering cross section in iron and lead. These numerical results have been compared with the available experimental results. The ratios $\frac{2F_i^{Pb}}{208F_i^{D}}$, $\frac{12F_i^{Pb}}{208F_i^{C}}$, and $\frac{56F_i^{Pb}}{208F_i^{Fe}}$ (i=2,3), have been calculated using the results from Ref. [20] in the case of carbon and iron, where for deuteron we have used the same formulae as for ^{12}C in Ref.[20], but performing the convolution with the deuteron wave function squared instead of the spectral function. We find that the nuclear-medium effects are also quite important in the case of deep inelastic scattering and the effect of nuclear medium in $F_2^A(x)$ and $F_3^A(x)$ structure functions is of different nature. Furthermore, we observe that the ratio of the structure functions for the different nuclei are not the same.

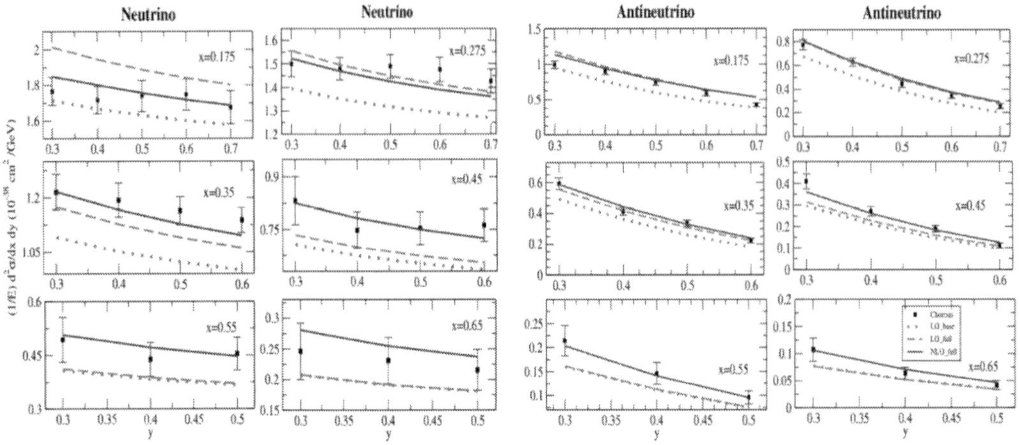

Figure 6. (Color online) $\frac{1}{E}\frac{d^2\sigma}{dxdy}$ vs y(left panel is for neutrino and right panel is for antineutrino) at different x for $\nu_\mu(E_{\nu_\mu} = 25$ GeV) induced reaction in ^{208}Pb. Dotted line is the results using Eq.38 with TMC. Solid (Dashed line) line is full calculation at NLO (LO). The experimental points are from CHORUS [11].

6. References

[1] G. P. Zeller et al., *Phys. Rev. Lett.* **88**, 091802 (2002).
[2] D. Allasia et al. Zeit. Phys.C **28**, 321 (1991).
[3] J. P. Berge et al. Zeit. Phys. C **49**, 187 (1991).
[4] K. Varvell et al. Zeit. Phys. C **36**, 1 (1991).
[5] E. Oltman et al. Zeit. Phys. C **53**, 51 (1992).
[6] W. G. Seligman et al. Phys. Rev. Lett. **79**, 1213 (1997).
[7] A. V. Sidorov et al. Eur. Phys. J. C **10**, 405 (1999).
[8] Bonnie T. Fleming et al. Phys. Rev. Lett. **86**, 5430 (2001).
[9] M. Tzanov et al., Phys. Rev. D **74**, 012008 (2006).
[10] G. Onengut et al., Phys. Lett. B **632**, 65 (2006).
[11] http://choruswww.cern.ch/Publications/DIS-data.
[12] R. Petti, Private Communication.
[13] T. Adams et al. [NuSOnG collaboration], Int. J. Mod. Phys. A **24**, 671 (2009).
[14] N. Agafonova et al. (The OPERA collaboration) New J. Phys. **13**, 053051 (2011).
[15] M. Hirai, S. Kumano, and T. H. Nagai, Phys. Rev. C **76**, 065207 (2007).
[16] K. J. Eskola H. Paukkunen and C. A. Salgado, JHEP **0807**, 102 (2008).
[17] I. Schienbein et al. Phys. Rev. D **80**, 094004 (2009).
[18] S. A. Kulagin and R. Petti, Phys. Rev. D **76**, 094023 (2007).
[19] M. Sajjad Athar, S. K. Singh and M. J. Vicente Vacas, Phys. Lett. B **668**, 133 (2008).
[20] H. Haider, I. Ruiz Simo, M. Sajjad Athar and M.J. Vicente Vacas, Phys. Rev. C **84**, 054610 (2011).
[21] P. Fernandez de Cordoba and E. Oset, Phys. Rev. C **46**, 1697 (1992).
[22] E. Marco, E. Oset and P. Fernandez de Cordoba, Nucl. Phys. A **611**, 484 (1996).
[23] C. Garcia-Recio, J. Nieves and E. Oset, Phys. Rev. C **51**, 237 (1995).
[24] I. Schienbein et al., J. Phys. G **35**, 053101 (2008).
[25] S. A. Kulagin and R. Petti, Nucl. Phys. A **765**, 126 (2006).
[26] S. Moch, J. A. M. Vermaseren and A. Vogt, Nucl. Phys. B **813**, 220 (2009).
[27] C. G. Callan, Jr. and D. J. Gross, Phys. Rev. Lett. **22**, 156 (1969).
[28] Pavel M. Nadolsky et al., Phys. Rev. D **78**, 013004 (2008); http://hep.pa.msu.edu/cteq/public
[29] J. Nieves, M. Valverde and M.J. Vicente Vacas, Phys. Rev.C **73**, 025504 (2006).
[30] H. de Vries, C.W. de Jager and C. de Vries, At. Data Nucl. Data Tables **36**, 583 (1971).
[31] M. Gluck, E. Reya and A. Vogt, Z. Phys. C **53**, 651 (1992).
[32] M. Gluck, E. Reya and I. Schienbein, Eur. Phys. J. C **10**, 313 (1999).
[33] M. Lacombe et al., Phys. Lett. B **101**, 139 (1981).

5th DAE-BRNS Workshop on Hadron Physics (Hadron 2011) IOP Publishing
Journal of Physics: Conference Series **374** (2012) 012021 doi:10.1088/1742-6596/374/1/012021

Probing strongly interacting matter with heavy resonances in Pb+Pb collisions at LHC energies

Prashant Shukla and Abdulla Abdulsalam

Nuclear Physics Division, Bhabha Atomic Research Centre, Mumbai - 400085, INDIA

E-mail: pshukla@barc.gov.in

Abstract.

With the recent Pb+Pb collisions experiments at LHC, a new era has began in the field of heavy ion collisions. The heavy hadronic resonances such as J/ψ, Υ and their states are produced in plenty at LHC and can be tackled in perturbative QCD. These resonances along with other low mass resonances can be very well reconstructed with advanced detector technologies and are used to study the detailed tomography of the matter which would not have been possible at lower energies. In addition to the hot matter effects one also has to address the questions such as modification of parton distributions of protons when placed inside a nucleus. This is known as initial state effect and can be well addressed at LHC by measuring the Z boson. In addition, there are (cold) nuclear matter effects which can modify the probes and are planned to be studied in future p+Pb collisions. In this paper, we review the recent LHC results with emphasis on quarkonia. We devote a section on modeling of quarkonia suppression. The new calculation done in the case of Υ and its states are given.

1. Introduction

The heavy ion collisions are performed to study the interaction of matter at extreme temperatures and densities where it is expected to be in the form of Quark Gluon Plasma (QGP), in which the quarks and gluons can move far beyond the size of a nucleon making color degrees of freedom dominant in the medium. The experimental effort in this direction started with low energy CERN accelerator SPS and evolved through voluminous results [1] from heavy ion collision at Relativistic Heavy Ion Collider (RHIC) in Brookhaven National Lab in recent years. The RHIC experiments at the energy $\sqrt{s}_{NN}$ = 200 GeV indicate that such matter is strongly coupled and can consist of bound states which carry net color.

One of the most striking QGP signals is quarkonium suppression [2]. Quarkonia are identified by their reconstructed mass peaks in the dilepton invariant mass distribution. At both SPS and RHIC energies, the suppression of the J/ψ resonance suggests that a very high temperature system was created [3, 4]. With the start of Pb+Pb collisions at per nucleon center of mass energy $\sqrt{s_{NN}}$ = 2.76 TeV at LHC, there is renewed excitement in this field. There are number of light and heavy resonances measured appear in dilepton invariant mass distribution. Below $\sim$ 12 GeV/c^2, it includes a number of resonance peaks: ρ, ω and ϕ at low masses and the ψ and Υ states at higher masses. At 91 GeV/c^2, the $Z^0 \rightarrow l^+l^-$ peak appears. The first measurements of the dilepton spectra at the LHC have recently been reported [5, 6, 7]. The CMS experiment reported the first measurements of the Z^0 mass region in Pb+Pb collisions [5] as well as measurements of the full dimuon distribution, including quarkonia [6]. ATLAS has also reported J/ψ and Z^0 measurements in the dimuon channel [7]. The second LHC Pb+Pb run, at much higher luminosity, will provide higher statistics measurements of the dilepton spectra over the full available phase space.

In this paper, we review the results from LHC starting with some global features then quarkonia and heavy flavour. We devote a section on modeling of quarkonia suppression. The new calculation done in the case of Υ states and their ratios as a function of transverse momentum are given.

2. Global features of heavy ion collisions at LHC as compared to RHIC

The multiplicity of produced particles is probably the most basic of observables. It provides information on the energy density achieved in the collisions and constitutes a primary input for most model calculations. For central Pb-Pb collisions at the LHC (at $\sqrt{s_{NN}} = 2.76$ TeV), the produced multiplicity is about twice as at RHIC [8, 9, 10]. As function of collision energy the charged particle multiplicity measured in central collisions at LHC [11] is larger than most of the predictions and about 50 % more than expected from simple phenomenological extrapolations from RHIC energy. The centrality dependence of the charged particle multiplicity is rather mild, favouring models incorporating some mechanism (such as parton saturation) moderating the increase with centrality of the average multiplicity per participant pair [12, 13]. Once the multiplicities are rescaled to account for the difference in the central values, the centrality dependences at the LHC and at RHIC are remarkably similar.

The average amount of transverse energy produced per unit of pseudorapidity per participant pair in central collisions is about 9 GeV, or about a factor 2.7 larger than at RHIC (the larger multiplicity at LHC being accompanied by an increase in the average transverse momentum of the produced particles), corresponding to an energy density of about 16 GeV/fm^3 (taking the conventional value of 1 fm/c for the plasma formation time). Hanbury Brown – Twiss (HBT) interferometry [14, 15] exploits the quantum interference between identical bosons (e.g. charged pions) to evaluate the size of the system at the time of decoupling. Compared to RHIC, the ALICE experiment finds an increase in the dimensions of the system in all three components [16, 17] (including, finally, also the sidewards component R_{side}). The system expands significantly more than at RHIC, with an estimated increase by about a factor 2 of the homogeneity volume for central collisions.

3. Quarkonia as probes of QGP

One of the most talked about signal of QGP is the suppression of quarkonium states, both of the charmonium (J/ψ, ψ',χ_c, etc) and the bottomonium ($\Upsilon(1S)$, $\Upsilon(2S)$, χ_b, etc) families. This is thought to be a direct effect of deconfinement, when the binding potential between the constituents of a quarkonium state, a heavy quark and its antiquark, is screened by the colour charges of the surrounding light quarks and gluons. The melting temperature depends on the binding energy of the quarkonium state. The ground states, J/ψ and $\Upsilon(1S)$ are expected to dissolve at significantly higher temperatures than the more loosely bound excited states. The masses and the radii of quarkonia states are given in Table 1 [18]. Since the $\Upsilon(1S)$ is the most tightly bound state among all quarkonia, it is expected to be the one with the highest dissociation temperature. Examples of dissociation temperatures are $T_{diss} \sim 1T_c$, $1.2T_c$, and $2T_c$ for the $\Upsilon(3S)$, $\Upsilon(2S)$ and $\Upsilon(1S)$ respectively. Similarly, in the charmonium family the dissociation temperatures are $\leq 1\ T_c$ and $1.2\ T_c$ for the ψ' and J/ψ respectively [6].

However, there are other possibilities that would affect the quarkonium production in heavy-ion collisions. The modifications to the parton distribution functions inside the nucleus (shadowing) and other cold-nuclear-matter effects can reduce the production of quarkonia without the presence of a QGP [9, 10]. Also, the large number of heavy quarks produced

Table 1. Quarkonium Spectroscopy from Non-Relativistic Potential Theory

state	J/ψ	χ_c	ψ'	Υ	χ_b	Υ'	χ_b'	Υ''
mass [GeV]	3.10	3.53	3.68	9.46	9.99	10.02	10.26	10.36
r_0 [fm]	0.50	0.72	0.90	0.28	0.44	0.56	0.68	0.78

in heavy-ion collisions, in particular at the energies accessible by the Large Hadron Collider (LHC), could lead to an increased production of quarkonia via statistical recombination.

Charmonium studies in heavy-ion collisions have been carried out first at the Super Proton Synchrotron (SPS) by fixed-target experiments at 17.3-19.3 GeV centre-of-mass energy per nucleon pair, then at the RHIC experiment and now at LHC with higher energy and luminosity. In all cases, J/ψ suppression was observed in the most central collisions[6]. At the SPS, the suppression of the ψ' meson was also measured. At RHIC, the reference was provided by the properly scaled yield measured in pp collisions. The PHENIX collaboration has updated the results on J/ψ suppression using the 2007 data sample [19, 20]. The previous results, and in particular the stronger suppression at forward than at central rapidity, are confirmed, with reduced statistical and systematic uncertainties.

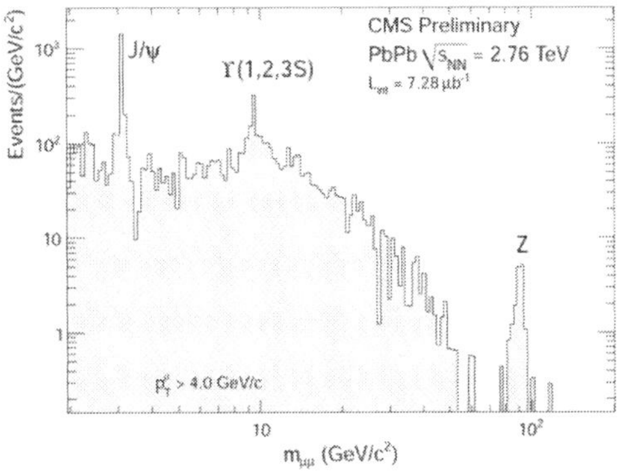

Figure 1. Dimuon mass spectrum with first year PbPb collision at $\sqrt{s_{NN}} = 2.76$ TeV at CMS.

The CMS collaboration obtained production rates of J/ψ mesons and of the Υ family in dimuom channel [6, 21]. Figure 1 shows the dimuon mass reconstructed in PbPb collisions at $\sqrt{s_{NN}} = 2.76$. Non-prompt J/ψs (those produced from B-meson decays) could be identified by their displaced decay vertex. The suppressions of prompt and non-prompt J/ψ particles were measured separately. Figure 2 gives an idea of R_{AA} of prompt J/ψ as function of centrality measured by CMS. The suppression pattern with CMS is very similar to PHENIX. One can easily see that a suppression of $\sim$5 is observed in the 10% most central PbPb collisions with respect to pp. This suppression is reduced in more peripheral collisions, reaching a factor of $\sim$1.6 in the 50-100% centrality bin. The R_{AA} as function of the number of participants N_{part} indicates that high p_T J/ψs are strongly suppressed as low as 0.2 for central collisions as shown in Fig. 2.

ATLAS also observes strong centrality dependence in the central-to-peripheral nuclear modification factor $R_{cp}^{J/\psi}$ at high p_T [7]. For $p_T > 0$ and $2.5 < y < 4$, ALICE observes a J/ψ nuclear modification factor $R_{AA}^{J/\psi}$ of about 0.5, practically independent of centrality [22].

CMS reported suppression for the Υ(1S) (around 0.6, with a weak energy dependence) and further suppression by about a factor 3 relative to the Υ(1S) for the excited states Υ(2S+3S)

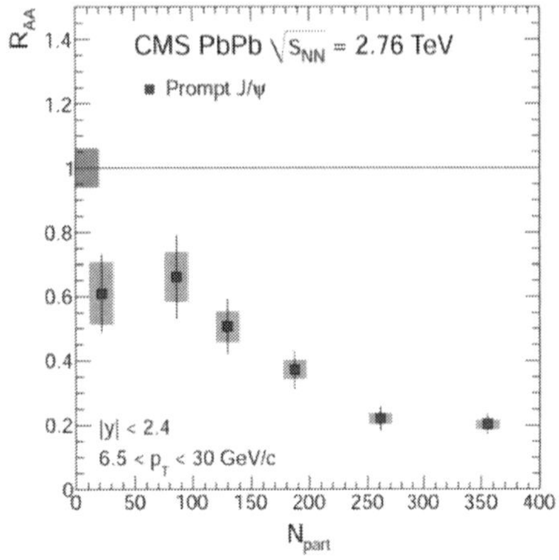

Figure 2. Measured R_{AA} as a function of centrality, for prompt J/ψ.

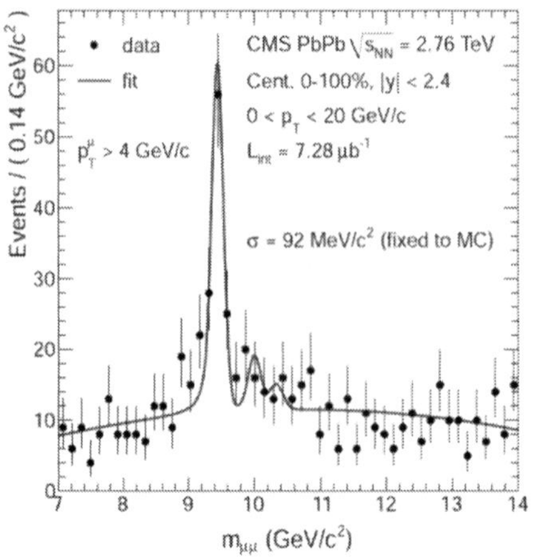

Figure 3. The upsilon states measured in dimuon invariant mass distribution.

[21]. Fig. 3 shows that the excited states, $\Upsilon(2S)$ and $\Upsilon(3S)$, are suppressed as compared to the $\Upsilon(1S)$. This is compatible with differential melting of quarkonia states in the high temperatures produced by PbPb collisions.

The detailed pattern of quarkonia suppression will emerge with high statistics measurements. Establishing a good pA reference will also be essential in order to disentangle the contributions from cold nuclear effects.

4. Z boson in heavy ion collisions

To study QGP effects the measurements in heavy-ion (AA) collisions are compared to scaled measurements in proton-proton (pp) collisions. But scaling pp to PbPb is not sufficient as the production of particles is different due to effects such as shadowing (initial state effects). So we use the baseline, Z^0 to compare in the same AA sample the yields of particles that are modified by the QGP to those of unmodified reference particles.

In AA collisions, Z boson production can be affected by various initial-state effects, though predictions indicate that these contributions are rather small [5]. Firstly, the mix of protons and neutrons in AA collisions (the so-called isospin effect) is estimated to modify the Z yield by less than 3% compared to pp collisions. Secondly, energy loss and multiple scattering of the initial partons can also alter the Z production, by about 3%. The parton distributions functions (PDFs) however are modified in nuclei and a depletion (shadowing) is expected for Z bosons at the LHC, modifying their yield by as much as 20%. Precise measurements of Z production in heavy-ion collisions can therefore help to constrain nuclear PDFs. The Z bosons are therefore ideally suited to serve as a standard candle of the initial state in PbPb collisions at the LHC energies. CMS [5] has shown that Z yield does not vary in different centrality giving an impression that no centrality dependence of the binary-scaled Z yields is observed in data.

5. Open charm and beauty mesons

The heavy flavour such as D and B probe the medium properties. At RHIC it has been observed that heavy quarks loose substantial energy in the medium. Both STAR and PHENIX are planning to upgrade the apparatus with the addition of vertex detectors, which should allow them to separate the vertices of the weak decays of heavy flavour particles from the primary vertex. Heavy flavour vertexing is already available in the LHC experiments. Reconstructed D^0

and D^+ signals have been presented by ALICE [24, 25], who observe substantial suppression of the production of D mesons over the currently available transverse momentum range (out to 12 GeV/c) in both central (0-20%) and peripheral (40-80%) Pb–Pb collisions. In BDMPS energy loss calculations for LHC, D are expected to be suppressed about a factor 2 less than pions for $p_T \sim 8$ GeV/c [26]). Could charm quarks be essentially thermalized in the QGP, and thereby "lose memory" of their energy loss history? CMS have presented results on the nuclear modification factor of J/ψ originating in the decay of B mesons for $p_T^{(J/\psi)} > 6.5$ GeV/c [21]. The suppression is again about a factor 3, with very little – if any – centrality dependence.

The future will bring high statistics D and B measurements, allowing us to establish the mass and colour charge dependence of the parton energy loss.

6. Ratios of upsilon states as probe of QGP

The ratio of upsilon states has been measured in Pb+Pb collisions at $\sqrt{s_{NN}} = 2.76$ TeV [23] which looks a more promising signal of QGP as the nuclear matter effects cancel out in the ratio. It was shown many years back [27, 28] that the p_T dependence of such ratio would show large variation and would be a direct probe of the QGP. In this work, we calculate the ratios of upsilon states using essentially same model with the parameters relevant for LHC.

6.1. Radius and formation time of quarkonia

Radius and formation time of quarkonia are calculated in potential models which are based on the assumption that the interaction between the heavy quark and its antiquark inside the quarkonium can be described by a potential. Due to the largeness of the heavy quark mass and the smallness of the heavy quark velocity, v<< 1, one can treat the quark-antiquark system nonrelativistically and solve the Schrodinger equation to obtain the bound state properties. Following Ref [29, 30], the interquark potential for non-relativistically bound quarkonium at zero temperature is

$$V(r,0) = \sigma r - \frac{\alpha}{r}, \tag{1}$$

where r is the separation between Q and $\overline{Q}$ with σ=0.192 GeV^2 and α=0.471. At a finite temperature T, the quark binding is modified by the color screening and the potential becomes

$$V(r,T) = \frac{\sigma}{\mu(T)}(1 - e^{-\mu(T)r}) - \frac{\alpha}{r}\, e^{-\mu(T)r}, \tag{2}$$

where $\mu(T)$ is called screening mass. For μ=0, we get the confining potential equation (1). Since the range of the binding force decreases exponentially with the screening mass, $\mu(T)$ the temperature dependence of the potential can be included in μ [29]. Screening effect can be evaluated with the semi-classical approximation,

$$E(r,\mu) = 2m + \frac{c}{mr^2} + V(r,\mu). \tag{3}$$

where m is mass of quark and c is obtained from the uncertainty relation $c = <p^2><r^2>$. Minimizing the $E(r,\mu)$ with respect to r gives us the temperature dependence of the lowest bound state radius. Each bound state has definite formation time τ_F, the time needed to form a bound state after the production of heavy quarks

$$\tau_F = m\, r/p, \tag{4}$$

where p is the momentum of each quark inside the states. The values obtained from the non-relativistic Schrodinger equation [29] are given in Table 2. In the rest-frame of medium, the time τ_F is obtained by

$$t_F = \tau_F \left(1 + \frac{P^2}{M^2}\right)^{1/2}. \tag{5}$$

with P and M denoting the momentum and mass of the pairs respectively.

6.2. Dissociation temperature

In order to get quantitative idea about how the bound state properties change with temperature one should derive this directly from QCD as in zero temperature potential [31]. One of the common features of potential models is that there is strong decrease of binding energy with increasing temperature. The binding energy is defined as the distance between the peak position and the continuum threshold [32]

$$E_{bin} = 2m + V_\infty - M, \tag{6}$$

where $V_\infty = \sigma/\mu$ in the case of screened Cornell potential [30]. When the binding energy of a resonance drops below a particular temperature called dissociation temperature T_D, the state is weakly bound, and thermal fluctuation can destroy it by transferring energy and exciting the quark anti-quark pair into the continuum. Above this temperature no resonance structure can be seen. The rate of this excitation or the width of the states is determined by the binding energy. Due to the uncertainty in potential one cannot determine the binding energy exactly but set an upper limit for it.

At a temperature where the width $\Gamma(T)$ is greater than the binding energy the state will likely to be dissociated. The upper limit for the dissociation temperatures of the quarkonium states is determined by the applying conservative quantitative condition $\Gamma(T) \geq 2E_{bin}(T)$. The corresponding dissociation temperatures are given in Table 2 [32].

Table 2. Formation time and dissociation temperature.

	Υ	Υ'	χ_b
τ_F [fm]	0.76	1.9	2.6
T_D [GeV]	$2T_c$	$1.2T_c$	$1.3T_c$

6.3. Quarkonia suppression model

For a screening mass μ(T) above a certain critical value, the resonance no longer forms in the plasma. The temperature T_D at which this happens is break up temperature for each bound state. The essential of this model can be found in Ref. [33]. The time at which the temperature drops to T_D can be obtained using Bjorken scenario [34] as

$$t_D = t_0 \left(\frac{T_0}{T_D}\right)^3, \tag{7}$$

where t_0 is formation time and T_0 is initial temperature of the QGP.

The p_T suppression pattern of a resonance is the consequence of the competition between the resonance formation time, τ_F and temperature, lifetime and spatial extent of the QGP. As long as $t_D/\tau_F > 1$, quarkonium formation will be suppressed. The maximum p_T below which suppression occurs is

$$p_{(T,max)} = M\sqrt{\frac{t_D}{\tau_F} - 1}. \tag{8}$$

At any time t, during the evolution of the QGP, the spatial boundary of the screening region is located at the transverse radius r_S given by $t_D(r_S) = t$,

$$r_S = R\left(1 - \left(\frac{\gamma\tau_F}{t_D(0)}\right)^{1/\beta}\right)^{1/2}, \tag{9}$$

where $t_D(0)$ is the value of t_D calculated by equation (7) for resonances produced in the center of the system and $\gamma = \sqrt{1 + (p_T/M)^2}$. The $Q\overline{Q}$ can escape the QGP before the τ_F for a range of angles between r and p_T provided $((r + \tau_F p_T/M)) > r_S$.

The probability $S(p_T)$ that the $Q\overline{Q}$ pair survives and is able to form an onium state is the ratio of the number of bound states produced by those pairs that escape the plasma to the maximum possible number of onium bound states that would be formed at the given p_T in the absence of the QGP:

$$S(p_T) - \frac{\int_0^R drr\rho(r)\theta(r, p_T)}{\pi \int_0^R drr\rho(r)}, \tag{10}$$

where $\rho(r)$ and $\theta(r, p_T)$ are

$$\rho(r) = \left(1 - \left(\frac{r}{R}\right)^2\right)^{\alpha} \tag{11}$$

and

$$\begin{aligned} \theta(r, p_T) &= \pi \quad z \leq -1 \\ &= \cos^{-1} z \quad |z| < 1 \\ &= 0 \quad z \geq 1, \end{aligned}$$

where z

$$z = \frac{r_S^2 - r^2 - (\tau_F p_T/M)^2}{2r\tau_F p_T/M}. \tag{12}$$

The parameters $\beta = 1/3$ and $\alpha = 1/2$.

Table 3. Screening temperature, screening time and $p_{(T,max)}$

	Υ	Υ'	Υ''	χ_b
		Set1		
T_D (GeV)	$2T_c$	$1.2T_c$	$1T_c$	$1.3T_c$
t_D (fm/c)	0.785	3.63	6.28	2.85
$p_{(T,max)}$	2.45	16.3	25.05	4.50
		Set2		
T_D (GeV)	$1.7T_c$	$1.1T_c$	$0.9T_c$	$1T_c$
t_D (fm/c)	1.28	4.72	8.62	6.28
$p_{(T,max)}$	12.79	22.78	35.72	21.68

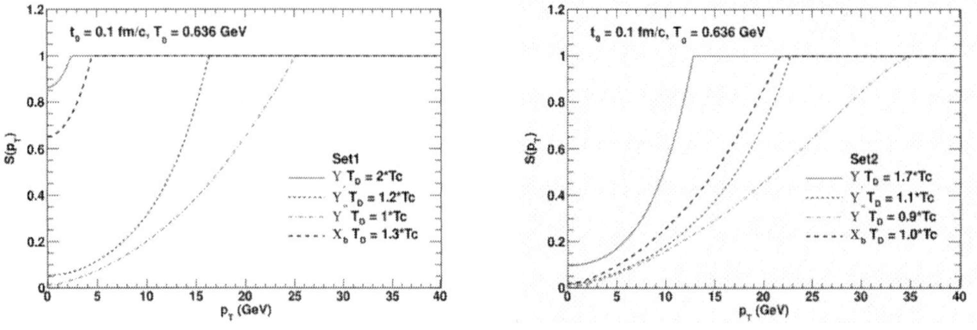

Figure 4. Survival probability for different bottomonium states versus p_T for two sets of dissociation temperatures.

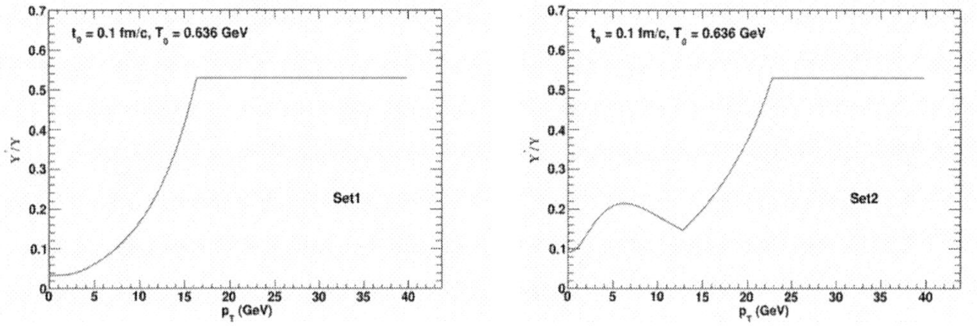

Figure 5. The ratio Υ'/Υ versus p_T for two sets of dissociation temperatures.

The initial temperature T_0 for a given initial time t_0 is obtained by

$$T_0^3 t_0 = \frac{3.6}{4a_q \pi R^2}\frac{dN}{dy} \tag{13}$$

$dN/dy = 1.5\times 1600$, charge multiplicity measured in Pb+Pb collisions at 2.76 TeV and $a_q = 37\pi^2/90$ is the degrees of freedom in quark phase. R is the radius of the nucleus.

Figure 4 shows the survival probability versus p_T for Υ, Υ', Υ'', χ_b with initial conditions $T_0 = 0.636$ GeV, $t_0 = 0.1$ fm/c, $T_c = 0.160$ GeV. Taking $R = 7.1$ fm, equivalent to the radius of Pb nucleus we can produce medium effect in the calculation of survival probability. The two sets of dissociation temperatures used are shown in Table 3. One can observe that all resonance states are suppressed in the medium. At small p_T, $S(p_T) \sim 0$ corresponding to total suppression. The maximum transverse momentum $p_{T,\max}$ below which the upsilon is suppressed is different for different states. If the QGP is of smaller spatial extent, the survival probability can become unity even for a smaller p_T.

Figure 5 shows ratio Υ'/Υ again for two sets of dissociation temperatures. We assumed that the p_T distribution of the Υ states are similar and we have taken $\Upsilon'/\Upsilon = 0.53$ and $\Upsilon''/\Upsilon = 0.17$ corresponding to the measurement at Tevatron [35]. We used these values for calculating the ratio of different states. Only directly produced Υ' and Υ are included in the ratio. Due to different suppression patterns of the states, the ratio is small at low p_T with some fluctuating behavior and will become constant at high p_T.

7. Conclusions

With the start of LHC with its high tech detectors high energy nucleus-nucleus collisions are entering an era of precision measurements that should allow us to impose tight constraints on the properties of the medium. The multiplicity and transverse energy measured at LHC are larger than most of the predictions. With CMS, J/ψ suppression of ~5 is observed in the 10% most central PbPb collisions with respect to pp. This suppression is reduced in more peripheral collisions, reaching a factor of ~1.6 in the 50-100% centrality bin. High p_T J/ψs are strongly suppressed as low as 0.2 for central collisions. CMS shows that the excited states, $\Upsilon(2S)$ and $\Upsilon(3S)$, are suppressed as compared to the $\Upsilon(1S)$. This is compatible with differential melting of quarkonia states in the high temperatures produced by PbPb collisions. The Z boson yield in PbPb collisions at $\sqrt{s_{NN}} = 2.76$ TeV has been measured inclusively and as a function of rapidity, transverse momentum, and centrality. No centrality dependence of the binary-scaled Z yields is observed in data. With higher luminosity data, the Z boson can be a powerful reference tool for final-state heavy-ion related signatures as well as providing a means to study the modifications of the parton distribution functions.

The ratio of upsilon states looks a more promising signal of QGP as the nuclear matter effects cancel out in the ratio. The p_T dependence of such ratio would show large variation and would be a direct probe of the QGP. In this work, we calculate the ratios of upsilon states with the parameters relevant for LHC. One can observe that all resonance states are suppressed in the medium. At small p_T, $S(p_T) \sim 0$ corresponding to total suppression. The maximum transverse momentum $p_{(T,\max)}$ below which the upsilon is suppressed is different for different states. Due to different suppression patterns of the states, the ratio is small at low p_T with some fluctuating behavior and will become constant at high p_T.

With a high luminosity LHC Pb–Pb run in 2011 (20 times the 2010 luminosity) and possibly a p–Pb run as early as next year, we could be in for some paradigm shifting in this field.

8. References

[1] Arsene I *et al* (BRAHMS Collaboration) 2005 *Nucl. Phys.* A **757** 1 ; Back B B *et al* (PHOBOS Collaboration) 2005 *Nucl. Phys.* A **757** 28; Adams J *et al* (STAR Collaboration) 2005 *Nucl. Phys.* A **757** 10 ; Adcox K *et al* (PHENIX Collaboration) 2005 *Nucl. Phys.* A **757** 184
[2] Matsui T and Satz H 1986 *Phys. Lett.* B**416** 178
[3] Baglin C *et al* (NA38 Collaboration) 1991 *Phys. Lett.* B**255** 459
[4] Adare A *et al* (PHENIX Collaboration) 2007 *Phys. Rev. Lett.* **98** 232301
[5] Chatrchyan S *et al* (CMS) 2011 *Phys. Rev. Lett.* **106** 212301
[6] Chatrchyan S *et al* (CMS) 2012 Suppression of non-prompt J/ψ, prompt J/ψ, and $\Upsilon(1S)$ in PbPb collisions at $\sqrt{s_{NN}} = 2.76$ TeV (arXiv:1201.5069v1)
[7] Aad G *et al* (ATLAS) 2011 *Phys. Lett.* B **697** 294
[8] Krisztian Krajczar *et al* (CMS) 2011 *J. Phys.* G**38** 124041
[9] Aamodt K *et al* (ALICE) 2010 *Phys. Rev. Lett.* **105** 252301
[10] Toia A *et al* (ALICE) 2011 *J. Phys.* G **38**, 124007 (arXiv:1107.1973)
[11] Steinberg P *et al* (ATLAS) 2011 *J. Phys.* G **38** 124004 (arXiv:1107.2182)
[12] Aamodt K *et al* (ALICE) 2011 *Phys. Rev. Lett.* **106** 032301
[13] Loizides C *et al* (ALICE) 2011 *J. Phys.* G **38** 124040 (arXiv:1106.6324)
[14] Brown R H and Twiss R Q 1954 *Phil. Mag.* **45** 663
[15] Brown R H and Twiss R Q 1956 *Nature* **178** 1046
[16] Aamodt K *et al* (ALICE) 2011 *Phys. Lett.* B **696** 328
[17] Kisiel A *et al* (ALICE) 2011 *J. Phys.* G **38** 124008
[18] Satz H 2006 *J Phys* G **32** R 25
[19] Adare A *et al* (PHENIX) (arXiv:1103.6269 [nucl-ex])
[20] da Silva C *et al* (PHENIX) 2011 *J. Phys.* G **38** 124031

[21] Silvestre C *et al* (CMS) 2011 *J. Phys.* G **38** 124033 (arXiv:1108.5077)
[22] Martínez García G *et al* (ALICE) 2011 *J. Phys.* G **38** 124034
[23] Chatrchyan S *et al* (CMS) 2011 *Phys. Rev. Lett.* **107**, 052302
[24] Dainese A *et al* (ALICE) 2011 *J. Phys.* G **38** 124032 (arXiv:1106.4042)
[25] Rossi A *et al* (ALICE) 2011 *J. Phys.* G **38** 124139
[26] Armesto N *et al* 2005 *Phys. Rev.* D **71** 054027
[27] Blaizot J P and Ollitrault J Y 1987 *Phys. Lett.* B **199** 499
[28] Gunion J F and Vogt R 1997 *Nucl. Phys.* B **492** 301
[29] Karsch F, Mehr M T, Satz H 1988 *Z. Phys. C Particles and Fields* **37** 617
[30] Mocsy A and Petreczky P 2008 *Phys. Rev.* D **77** 014501
[31] Mocsy A 2009 *Eur. Phys. J.* C **61**
[32] Mocsy A and Petreczky P 2007 *Phys. Rev. Lett.* **99** 211602 (arXiv:0706.2183 [hep-ph])
[33] Chu M C and Matsui T 1988 *Phsyical Review* C **37** 1851
[34] Bjorken J D 1983 *Phys. Rev.* D **27** 140
[35] Abe F *et al* 1995 Fermilab-Pub-95/271-E

Institute of Physics
Dirac House, Temple Back
Bristol BS1 6BE UK

CURRAN ASSOCIATES INC
proceedings.com